Randolf Rausch, Christoph Schüth, Thomas Himmelsbach (eds.)

Hydrogeology of Arid Environments

Proceedings

Borntraeger Science Publishers

Hydrogeology of Arid Environments
Proceedings

Editors: Prof. Dr. Randolf Rausch
Technical Director – Water Resources Studies, giz International Services / Dornier Consulting, P.O. Box 2730, Riyadh 11461, Kingdom of Saudi Arabia

Prof. Dr. Christoph Schüth
Institute of Applied Geosciences, Technische Universität Darmstadt, Schnittspahnstraße 9, 64287 Darmstadt, Germany

Prof. Dr. Thomas Himmelsbach
Head of Sub-Department Groundwater Resources – Quality and Dynamics, Federal Institute for Geosciences and Natural Resources (BGR), Stilleweg 2, 30655 Hannover, Germany

Front cover: A water jet from a pumping test in the desert of Abu Dhabi. Water and desert illustrate the topic of the conference "Hydrogeology of Arid Environments". Photo by Georg Koziorowski.
Back cover: Regional distribution of deserts and savannas on earth (modified from Köppen-Geiger).

ISBN 978-3-443-01070-6
Information on this title: www.borntraeger-cramer.com/9783443010706

Publisher: Gebr. Borntraeger Verlagsbuchhandlung
Johannesstr. 3A, 70176 Stuttgart, Germany
mail@borntraeger-cramer.de, www.borntraeger-cramer.de
♾ Printed on permanent paper conforming to ISO 9706-1994

Layout: DTP + TEXT Eva Burri, Stuttgart
Printed in Germany by Gulde Druck, Tübingen

Organizing Commitee

Fachsektion Hydrogeologie der Deutschen Gesellschaft für Geowissenschaften – Working Group "Hydrogeology of Arid Environments" (FH-DGG)

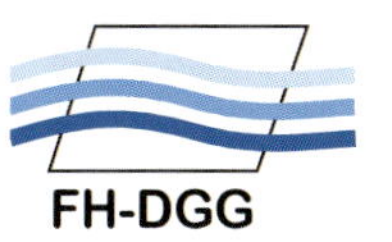

Federal Institute for Geosciences and Natural Resources (BGR)

Technische Universität Darmstadt – Institute of Applied Geosciences (IAG)

Advisory Commitee

Klaus Mayer

Klaus Schelkes

Anke Steinel

Leonard Stöckl

Markus Zaepke

BGR, Hannover

Supporters

German Water Partnership e.V.

Helmholtz Centre for Environmental Research (UFZ)

International Association of Hydrogeologists (IAH)

Ministry of Water & Electricity, Kingdom of Saudi Arabia (MoWE)

National Technical University of Athens (NTUA)

Sponsors

Abu Degen Drilling Company Ltd. (ADC)

Büro für Geohydrologie und Umweltinformationssysteme
Dr. Brehm & Grünz GbR – Diplom Geologen (BGU)

Dornier Consulting (DCo)

Eijkelkamp Agrisearch Equipment

Deutsche Gesellschaft für Internationale Zusammenarbeit (GIZ) GmbH

GWE pumpenboese GmbH

Hajjan Drilling Est. Water Well Drilling Contractor

Hydrosolutions GmbH

Ingenieurgesellschaft Prof. Kobus und Partner GmbH (kup)

OTT Hydromet GmbH

Schlumberger Water Services (SWS)

Seba Hydrometrie GmbH

United Nations Educational, Scientific and Cultural Organization –
International Hydrological Programme (UNESCO)

Preface

Most of the pressing future challenges of our society are related to geosciences. In particular, the limited availability of geo resources, such as water, soils, fossil and renewable energies, or raw materials is of overriding concern. This has been clearly emphasized by the 'geo-commission' of the German Research Foundation in their recently published strategic paper on future challenges in the geosciences.

Water is an essential resource. It is forecasted that the global water consumption will double by the year 2025, in comparison to the 60ies of the last century. However, especially in arid areas, that make up about 30% of the earth's land mass, water resources are dramatically overexploited. Considering this it is obvious that in arid regions the sustainable use of water resources is difficult or even unrealistic. We have to face the fact that water, in the same way as oil, is withdrawn from non renewable supplies, i.e. consumed.

Smart and efficient concepts are therefore needed to manage these resources to the greatest possible benefit for human society. This requires a fundamental understanding of hydrodynamic processes, of the local and regional hydrogeology, as well as a precise quantification of the available resources and water cycles. A sound management can be based only on reliable data rather than on broad estimates.

For scientists, especially hydrogeologists, it is a great challenge and at the same time a responsibility and chance. We therefore initiated the working group 'Hydrogeology of Arid Environments' within the framework of the hydrogeology chapter of the German Society for Geosciences (FH-DGG). It is the aim of this initiative to focusing the expertise on this topic that exists in Germany and in other countries all over the world.

The current conference was organized in order to understand the dynamics of groundwater resources under changing climatic and hydrological conditions as well as to address the challenges in groundwater use. The conference will serve as a platform for the exchange of experiences, ideas and concepts on the hydrogeology of arid environments. It focuses on the assessment of the water cycle and its interactions, the assessment of groundwater resources, aspects of water resources engineering and management, as well as on economic factors.

Riyadh, Darmstadt and Hannover, March 2012

Randolf Rausch
Christoph Schüth
Thomas Himmelsbach

Contents

Poster Abstracts

Proceedings "Hydrogeology of Arid Environments", p. 1–6
Stuttgart, March 2012

Key Note

Integrated Groundwater Management in the Kingdom of Saudi Arabia

Mohammed Al Saud[1], Randolf Rausch[2]

[1] Ministry of Water & Electricity, Saud Mall Center, Riyadh 11233, Kingdom of Saudi Arabia, malsaud@mowe.gov.sa.
[2] GIZ IS, P.O. Box 2730, Riyadh 11461, Kingdom of Saudi Arabia, Randolf.Rausch@gizdco.com.

Key words: integrated groundwater management, assessment of water resources, groundwater budget, groundwater in storage, water strategy, Saudi Arabia

Introduction

The Kingdom of Saudi Arabia is known as one of the most water-scarce countries in the world. Meeting the growing water demand and managing the limited water resources has become a major challenge. The successful economic progress and population growth within the last years have triggered a rapid development of the water sector. That means, that the existing water resources should be conserved and their utilization rationalized.

Water Consumption

The water resources of Saudi Arabia are made up of non-renewable groundwater, renewable groundwater, surface water, desalinated seawater, and treated wastewater. The availability of these resources varies throughout the country.

The total water consumption for 2010 was about 17.4 BCM/a (552 m^3/s). About 14.4 BCM/a (457 m^3/s) were taken from non-renewable groundwater resources, 1.8 BCM/a (57 m^3/s) from renewable water resources, and 1.2 BCM/a (38 m^3/s) from desalinated seawater. About 83% of the water is used for agriculture, 12% for municipal use, 4% for industrial use, and 1% for use by the oil and mining industry.

Geological and Hydrogeological Settings

The Kingdom of Saudi Arabia is made up of two distinct geologic regions. The oldest is the Arabian Shield, on the western side of the Arabian Peninsula. The youngest is the Arabian Platform or Shelf, located in the middle and eastern side of the peninsula (Figure 1).

Arabian Shield: It comprises of igneous and metamorphic rocks of Precambrian age and volcanic rocks (harrats) of Tertiary and younger age. In the valleys alluvial wadi fillings exist. On the Arabian Shield only secondary aquifers can be found. The igneous and metamorphic bedrocks have in general a low yield. Moderate yields can be found in the harrats and in the alluvial wadi fillings. However, because of the limited extent of these aquifers they are only of local importance.

Arabian Platform: The geology is characterized by a thick sedimentary succession which is lying on the basement rocks. The sediments range from Cambrian age to recent. They consist of carbonates, sulphates, shales, marls, and sandstones. Deposition occurred during several transgressive-regressive cycles. In the lower part of the succession clastic sediments like sandstones dominate while in the upper part limestones, dolomites and sulphates prevail.

The sediments form a big aquifer system with several principal aquifers which are imperfectly hydraulically connected to each other. Water-bearing sandstone and limestone layers (aquifers) alternate with low permeability shale and sulphates layers. In general the aquifers have a high yield and a good water quality. The aquifers are of regional importance. Most of the groundwater stored in these aquifers is non-renewable. The aquifers were recharged many thousands years ago, during wet climatic periods. Recent groundwater recharge is low.

www.borntraeger-cramer.de
2012/0001/$ 1.50

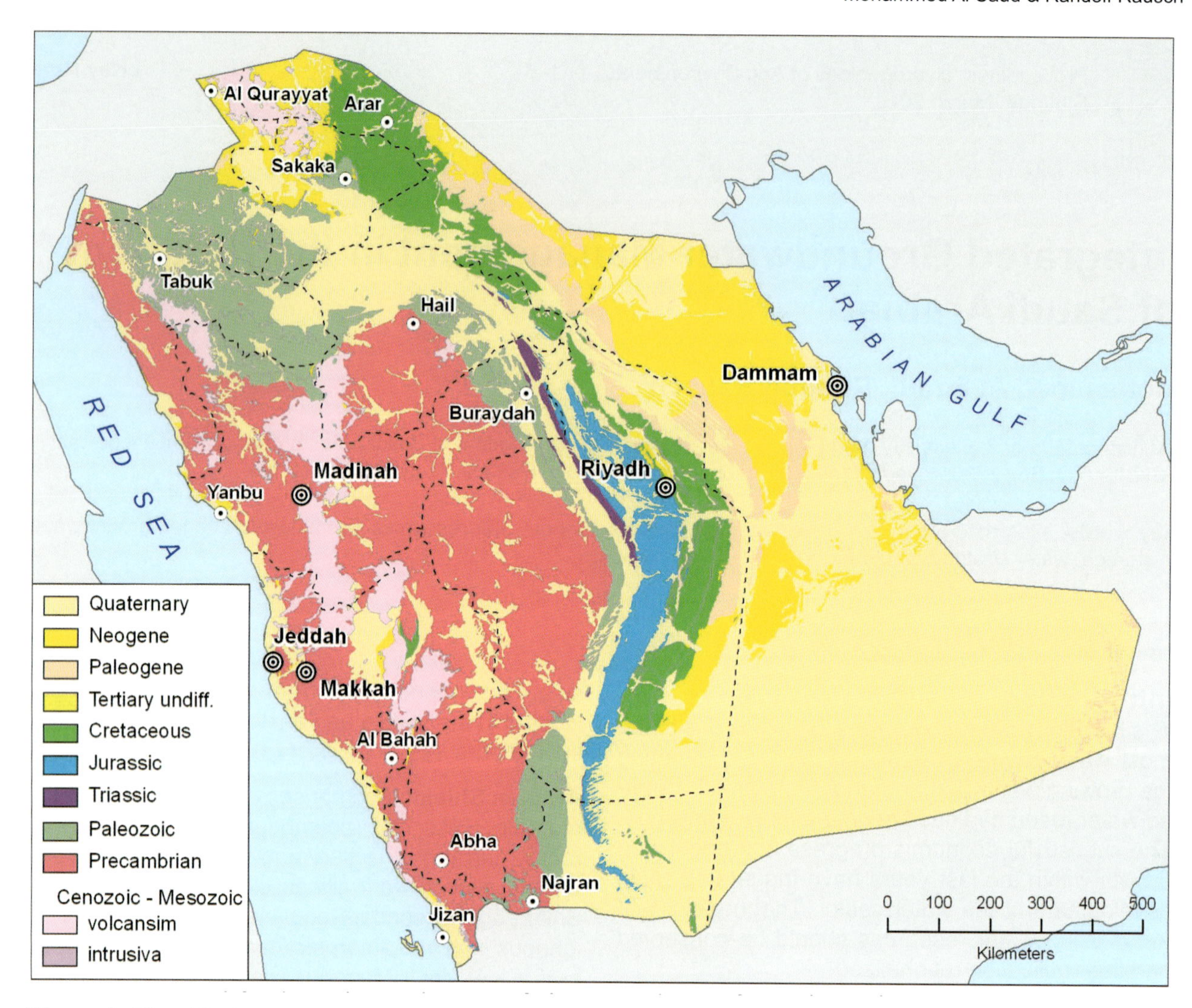

Figure 1: Simplified geological map of the Kingdom of Saudi Arabia.

Water Provinces in the Kingdom of Saudi Arabia

Depending on the geological, morphological, and climatological settings the Kingdom of Saudi Arabia can be divided into five major water provinces (Figure 2):

Red Sea Coast: The water resources of this region are made up by renewable water resources and desalinated seawater.

Northern Arabian Shield: Renewable water resources are the only water resource in this region. However, the amount of renewable resources is low compared to the Southern Shield, because the rainfall is lower.

Southern Arabian Shield: Significant amounts of rainfall (up to 500 mm/a) enable a water supply that is based solely on renewable water resources, if a good water management is implemented. This is the only region in Saudi Arabia, where a sustainable use of the water resources is possible.

Arabian Platform: In this region, only non-renewable groundwater resources exist. Some remote supply from desalinated seawater from the East Coast takes place.

East Coast: Desalinated seawater and, to much lesser extent, non-renewable groundwater resources make up the water resources of this region.

Assessment of Water Resources

In the seventies and eighties of the last century a first countrywide assessment of the water resources was carried out, which was the basis for the National Water Master Plan of the Kingdom. With grow-

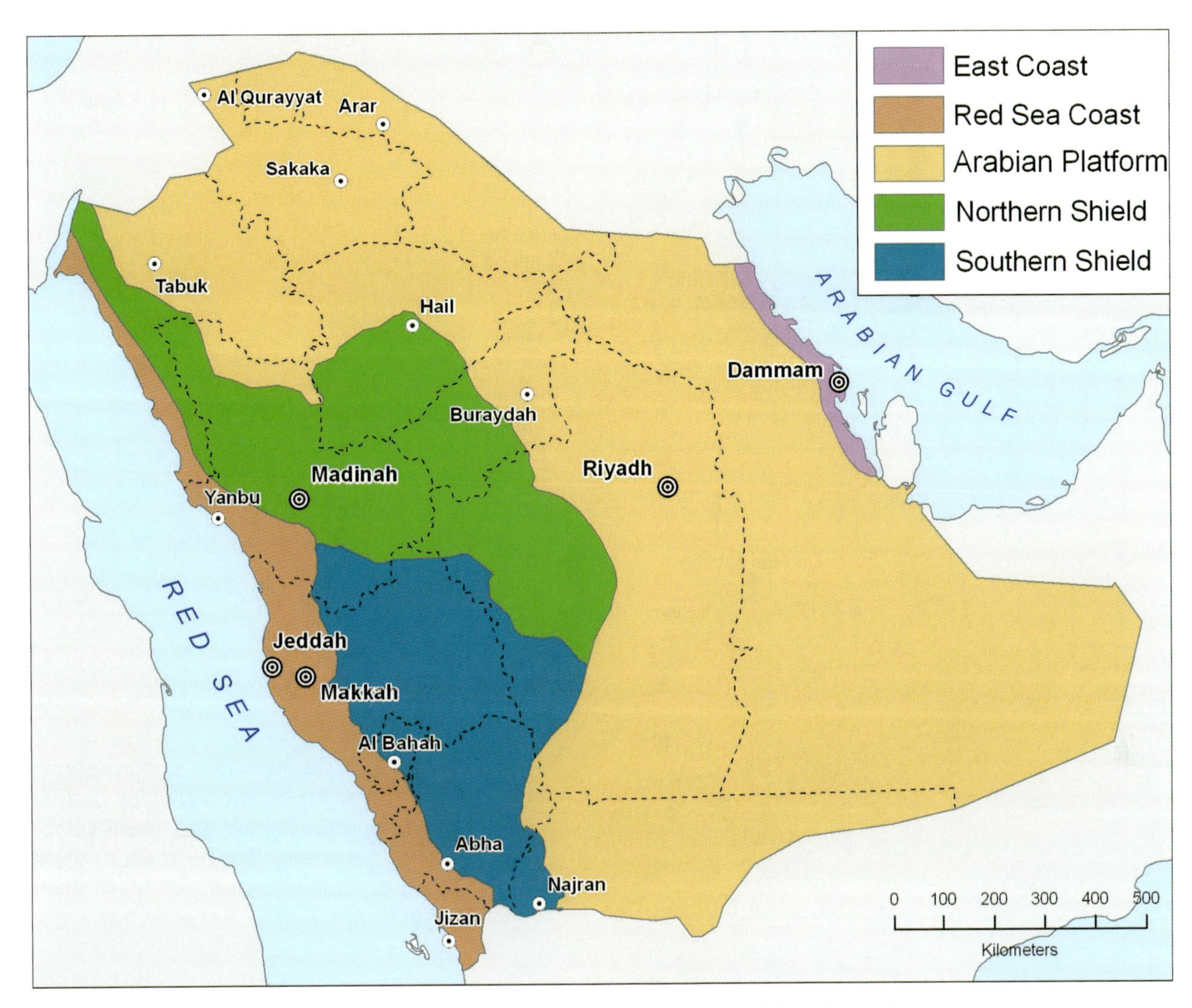

Figure 2: Map showing the major water provinces in the Kingdom of Saudi Arabia.

ing water demand and knowledge, a reassessment was necessary. Therefore, in 2003 the Ministry of Water & Electricity launched a project to investigate all aquifers in the Kingdom. Since then several studies were carried out or are in progress. The main objectives of these studies are:

- Assessment of groundwater budget. The question is what are the in- and outflows to the aquifer systems?
- Assessment of groundwater resources. This looks at the question of how much groundwater is – or remains – available.
- Managing of groundwater resources. How can we make best use of the groundwater resource? Is sustainable groundwater management possible?

To answer these questions, robust and reliable data are needed, which can only achieved by applying the latest technologies in groundwater sciences. Therefore, research cooperation between the Helmholtz Centre for Environmental Research (UFZ), Technische Universität Darmstadt (TUD), German International Cooperation (GIZ IS), Dornier Consulting (DCo) and the Ministry of Water & Electricity was founded. The research topics focus on estimation of groundwater recharge from precipitation, large scale groundwater modeling, concepts for smart groundwater mining, and on groundwater quality.

Groundwater Budget: For the assessment of the groundwater budget in arid areas two states must be considered: the predevelopment and the present state, where the predevelopment state is defined by a site which is in its natural condition prior to any major human activity.

For each state the identification and the quantification of the water budget components are needed. An example is given in Figure 3. It shows a schematic sketch of the groundwater budget components

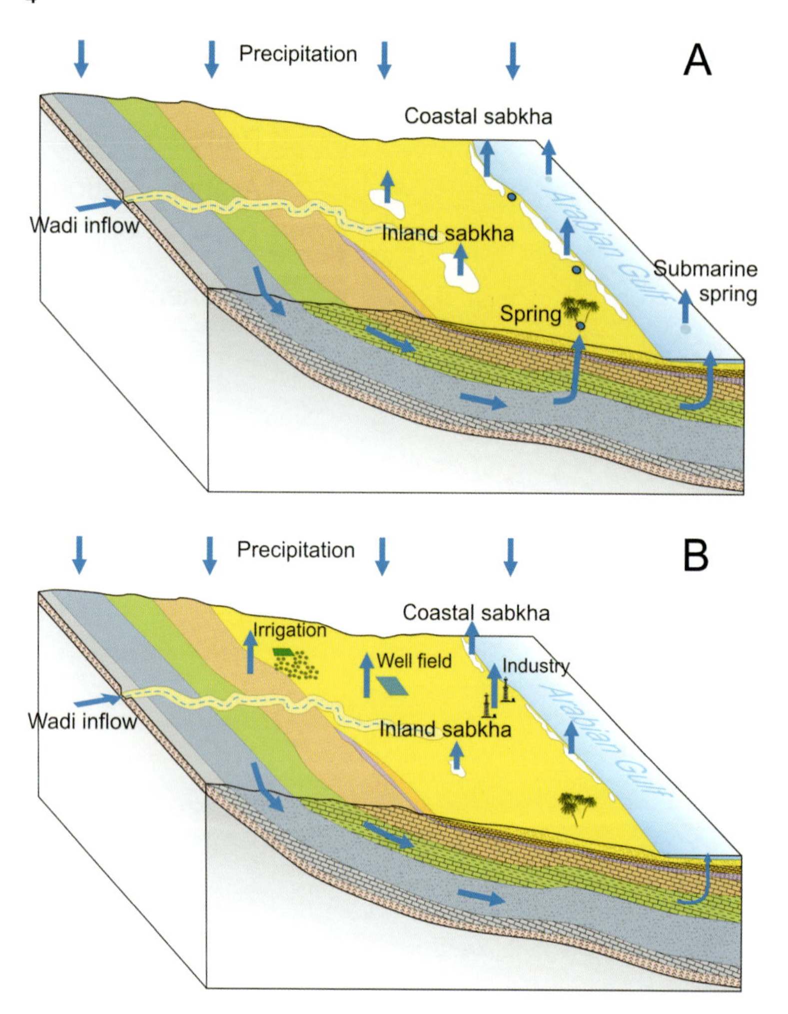

Figure 3: Schematic sketch showing the different groundwater budget components for (A) the predevelopment state and (B) the present state for the aquifer system on the Arabian Platform.

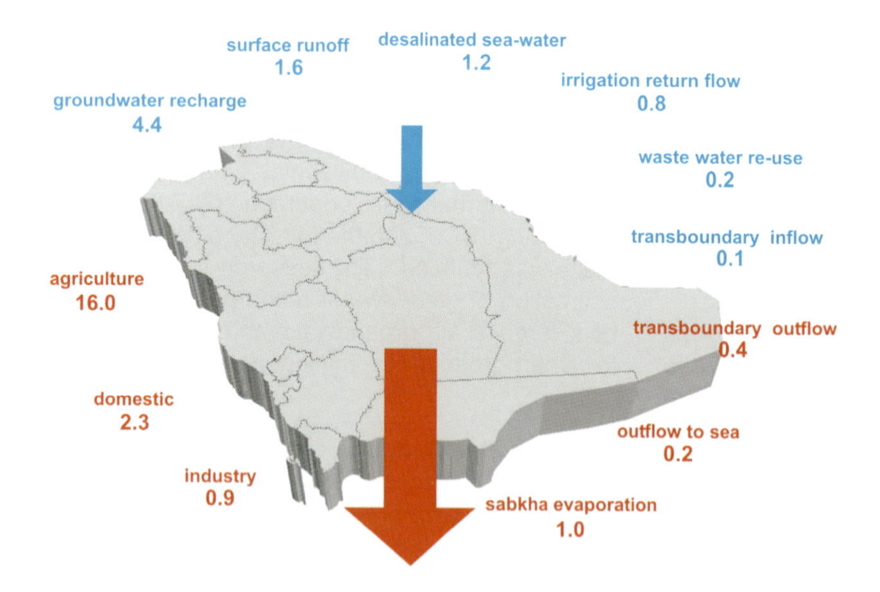

Figure 4: Water budget for the Kingdom of Saudi Arabia for the year 2009. BCM/a is the abbreviation for Billion Cubic Meters per year.

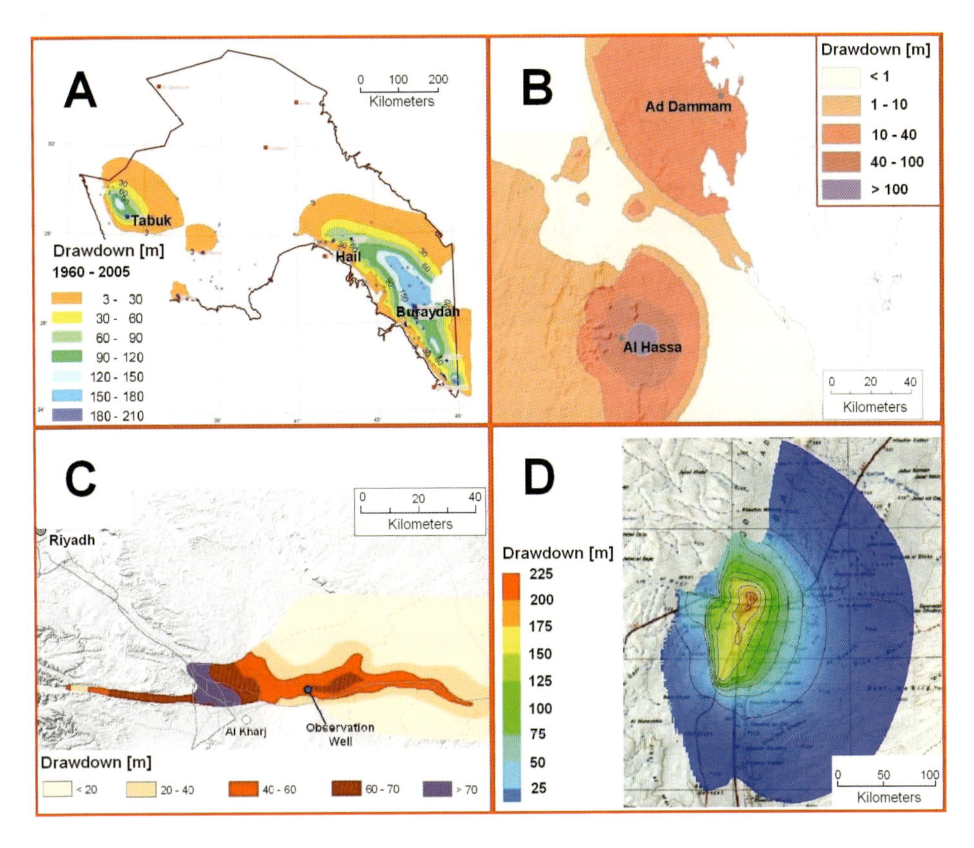

Figure 5: Groundwater drawdown and cones of depression from agricultural abstraction in the Kingdom of Saudi Arabia. A) Saq aquifer near Tabuk, Hail, Buraydah. B) Umm Er Radhuma aquifer near Al Hassa. C) Wasia-Biyadh aquifer near Kharj. D) Wajid aquifer near Wadi Ad Dawasir.

for the aquifer system on the Arabian Platform. The inflow components are groundwater recharge from precipitation and inflow through wadi channels. The outflow components are natural discharge to springs, sabkhas, and to the Arabian Gulf and for the present state additionally well abstraction for agricultural, industrial, and domestic water use.

The quantification of the water budget showed that for the predevelopment state natural inflow equals about natural outflow. Natural depletion is small. That means that the system during these times was in equilibrium and the groundwater use was sustainable.

For the present state the natural inflow is more or less the same like in the predevelopment state but the outflow increased dramatically by groundwater abstractions. Today, the groundwater system is no longer in equilibrium, the groundwater use is non sustainable and groundwater mining takes place. This situation is presented in Figure 4 which shows the total water budget for the Kingdom of Saudi Arabia for the year 2009. The inflow is 8.3 BCM/a and the outflow is 20.8 BCM/a. The biggest part is by agriculture which takes 16.0 BCM/a for irrigation. It entails that 12.5 BCM/a are taken from storage. The consequences of this high groundwater abstraction are:

- Declining groundwater levels / large groundwater drawdown
- Deterioration of groundwater quality
- Land subsidence
- Increase in production costs
- Conflict between users: agriculture, industry, domestic water use and current generation, future generation
- Destruction of the environment - loss in quality of life

Some examples for large groundwater drawdown are presented in Figure 5 for four different locations in the Kingdom. The locations are characterized by big groundwater abstractions for agricultural use.

Groundwater in Storage: The assessment of the total water volume in storage is academic because only a part of this water can be abstracted considering technical and financial criteria. Therefore, the assessment of groundwater resources in storage is calculated by taking into consideration exploitable criteria like drilling depth (< 2,000 m), pumping height (< 300 m), and salinity (< 2,000 mg/l). The aquifer studies are still ongoing. Preliminary estimations of total exploitable water volume show that there are big non-renewable resources. However, the spatial distribution is irregular over the country. The biggest water resources are located in the eastern part of the Kingdom on the Arabian Platform. In many cases agricultural centers coincide with areas of high population which leads to a conflict between agricultural and municipal use. In the future these resources will be conserved only for the domestic water supply of the cities and towns.

Groundwater Strategy

The results and data of the studies are the basis for a smart and efficient management of the groundwater resources and are central part of the future water strategy of the Kingdom of Saudi Arabia. The main result of the studies carried out so far is that the biggest part of groundwater resources in the Kingdom is non renewable and finite. That means it must be treated as a natural resource, like oil or minerals, and can be used to support economic growth. The consumption of these resources for agricultural mass production is uneconomically. These resources must be conserved for the municipal use. To achieve theses goals steps are taken by the Kingdom. The most important indicators for change in water strategy are:

- 2003: Establishment of the Ministry of Water & Electricity (MoWE). Separation of the agricultural sector from the water sector.
- 2003: Launch of reassessment of the water resources of the Kingdom. Studies will be finished until 2014.
- 2008: Establishment of National Water Company (NWC).
- 2008: Phasing out wheat production until 2016 (Royal Decree 335).
- 2009: Phasing out the production of water intensive crops.
- 2010: Introduction of restricted areas for domestic groundwater use.
- 2010: Intensive development of groundwater resources for municipal water supply.
- 2010: Preparation of legal framework.

As conclusion it can be stated that the water sector in the Kingdom of Saudi Arabia is on the right track. The implementation of Integrated Water Management leads to an optimized management for the benefit of the people in the Kingdom of Saudi Arabia.

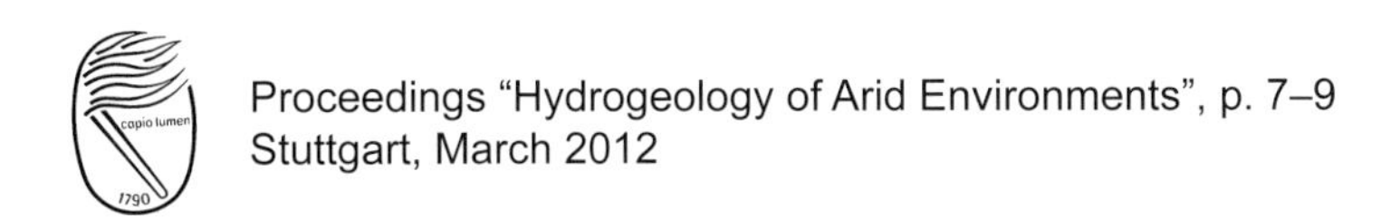

Proceedings “Hydrogeology of Arid Environments”, p. 7–9
Stuttgart, March 2012

Key Note

Agricultural Irrigation – A Critical Nexus for Groundwater Resources in More Arid Climates

Stephen Foster

Global Water Partnership (GWP)-Senior Adviser, GWMATEfoster@aol.com & International Association of Hydrogeologists (IAH), Immediate Past President, c/o PO Box 4130, Goring RG8-6BJ, United Kingdom,
email: IAHfoster@aol.com

Key words: groundwater resources, arid-zone hydrogeology, conjunctive use, groundwater irrigation, agricultural water-use, irrigation efficency

General Background

The last 20–40 years have witnessed a global boom in groundwater use for irrigation in the world's more arid and/or drought-prone areas (except as yet in Sub-Saharan Africa). Today irrigated agriculture is by far the largest abstractor and consumer of groundwater – with some 38% of all cultivated land under irrigation being waterwell equipped and global consumptive groundwater use put at 545 km^3/a (2010 data). Large groundwater-dependent agro-economies already exist in South & East Asia and there are many examples of groundwater-based commercial agriculture in Latin America, the Middle East & North Africa.

From the outset it must be emphasised that access to groundwater has contributed greatly to increasing food security – principally by ensuring water availability at critical times in the crop-growth cycle and by mitigating the devastating effects of surface-water drought on crop yields. Groundwater is thus a ‘popular commodity’ with most farmers. But associated resource depletion, often accompanied by counterproductive competition between irrigation users and conflicts with drinking-water provision, and sometimes by incipient aquifer salinisation (through various mechanisms) and degradation of important groundwater-dependent aquatic ecosystems, is occurring rather widely.

Much more attention to groundwater resource management is thus needed. And whilst It is recognised that the ‘institutional landscape’ and ‘policy framework for water, food and energy’ (involving stakeholder participation, regulatory capacity, economic instruments and macro-economic provisions) are critical for the implementation of management measures, the focus here will be more upon the conceptual understanding and practical interventions required to promote more sustainable groundwater irrigation.

Scope & Approach of Paper

The paper will introduce numerous critical issues for sustainable groundwater irrigation use. It will be illustrated by actual case examples, taken mainly from GW-MATE operations for World Bank-financed projects during 2001–2010, whose results are now being disseminated with support from the Global Water Partnership. The areas concerned include the Central Punjab of India, Guantao County on the North China Plain, the Silao-Romita Aquifer of Mexico, the Ica Valley in Peru, the Carrizal Valley of Mendoza-Argentina, the Hard-Rock Aquifers of Peninsular India, and wastewater reuse areas of Lima and Mexico DF.

Some discussion of the concepts of resource ‘sustainability’ and ‘overexploitation’ might be expected here – but is not proposed to get hung-up over semantics! Essentially when trying to promote a more balanced approach to groundwater use in irrigated agriculture (which also values its other roles) public administrations should bear in mind that:

- undesirable side-effects from resource exploitation can sometime commence well before groundwater abstraction exceeds average replenishment – natural susceptibility to irreversible degradation varying considerably with hydrogeological setting

www.borntraeger-cramer.de
2012/0007/$ 0.75

- maintaining groundwater stocks against all depletion is rarely appropriate (given the long periodicity of major arid-zone recharge events), since groundwater storage is important for mitigating drought impacts on surface-water and to provide time for transition to a lower water-use economy.

Intensification of agricultural practice and changes in irrigation management have a major impact on groundwater recharge (not to mention quality) and it will be essential to exert control over irrigation use to achieve resource sustainability. Such is the interrelationship that harmonised irrigation and groundwater accounting with improved field-level metering and monitoring, together with cross-sector dialogue and 'integrated vision', will be required to promote sustainable land and water management. The nexus between agricultural irrigation and groundwater resources is thus especially significant in the more arid environments, with two rather distinct cases being recognised – 'groundwater-only irrigation areas' and 'conjunctive use' (with surface-water or wastewater).

Key Issues in Groundwater – Only Irrigation Areas

In this class of irrigation area under arid climate conditions, it will be essential to focus resource management on controlling groundwater consumptive use, since progressive changes from gravity (flood) irrigation to pressurised (drip) irrigation inevitably result in substantial increases, even where actual abstraction is successfully capped. The most direct approach is to effect reductions in irrigated area – but without parallel action to sustain farmer incomes (eg. by increasing water productivity through improving crop yields or cultivating higher-value crops) such policies can prove difficult to sustain. Other possibilities include:

- community bans on cultivation of highly water-intensive crops (like sugarcane, paddy rice, alfalfa or maize as animal feed, etc) or nationally phasing-out guarantee prices or subsidies for such crops
- statutory control over the date of planting-out paddy rice to reduce non-beneficial evaporation.

Managing the consumptive use of groundwater will also require much better hydrogeological accounting, so as to describe in detail the relationship the soil-water components of irrigated cultivation and the factors on which different recharge components depend. This is vital for the sound quantification of groundwater resources and their integrated management with irrigation use.

Moreover, a number of other key technical considerations will need to inform the approach to groundwater resources management:

- *Improving in Irrigation Water-Use Efficiency*: can be the key to increasing water-use productivity and reducing unit energy consumption of irrigated agriculture, but such improvements do not necessarily equate to 'real water resource savings' because a substantial proportion of the so-called 'losses' associated with 'inefficient irrigation' are in fact returns to groundwater
- *Relative Potential of Demand-Side versus Supply-Side Measures*: whilst 'managed aquifer recharge' should be encouraged, it is not usually the solution to groundwater resource imbalance unless pursued as a complement to demand-side management measures
- *Harsh Reality of Weakly-Recharged Aquifers*: must be recognised where average rainfall is low (<500 mm/a) with little diffuse recharge to shallow aquifers and more generally for deeper confined aquifers due to physical isolation from the land surface – in such cases groundwater-irrigated agriculture may have developed despite limited contemporary recharge or even 'non-renewable groundwater resources' and there is need for public administrations and private irrigators to come to terms with its implications.

Improving Conjunctive Resource Use for Irrigation

The spontaneous drilling of waterwells by farmers, in and around major irrigation-canal commands on extensive alluvial plains, has occurred widely as a coping strategy in face of inadequate canal-water service levels, for a variety of reasons. In effect conjunctive use of groundwater and surface water is capable of achieving, with varying degrees of effectiveness:

- greater water-supply reliability and security – by using natural aquifer storage to 'buffer' the temporal and spatial variability of canal-water
- reduced environmental impact – by counteracting land waterlogging and soil salinisation.

Moreover, groundwater use in such areas is often characterised by higher water productivity (kg crop or US$ profit per ha/m3), despite (or perhaps because of) the fact that the unit cost of this water-supply to the user is much higher.

Spontaneous (unplanned and unmanaged) groundwater resource use sometimes results in aquifer depletion to water-table levels that complicate the deployment of low-cost ground-level lift pumps and/or that induce saline groundwater encroachment (although a number of distinct mechanisms can be responsible for the latter and careful hydrogeological diagnosis is required). But If conjunctive use can be better planned, it offers a major opportunity of increasing agricultural production (through improvements in cropping intensity and water productivity), usually without compromising groundwater sustainability, and is a preferred strategy for climate-change adaptation – although serious institutional impediments often have to be overcome.

The technical assessment of conjunctive resource use for irrigated agriculture, (especially in alluvial environments) requires sound characterisation of groundwater-surface water connectivity, so as to avoid the ever-present risks of 'double-resource accounting' and groundwater salinisation. Decisions on lining of primary/secondary irrigation-canals will often be complex but critical since:

- it will be a high priority if the phreatic aquifer is naturally saline (with fresh groundwater confined at depth), since canal seepage will then be a 'non-recoverable loss' causing rising water-table and possible soil salinisation
- in sharp contrast although unlined secondary canals on highly-permeable alluvial terraces in arid climate often carry water for relatively few weeks per year (and the majority of irrigation users depend entirely on waterwells) seepage from these canals is often responsible for major aquifer recharge.

Integrated numerical modelling of irrigation canal flows, groundwater use and aquifer response, soil-water status and crop water-use is a great aid to evaluating the potential benefits of varying the spatial and temporal use of groundwater and distribution of surface water, and thus of improving conjunctive use efficiency and sustainability.

Future Vision

Increasing farmer incomes from smaller irrigated areas is an attractive option in the quest for groundwater resource sustainability in more arid climatic areas – and the rising demand for 'precision irrigation' with pressurised systems offers an adaptable platform for conversion to the intensive cultivation of higher-value crops. But whether this trend follows a 'sustainable path' will depend on the detail of irrigation-water management and whether 'real water-resource savings' are pursued and groundwater use rights or allocations are capped in consumptive use terms.

There will, however, be inevitable market-related and risk-defined limits on the scope for conversion to high-value cropping, and the production of staple-crops (wheat, maize, rice, etc) will remain a very important component of groundwater irrigation in most developing nations. In most cases there exists great need to increase crop yields through improving soil management, seed-density and type, fertilizer and pesticide use to eliminate nutrient constraints or pest impacts on crop growth – but without adequate management this could impact on groundwater recharge (quantity and quality) through increasing unit rates of both consumptive use and nutrient or pesticide leaching. The increased cultivation of biofuels could represent an added complication.

General References

Foster, S., Steenbergen, F. van, Zuleta, J. & Garduño, H., 2010: Conjunctive use of groundwater and surface water–from spontaneous coping strategy to adaptive resource management. – GW·MATE Strategic Overview Series 2. World Bank (Washington DC). http://www.worldbank.org/gwmate.

Garduño, H. & Foster, S., 2010: Sustainable groundwater irrigation – approaches to reconciling demand with resources. GW·MATE Strategic Overview Series 6. World Bank (Washington DC) http://www.worldbank.org/gwmate.

Siebert, S., Burke, J., Faures, J. M., Frenken, K., Hoogeveen, J., Doell, P. & Portman, F. T., 2010: Groundwater use for irrigation – a global inventory. – Hydrology & Earth System Science **14**: 1863–1880.

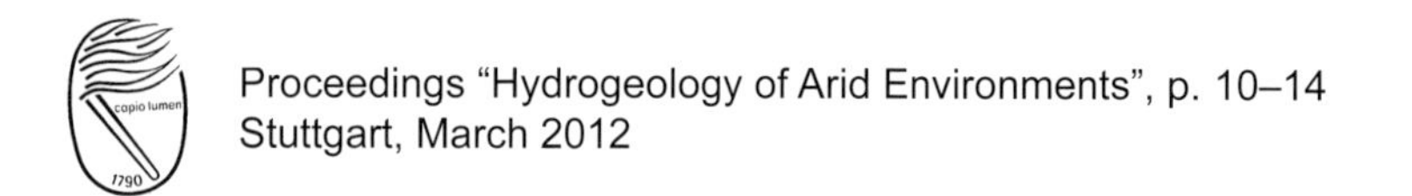

Proceedings "Hydrogeology of Arid Environments", p. 10–14
Stuttgart, March 2012

Key Note

Climatic Caused Variations of Groundwater Recharge in the Middle East and its Consequences for the Future Water Management

Heinz Hötzl

Karlsruhe Institute of Technology (KIT), Adenauerring 20 b, 76128 Karlsruhe, Germany, heinz.hoetzl@kit.edu

Key words: arid recharge conditions, Quaternary climate changes, not-renewable groundwater, water management

Introduction

With regard to the millennium goals of the United Nations to supply sufficient high quality water to the human population arid and semiarid countries are generally from their natural climatic conditions amongst the last in the row with regard to the amount of the yearly available water per capita. Under the consideration of the defined climatic condition this doesn't amaze, surprising is more the fact that in spite of such conditions a growing proportion of the mankind is forced to live under such restriction due to the worldwide excess of population. Anyhow, it is the task of governments and authorities to supply also for these peoples an adequate amount of water by responsible water management to enable the basic requirements of life.

Referring to this task it is worth to remind oneself the limiting factors of arid and semiarid regions. Arid areas receive little or no rain and they are too dry or barren to support vegetation. In general the yearly rate of precipitation is less than 250 mm, in wide areas less than 100 mm. Semiarid areas rank between desert and humid climates with regard to ecological characteristics and agricultural potential. The yearly rate of precipitation is in general less than 500mm with dry frequently hot summers and short wetter winters. The wet period refers to a period during which rainfall is higher than evapotranspiration and is usually distributed over a 3-4 month period in a year. The vegetative cover is characterized mainly by shrubs, scrubs and grass.

The distribution of the main arid areas on the globe depends strongly on the atmospheric circulation systems. For the wide extended dry belts of the sub-tropics accounts mainly the tropical Headley cells. This circulation systems are created in the inner tropical convection zone (ITCZ), where the wet lower level air rises after cooling and rainfall moves from the equatorial zone both northward and southward in the sub-tropical region, warming by compression and forming the sub-tropical high-pressure belt between 20 and 40 N and S respectively.

On the northern hemisphere this dry subtropical zone includes with the Sahara the largest desert area with about 13.000 km². To the east this arid belt extends directly to the deserts of the Middle East with the Arabian Peninsula and their northern extension in Jordan, Syria and Iraq on which the focus of this report will be laid. Further to the east follow the deserts of Iran and Pakistan, all together they form the largest connected and closed arid area on the world.

In a certain way it seems strange that just the arid Middle East has undertaken in the last fifty years an extreme economical and technical development with a tremendous population increase due to the oil exploration. However this has only intensified the problems of water scarcity. Even though the income from oil selling helps to introduce and implement new technologies like seawater desalinization it remains still an enormous challenge to cover the increasing demand of high quality water. The fast increase of the population in the last decades together with the increased demand for urban water, water for food production and industrial water has led already to a partially complete exhaustion of the natural resources.

www.borntraeger-cramer.de
2012/0010/$ 1.25

Groundwater Recharge Conditions and Assessment Methods

As mentioned before surface water resources are generally scarce and highly unreliable in semiarid and arid regions, with the result that groundwater is the primary source of water in these regions. For the management of this resource the groundwater recharge rate has proved as one of the most important parameter in order to assess the amount of available and renewable water in a sustainable way. Unfortunately it needs rather extensive and combined methods to attain significant and reliable values. This follows on one hand from the partially strong spatial and time variation and on the other hand from manifold influencing factors. The most important are variable precipitation and evapotranspiration, topographic conditions like slope angels or surface discharge channels, rock types and soil depth, the thickness of unsaturated zone with variable properties as well as vegetation cover and land use.

Recharge studies has been carried out in arid and semiarid regions using a variety of techniques, including physical, chemical, isotopic, and modeling techniques. These techniques have been described in previous studies. Scanlon et al. (2006) gave a short synopsis of the most quoted publications, starting with Simmers (1988) who edited a volume on recharge that focuses primarily on techniques for estimating recharge in arid and semiarid regions. Papers from a symposium on groundwater recharge that was held in Australia in 1987 are compiled in an edited volume by Sharma (1989). The International Association of Hydrogeologists published a volume on recharge edited by Lerner et al. (1990) that includes descriptions of recharge in different hydrogeologic settings. A compilation of papers by IAEA (2001) providing valuable information on recharge in water-scarce areas. A special volume of the Hydrogeology Journal devoted to recharge includes papers on recharge processes and methods of estimating recharge (Scanlon & Cook 2002). Recharge issues related to the SW United States are described in Hogan et al. (2004).

Regional recharge estimation for water resources evaluation has relied mostly on groundwater-based approaches, which integrate over large spatial scales. The most widely used technique for estimating recharge was the chloride mass balance (CMB) approach. Other techniques are chloride and tritium profiles in a selected borehole, the penetration depth of bomb tritium and bomb chlorine-36, artificial tritium injection, different tracer approaches (tritium mass balance and peak penetration) as well as analytical model that takes into account the water-table rise and numerical models based on water and energy balance processes.

The results of the different studies show partly extreme variations especially under point or local aspects. To show one example with only varying soil conditions results are given from a study in the

Figure 1: Diffuse Groundwater recharge in the background and preferential flow in a karst shaft after a rain event in the desert NE of Riyadh, Saudi Arabia.

Ad Dhana Desert in Central Saudi Arabia east of Riyadh (Hötzl et al. 1990). The yearly average precipitation range varies there between 70 and 100 mm. The example refers to one precipitation event of about 10 minutes and total rate of 15 mm. Comparable measurements were carried out in adjacent areas, where locations with sand-dunes over limestones and on other hand small karstified catchments with barren rock surface (UER-Formation) and sinkholes were observed (Fig. 1). While the sand-dunes only show a wedding of the uppermost 35 cm (below they remained completely dry) from the small karstified catchments between 40–70 % of the total precipitation were discharging and percolating via the shafts and sinkholes in the deep underground. With these studies could be documented that in spite of the extreme arid condition, the areas with the outcropping karstified surface still contribute to the recharge of the UER-Aquifer, which supplies huge springs along the Gulf coast. In contrast the sand dune covered desert delivers only negligible amounts of recharge as was documented by a long-term analysis of bomb tritium in the sand column with about 2% of the yearly precipitation rate (Dincer et al. 1974).

This observation is confirmed by other regional studies in arid areas showing that rainfall below 200 mm usually results in negligible recharge. In Table 1 some results of regional studies from different regions are compiled showing the similar trends. In general the recharge rates encompass 1 to 10 mm/year under mainly piston flow conditions without dominant preferential flow. With regard to the yearly precipitation rate this amounts between 0.5 –5 %. This emphasis the problems of larger exploration rates, even under the availability of extended recharge area for the aquifers a sustainable management of these resources needs the observance of the restricted conditions.

Recharge Conditions Under Changing Climate in the Past

Under the widespread overexploitation of the groundwater resources, which has been started about fifty years ago, and the dry climate one may be surprised about the large amount of groundwater still stored in some of the aquifers. Though it could be proofed that even under the recent arid condition still natural recharge is taking place, the region live basically on replenishment of the aquifers during the more wet periods in the past. Like in the Sahara basin the Middle East received several times stronger precipitation due to climatic changes during the Quaternary, which let to temporarily semiarid conditions even the core deserts which are now under hyper-arid conditions.

Climatic changes in the recent desert areas during the Quaternary period were recognized in early geologic studies from morphologic features and the occurrence of young fluviatile and limnic sediments, last but not least also from the rock drawings of the early men. The comparison with the climate change in the northern hemisphere led at first to an assumption that parallel to the glaciations the cooling of the earth caused pluvial conditions in the latitudes of the sub-tropics. The observation of obviously repeated wetter periods gave reason to compare these with the four main glacial periods. Modern age determinations with radiocarbon method starting from the sixties of the last century as well as comparable studies in ocean sediments

Table 1: Groundwater recharge rates from different arid and semiarid regions (after Scanlon et al. 2006); CMB-Chloride mass balance.

Region	Soil/Rock	Precipitation yearly average in mm	Recharge yearly rate in mm	Method recharge determin.
Central Saudi Arabia	sand dunes	70–100	2	^{3}H
NE-Jordan	desert pavement	100	1–6	water balance
Israel	fractured chalk	200	16–41	^{3}H
Tunisia	sands	100	1–3	CMB
Nigeria	inter-dune grassland	434	15–54	CMB
Botswana	sand savanna	420	1–10	CMB
India	calcrete gravel	240	9–18	CMB
South Australia	sand dune	300	13–14	CMB
USA Death Val	sands	171	1.7	water balance

Figure 2: Irrigation of alfalfa grass in the Wadi Disi with fossil water out of the Disi-Aquifer (Paleozoic sandstones), Southern Jordan. This water will be used for the supply of Amman in the future.

revealed a different picture of the climatic changes in the sub-tropics, cooler and with extreme hyper-arid conditions during the glacial maxima, but with frequent interruption by wetter phases.

The change of climate become noticeable by the shifting of the atmospheric front systems, either by the northward movement of the monsoonal front, bringing the southern and central part of the sub-tropics more rain, or by the southward movement of the polar front causing more rain in the northern part of this region. Intensity, coverage as well as frequency of these changes varied in a wide range. Larger effects are known from the Late Pleistocene as well as from the Early Holocene. Rather smaller effects are known from historic time and even from the last century like the change from drought and wet periods in the Sahel zone.

Impressive witnesses of these wetter periods for example are lake sediments indicating higher water levels like in the Dead Sea or the Lake Chad area or limnic sediments in the now dry basins of Damascus or in the Rub al Khali in the southern part of the Arabian Peninsula. Age determinations of the groundwater in this area indicate that parallel to the surface water rise the aquifers were filled up with fresh water and that the rising groundwater levels were aligned according the new base level of the rivers and lakes.

The storage and preservation of huge water masses needs ample aquifers as well as special stratification and tectonic settings. Fortunately for the Middle East the monocline or basin structure of the mighty sediment sequence on the Arabian plate favors the storage of water in the underground. As examples are mentioned the thick Paleozoic sandstones (Fig. 2) in the direct surrounding of the Arabian shield as well as the mighty Cretaceous and Tertiary carbonate rock sequences in the Arabian shelf platform. Similar conditions can be found in the thick Nubian sandstone sequences of the Sahara basin. The huge groundwater resources there form for wide areas in the Middle East the only direct available natural water resources. Due to their partially or fully non-rechargeable conditions particular careful usage with special management strategies are necessary, which will be stressed in more detail in the report.

Conclusions for Future Climate Changes

With regard to the announced climate changes within the current century one can estimate the possible losses of groundwater according to the knowledge from the past climatic changes. After the rather short phases of wetter conditions, like the Neolithic "pluvial" phase (8.000–6.000 B.P.) in the southern and central subtropics, the rivers and lakes dried up rapidly by evaporation and caused a parallel depletion of the groundwater levels. The large amount of water, still stored in many of the aquifers, are mainly due to the specific hydrogeologic conditions, where aquifers formed closed structures below the base levels. These resources are of course available for utilization, however the non-renewable condition under the recent arid conditions restrict a management under normal sustainable criteria. It is groundwater-mining which will end up with an empty reservoir.

With regard to still renewable aquifers the announced climate change with increased temperatures and reduced precipitation for the Middle East will cause a significant reduction of groundwater recharge. Recent studies in different arid regions have shown that the relationship between the amount of precipitation and recharge doesn't follow a linear function. Due to the increasing share of precipitation stored in the unsaturated upper soil layer with decreasing rain the recharge rate is reduced in an exponential way. For example in the northern more semiarid parts of the Middle East, where still notable national recharge occurs a loss of 10 percent of rain might implicate a reduction of 30–50 percent of the recharge as new calculation have shown. In addition the lowering of the base level during more dry conditions like the depletion of the Dead Sea will drain important parts of the stored water in the aquifers. These implicate a further escalation of the existing water crisis.

References

Dincer, T., Al-Mugrin, A., Zimmermann, U., 1974: Study of the infiltration and recharge through the sand dunes in arid zones with special reference to stable isotopes and thermonuclear tritium. – Journal of Hydrology **23**: 79–109.

Hötzl, H., Wohnlich, S., Zötl, J.G. & Benischke, R., 1993: Verkarstung und Grundwasser im As Summan Plateau (Saudi Arabien). – Steir. Beitr. Hydrogeologie **44**: 5–158, Graz.

Hogan, J.F., Phillips, F.M. & Scanlon, B. R., 2004: Groundwater Recharge in a Desert Environment: The Southwestern United States: Water Science. – Applications Series, Vol. 9. American Geophysical Union: Washington, DC.

IAEA, 2001: Isotope Based Assessment of Groundwater Renewal in Water Scarce Regions, IAEA TecDoc 1246. IAEA, Vienna. 273.

Lerner, D. N., Issar, A. S. & Simmers, I., 1990: Groundwater recharge, a guide to understanding and estimating natural recharge. – International Association of Hydrogeologists, Kenilworth, Rep 8, 345 pp.

Scanlon, B. R. & Cook, P.G., 2002: Theme issue on groundwater recharge. – Hydrogeology Journal **10**,n1, 237p.

Scanlon, B. R., Keese, K. E., Flint, A. L., Flint, L. E., Gaye, C. B., Edmunds, M. W. & Simmers, J. (2006): Global synthesis of groundwater recharge in semiarid and arid regions. – Hydrol. Process. **20**: 3335–3370, Published online in Wiley InterScience.

Sharma, M. L., 1989: Groundwater Recharge. – A.A. Balkema, Rotterdam, 323.

Simmers, I., 1988: Estimation of Natural Groundwater Recharge. – D. Reidel Publishing Co: Boston, MA, 510.

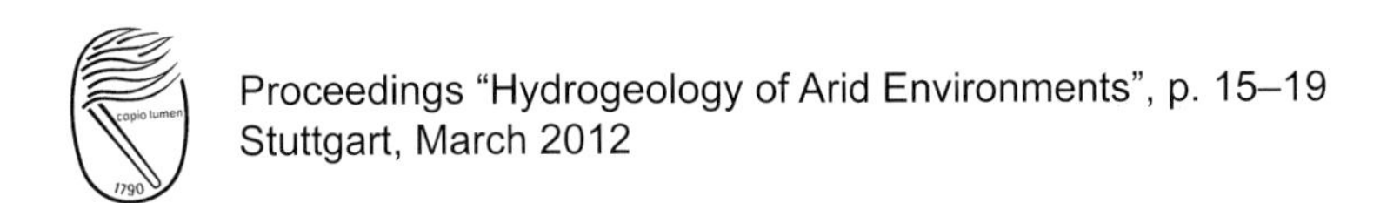
Proceedings "Hydrogeology of Arid Environments", p. 15–19
Stuttgart, March 2012

Key Note

Modelling of the Hydrology and Hydraulics of the Okavango Delta

Wolfgang Kinzelbach

Institute of Environmental Engineering, ETH Zurich, Zurich Switzerland, email: kinzelbach@ifu.baug.ethz.ch

Key words: wetland, salinity, Okavango, aerogeophysics, modelling

Introduction

The Okavango Delta is an inland delta of the Okavango River (Fig. 1). The river originates in the humid highlands of Angola, flows southwards crossing the Namibian Caprivi-Strip and finally enters Botswana in the so called panhandle. It forms an alluvial fan in the graben between Gomare and Thamalakane faults. The slopes on top of the fan are so small that the river spreads out into a delta shape, forming a huge wetland. The wetland is a freshwater anomaly in the otherwise semi arid Kalahari Desert, where high potential evapotranspiration rates cause over 95% of the incoming waters to be lost to the atmosphere together with the local precipitation in the Delta of 450 mm per year. Besides the presence of water as such it is the variability of the hydrological situation over the year, which makes these wetlands unique. The area of the wetland varies over the seasons between the area of the permanent swamps of about 2000 km^2 and the area of the seasonal swamps which can reach more than 10000 km^2 depending on the precipitation in the highlands. The waters of the rainy season arrive at the panhandle in April. But due to the small slope of the delta the flood reaches the distal part at Maun only in July. This implies that high water availability stretches 3 to 4 months into the dry season. The combination of a highly seasonal inflow and local dry and wet seasons result in an ever-changing flooding pattern. A multitude of different environments and ecological niches have developed accordingly. The extraordinary biodiversity found in the wetlands can be primarily attributed to the special hydrological setting. With upstream activities still minor, the Okavango presently remains one of the largest virtually pristine river systems of Africa. The Okavango Delta is included in the Ramsar List of Wetlands of International Importance. A number of future threats to the wetland are apparent. They are linked mainly to the development of the upstream basin and eventually climate change.

Phenomena and Drivers

The wetlands are formed by a multitude of drivers, from the forces of tectonics, the hydrology, the dust and sediment inputs, via the evapotranspiration to the animals like termites, hippos and elephants. The phenomena have been researched and described in detail by the group of Terence (Spike) McCarty at the University of the Witwatersrand, Johannesburg. Here we discuss only two aspects, the dynamics of the flooding and the distribution of salinity. The flooding is the result of the interaction of topography, inflow and precipitation including the coupling between surface water and groundwater. It has been modelled in a distributed deterministic way by (Bauer 2004). The model was developed further by Milzow (2008). The delta has shown to be robust against high variability of hydrological conditions in the past. The recent decades showed prolonged drought conditions which were only relieved in the last 3 years where flooding extent was much larger than in the 20 years before. Still, the main danger for the Delta results from diminishing inflows. These can be due to possible abstractions of water in the upstream or possibly lower inflows due to climate change. Reduction of inflows will reduce the flooded area and thus the vital space for plant and animal life. Changes in flooding frequency are reflected in the vegetation, as only land, which is flooding annually, will remain grassland providing forage for the herds of animals.

www.borntraeger-cramer.de
2012/0015/$ 1.25

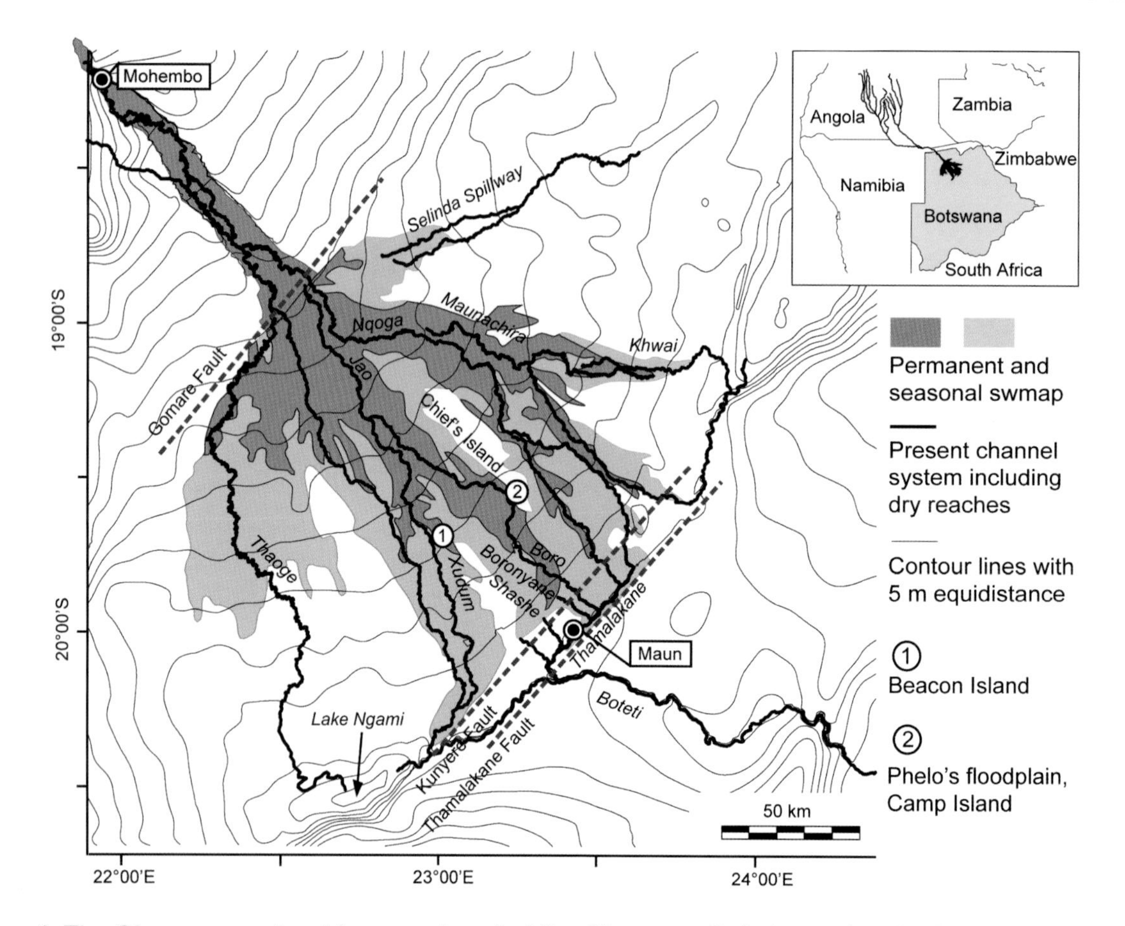

Figure 1: The Okavango wetland (commonly called the Okavango Delta) covering the Panhandle and parts of the Okavango fan. The channel network includes reaches that are not currently active. The wetland section to the northwest of Gomare Fault is referred to as the Panhandle (from Milzow et al. 2009).

Data

The hydrologic-hydraulic model requires data. The best data available are the inflow data measured at Mohembo in the panhandle on a daily basis. A few other stations are available in the delta as well as climatic data at few locations in the larger region. Besides these scarce data resources new sources of data can be accessed via remote sensing. The Famine Early Warning Systems provide close to real time data on the spatial distribution of precipitation and potential evapotranspiration for the whole of Southern Africa on an 8 km grid, based on METEOSAT 5 data. From multispectral satellite data (Aster or Landsat) the actual evaporation can be estimated in its spatial distribution by soil energy balance algorithms. Using the NDVI, the same data provide the distribution of flooded areas, at least for cloud free days. Water areas can also be obtained from ENVISAT SAR data, irrespective of clouds. The shuttle topography mission supplies a digital terrain model of the delta, which, however, is not sufficiently accurate for hydraulic modelling purposes. It has to be corrected with the observed flooding and the correlated vegetation/tree type distributions. Finally aerogeophysical data are a type of remote sensing data which are of high value for our purposes. From high resolution aerogeomagnetic data the thickness of the Kalahari sediment beds above the crystalline rock and thus the aquifer thickness can be estimated (Kgotlhang 2008). Another highly interesting data set is the airborne TEM (Time domain electromagnetic data) financed by the geological survey of Botswana, which allows to analyze the distribution of fresh and saline waters in the delta in three dimensions via the measurement of electrical conductivity (Kgotlhang 2008).

Hydrological scenarios

The hydrologic-hydraulic model allows to compute the development of the flooded area under management scenarios or hydrological conditions different from the today. In Figure 2 the model base run for the last 25 years is compared to a model run

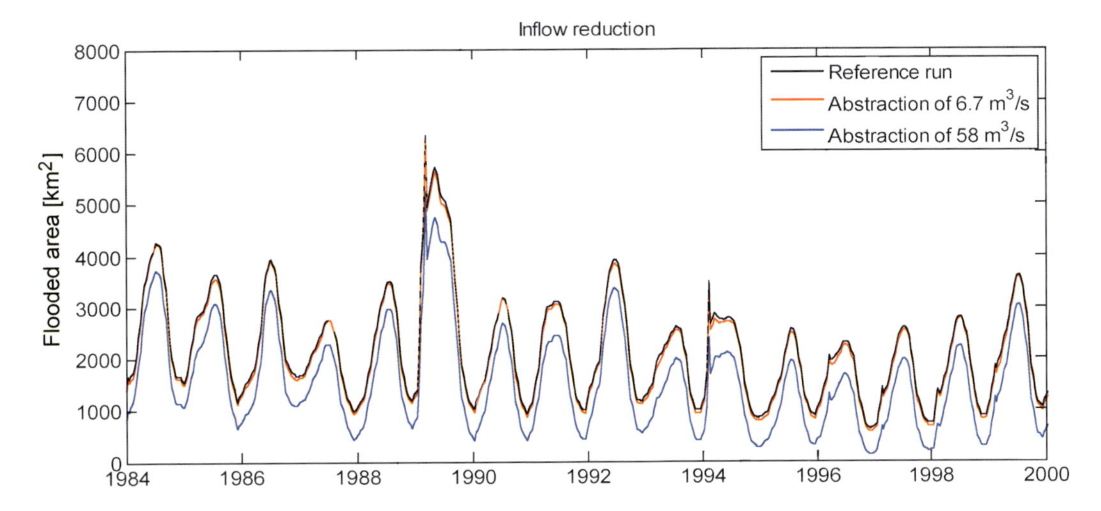

Figure 2: Flooded area of the delta as computed by the model for historical conditions of 1984–2000 without and with abstractions for agriculture (58 m^3/s) and households (6.7 m^3/s).

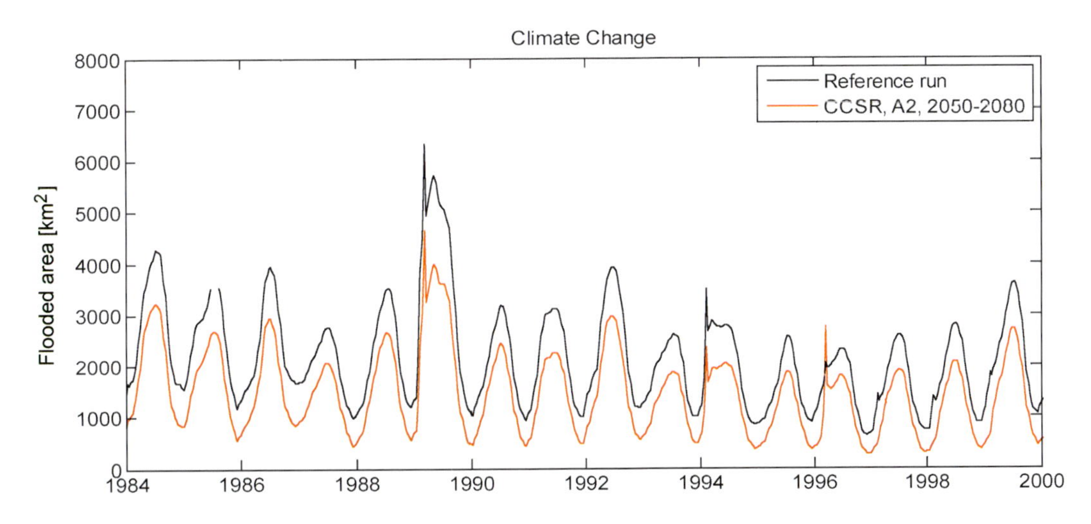

Figure 3: Flooded area of the delta as computed by the model for historical conditions and climate change conditions taken from CCSR model scenario 2050–2080 (Greenhouse gas scenario A2 of the IPCC).

for the same hydrological conditions except for the inflow, which is reduced by different amounts of abstraction. One is the amount of water needed to irrigate the agricultural lands in the basin suited for irrigation. The average flow required for that purpose would be 58 m^3/s, compared to the total average flow of the river of 300 m^3/s. The household consumption of 6.7 m3/s in the basin is virtually negligible vis-à-vis the agricultural demand. With agricultural abstraction the minimum extent of the flooded area would be reduced to less than one half of the minimum extent seen in the last 25 years.

The consequences of agricultural abstractions are comparable to consequences one obtains from a climate change scenario computed by the CCSR global circulation model of Tokyo University for the period 2050–2080, based on the greenhouse emission scenario A2 of the IPCC. (Fig. 3) The climate change predictions are highly uncertain if one compares the results of different global circulation models. Of five models compared one predicts wetter conditions while the others predict drier conditions to a strongly varying degree (Milzow et al. 2010).

Salinity and airborne TEM

In any highly evaporative environment the fate of salts has to be given special attention. The Okavango River is remarkably fresh. The total dissolved solids at Mohembo of 35 mg/l rise only to about

85 mg/l at the outflow from the delta. This sounds counterintuitive given the fact that 95% of all water entering the delta are lost by evaporation and transpiration. It can only be understood by considering that one third of the water is not evaporated from the water surfaces but rather through the vegetation. The coastline of islands is long enough to easily provide that amount of infiltration. Trees on islands act like pumps drawing down the extremely shallow groundwater table. Thus they draw water from the channels and floodplains. This loss does not influence the salinity of the remaining water, as water is taken away together with the dissolved salts. The evapoconcentration process of trees on the islands leads to an accumulation of salts. As a consequence trees in the middle of the islands die off. When the solubility limit is reached salt crusts and salt pans are deposited. Under islands the increased density of groundwater finally causes the formation of density fingers which dispose of saline water in the deeper and saline parts of the aquifer. The delta shows a remarkable organisation of fresh and saline areas on all scale levels. By storing salt, the saline parts keep the rest o the delta fresh. The fingering was proven by (Bauer et al. 2006) under a particular island, using surface to borehole geoelectrics. The TEM campaign of 2008 yielded data which showed even more impressively that all over the delta fingering occurs under surface salt accumulations visible as white spots from a plane. An example is shown in Fig. 4. Every year, the Okavango imports about 300'000 tons of dissolved solids into the delta. A preliminary estimate assigns about one half of it to precipitation while the other half is disposed off to the depth by salt fingering.

The decisive mechanism to keep the Delta waters fresh is the disposal of salt in islands. This requires a long coastline. Shrinking of the Delta reduces this coastline relatively more than it reduces the flooded area. Therefore a minimum size of delta is required to keep the process going.

Conclusions

The Okavango Delta is exposed to various dangers in the future. The worst scenario would be a reduction of inflows due to climate change superimposed with the maximum abstraction of water in the upstream for irrigation. Both dangers cannot be managed by Botswana alone as the causes are either located in the upstream or of global nature. The Okavango River Commission between Angola,

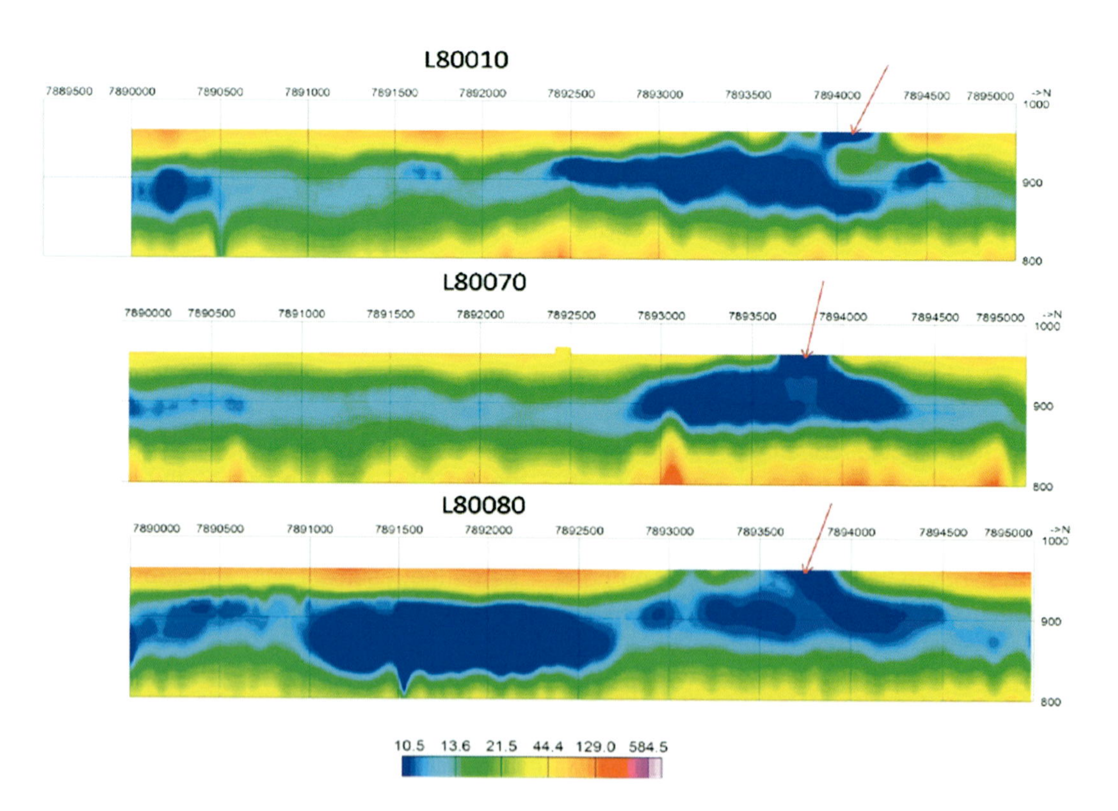

Figure 4: Salt fingers in vertical cross-sections as seen by airborne TEM. Blue indicates high electrical conductivity/high salinity. The origins of the salt plumes are indicated by the arrows. Vertical scale from 800 to 1000 m amsl. Horizontal scale UTM coordinates.

Namibia and Botswana is challenged to find ways – possibly with international help – to restrict abstractions to preferably less than 25 m^3/s. Such a restriction will also keep the size of the flooded area above values which are critical from the points of view of ecology and salinity.

Acknowledgement

The following Ph. D. and master students carried out the Okavango Research under my guidance at ETH Zurich over the last 11 years: Peter Bauer-Gottwein, Stefanie Zimmermann, Christian Milzow, Lesego Kgotlhang, Peter Meier, Vanessa Burg, Otlaathusa Tshekiso and Thomas Langer. I thank them for their enthusiasm and commitment. The financing of the research by ETH Zurich via its research commission and by the Swiss National Fund under several contracts is gratefully acknowledged. The Department of Water Affairs made the research possible by support with logistics in the field. The Geological survey of Botswana financed the airborne TEM campaign.

References

Bauer, P. (2004): Flooding and salt transport in the Okavango Delta, Botswana: key issues for sustainable wetland management. – Dissertation, ETH Zurich Nr. 15436

Bauer, P., Supper, R., Zimmermann, S. & Kinzelbach, W. (2006): Geoelectrical imaging of density fingers in the Okavango Delta, Botswana. – Journal of Applied Geophysics **60**(2): 126–141.

Bauer-Gottwein, P., Langer, T., Prommer, H., Wolski, P. & Kinzelbach, W. (2007): Okavango Delta Islands: Interaction between density-driven flow and geochemical reactions under evapo-concentration. – Journal of Hydrology **335** (3-4): 389–405:

Kgotlhang, L. (2008): Application of airborne geophysics in large scale hydrological mapping; Okavango Delta, Botswana. – Dissertation, ETH Zurich Nr. 18083.

Milzow, C. (2008): Hydrological and sedimentological modelling of the Okavango Delta Wetlands, Botswana. – Dissertation, ETH Zurich Nr. 18058.

Milzow, C., Kgotlhang, L., Bauer-Gottwein, P., Meier, P. & Kinzelbach, W. (2009): Regional Review: The hydrology of the Okavango Delta: Processes, data and modelling. – Hydrogeology Journal **17(6): 1297–1328.**

Milzow, C., Burg, V. & Kinzelbach, W. (2010): Estimating future ecoregion distributions within the Okavango Delta Wetlands based on hydrological simulations and future climate and development scenarios. – Journal of Hydrology **381**(1-2): 89–100.

Proceedings "Hydrogeology of Arid Environments", p. 21–24
Stuttgart, March 2012

Extended Abstract

Demand Management as a Potential Unconventional Source of Water, The West Bank Governorates

Muath Abusaada[1], Abdelrahman Tamimi[1], Martin Sauter[2]

[1] Palestinian Hydrology Group, P.O. Box 565, Al Masyoun, Ramallah, West Bank. Palestinian Authority, email: muath@phg.org
[2] Geoscientific Centre, University of Göttingen, Goldschmidtstr. 3, 37077 Göttingen, Germany

Key words: demand management, domestic demand, unmet demand

Introduction

In semi arid areas, water resources are usually limited. While the demand increases day by day, the available water from conventional resources decreases as a result of overexploitation, climate change and deterioration of its quality. Consequently, looking for new unconventional water sources has become more urgent. Using water in a more efficient way is considered as a new potential source of water, where the population improves their water utilization without affecting the water requirements for human demand (i.e. domestic use). These actions are known as Demand Management (DM) actions. In this paper, the DM actions adapted in the West Bank were assessed. Accordingly, six DM actions including water harvesting, water conservation devices, decrease of losses, reuse of gray water for gardening, use of water tankers and public awareness were studied.

Methodology

The West Bank governorates widely vary in terms of water utilization, water availability, unit price, existing water infrastructure and socio-economic characteristics. For this reason, the Water Evaluation and Planning (WEAP) model was developed for the West Bank governorates to assess the supply-demand quantities. The final results were aggregated to represent the demand, supply and the unmet demand under the two scenarios; the "Do Nothing (DN)" and DM scenarios. A comparison between the DN and the DM scenarios was conducted. In the DN scenario, the current practices were kept without change while under the DM scenario six demand management actions were introduced (i.e. water harvesting, water conservation devices, decrease of losses, reuse of gray water for gardening, use water of tankers and public awareness).

Calculation

The schematic diagram of the WEAP model is shown in Figure 1. In general, the types of water supply, types of demand and the domestic water utilization are different in each governorate. For this reason, the characteristics of each demand node (i.e. governorate) were studied and then simulated within the model. The supply sources are represented as seven supply nodes in order to simulate the six DM actions as well as the existing supply sources.

The quantity of water that could be saved through the DM scenario mainly relies on the amount of water delivered per capita from the main sources (i.e. the supply from network) as well as the type of the interventions under each of the DM actions. The characteristics of each DM action including the aim, the tool and methodology needed for implementation and the target group are summarized in Table 1.

The cost benefit analysis was also conducted by the WEAP model, therefore, the fixed and variable costs as well as fixed and variable benefits were defined as shown in Table 2.

Results

The results shows that the DM actions could save up to 90 Mm^3 by year 2050 which is equivalent to

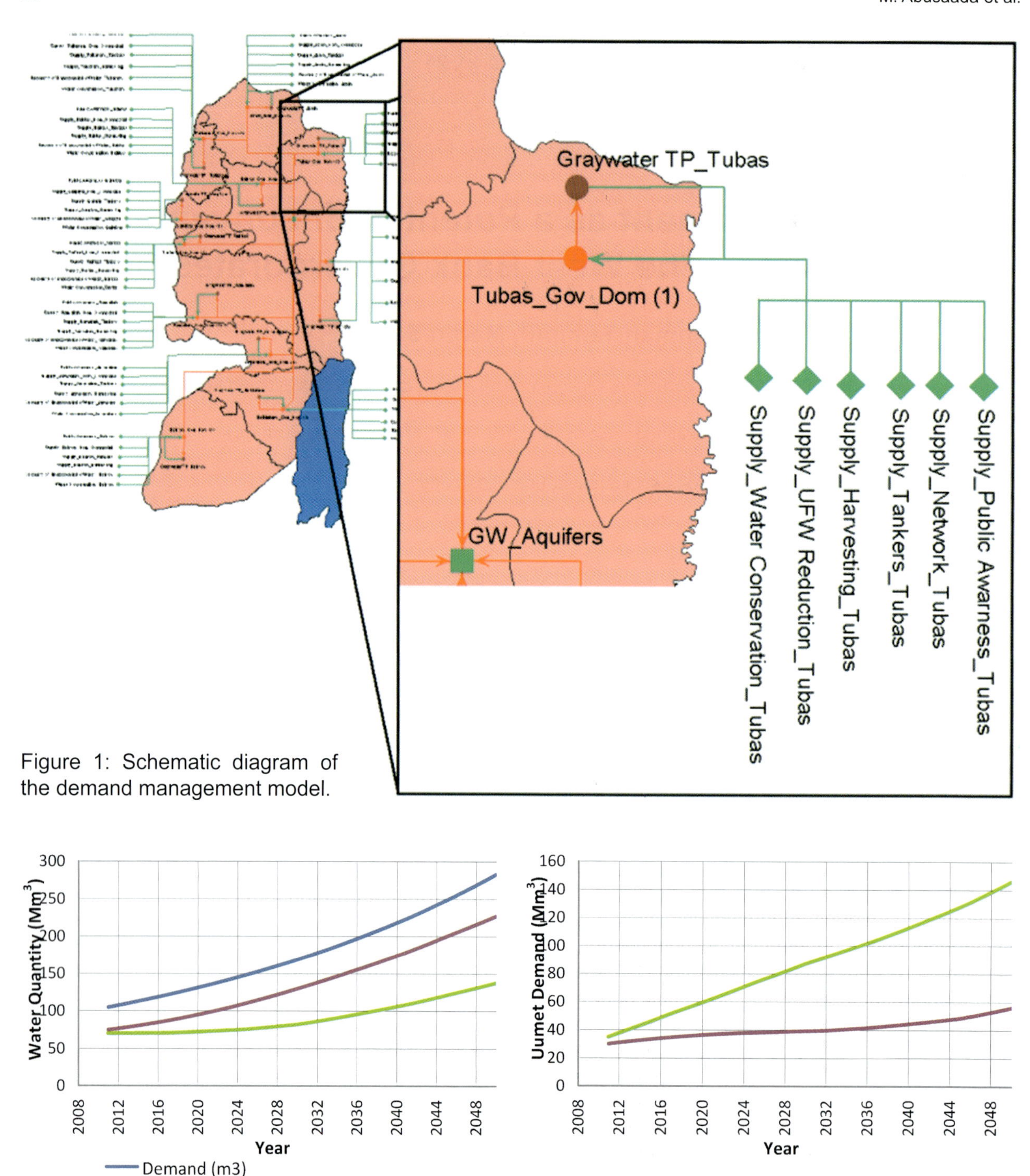

Figure 1: Schematic diagram of the demand management model.

Figure 2: Demand, supply and unmet demand under the two scenarios.

around 32% from the West Bank is domestic demand, Figure 2. This means that the unmet demand will be reduced by DM actions to 56 Mm^3 by the year 2050 compared with the DN scenario (i.e. 146 Mm^3). Consequently, the unmet demand by year 2050 will be reduced from 51.5% to 19.8% by implementing the DM actions. However, the unmet demand remains high, leaving the need to look for other water supply sources is needed.

The average cost of a cubic meter of water is calculated for both scenarios based on the capital and operational costs as well as quantitative benefits

Table 1: Description of demand management items.

Supply	Aim	Tools	Approach	Target Group
Public Awareness	Decrease water consumption	Training courses, workshops, booklets, leaflets, posters, short films, TV and radio	Train/teach people how to use water efficiently	Women and Children
Water Conservation Devices	Decrease the internal water losses	Conservation devices	Public awareness on the advantages of conservation devices /provide technical and financial support for their use	Households
Water Harvesting	Increase the storage capacity of water harvesting installations	Cisterns, Ponds	Provide families with technical and financial support	Households
Unaccounted for Water	Reduce the unaccounted for water in the distribution systems	Pipes, metering devices and other spare parts	Provide municipalities with technical and financial support	Municipalities
Treated Gray water	Increase the rate of water reuse/ decrease the pollution load/ decrease the cost of wastewater dumping	Gray water treatment plants	Provide households with technical and financial support/ Public awareness	Households
Supply by Tankers (this type is used as a supplementary source)	Reduce the use of tankers to decrease the price of water and for health purposes	–	Public awareness	Households

from DM interventions. The qualitative benefits (i.e. environmental and social benefits) were excluded from the cost-benefits analysis. The result shows that the cost of a cubic meter under the DM scenario will increase higher than the cost under the DN scenario during the beginning of the time management in order to cover the capital costs of the interventions before starting to decrease with time, Figure 3. The average cost by year 2050 will reach 1.82 $\$/m^3$ compared with 6.38 $\$/m^3$ under the DN scenario. Accordingly, the DM scenario should be considered as one of the potential sources of unconventional water resources in arid and semi arid areas.

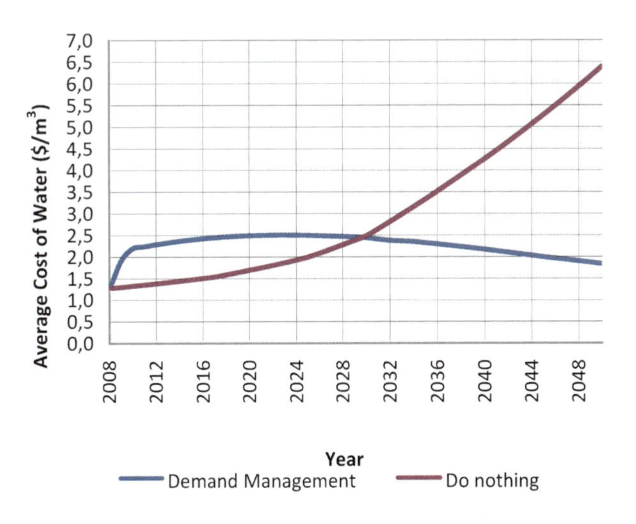

Figure 3: Average cost of water under the two scenarios.

Acknowledgments

We would like to thank the German Federal Ministry of Education and Research (BMBF) for financing this study as part of the GLOWA research initiative: Global Change in Hydrological Cycle.

Table 2: Description of cost-benefit analysis for demand management items.

Supply	Supply quantity	Fixed Cost	Variable Cost	Benefit
Public Awareness	Function of number of targeted population, the current water consumption and the percentage of demand reduction as a result of public awareness	Cost of training courses, workshops, booklets, leaflets, posters, short films, TV and radio	0	Quantity of saved water by public awareness (m^3) multiplied by the price of water from the domestic network ($/$m^3$)
Use of Water Conservation Devices (WCD)	Number of households using WCD multiplied by the percentage of supply reduction as a result of WCD multiplied by the current water consumption	Number of HH using WCD multiplied by the average number of WCD per HH multiplied by the average unit cost	0	Quantity of saved water WCD (m^3) multiplied by the price of domestic water network ($/$m^3$) assuming each unit will serve 10 years.
Water Harvesting (WH)	Number of HHs using WH multiplied by the average storage tank (e.g. 70 m^3/unit/yr)	Sum of the costs of cisterns and ponds	Negligible	The amount of harvested water (m^3/yr) multiplied by the price of water from the domestic network ($/$m^3$)
Unaccounted for Water (UFW)	Quantity supplied by distribution network multiplied by the percentage of UFW reduction	Cost of pipes, metering devices, other spare parts used and other costs related to the improvement of the network	–	The amount of water saved (m^3/yr) multiplied by the price of water from the domestic network ($/$m^3$)
Treated Gray water	Function of the number of HHs with gray water treatment units, average family size, average water consumption and the percentage of water returning to wastewater network	Cost of gray water treatment plant units	Negligible	The amount of treated water multiplied by the price of domestic network water + the amount of treated water multiplied by the cost of empting wastewater collection ponds
Supply by Tankers (this type is used as a supplementary source)	Function of the number of population supplied with less than 50 l/c/d	0	Cost of water delivered by tankers ($/$m^3$) multiplied by the total supplementary water	The amount of supplementary water reduction multiplied by the cost of water delivered by tankers ($/$m^3$)

References

Abusaada, M. & Tamimi, A., 2010: Demand Management and Dead-Red Project Assessments. – Report: Glowa Jordan River III.

Arlosoroff, S., 1997: Israel – A case study of Water Resources management. – Swiss Federal Institute for Environmental science and Technology(EAWAG) news 43E, PP 8-11.

Palestinian Hydrology Group (PHG) Database, 2011.

Palestinian Pater Authority (PWA) Database, 2011.

Ratna Reddy, V., 2009: Water Pricing as a Demand Management Option: Potentials, Problems and Prospects. – Strategic Analyses of the National River Linking Project (NRLP) of India: Series 3, PP 25-46.

Smout, I., Kayaga, S. & Muñoz-Trochez, C., 2008: Financial and Economic Aspects of Water Demand Management in the Context of Integrated Urban Water Management. – 3rd SWITCH Scientific Meeting, Belo Horizonte, Brazil.

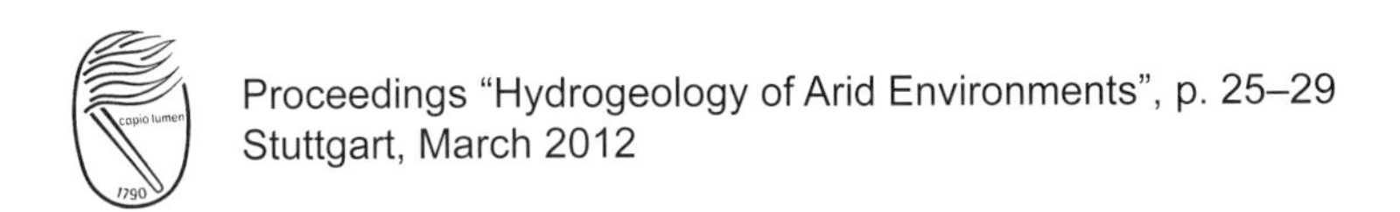
Proceedings "Hydrogeology of Arid Environments", p. 25–29
Stuttgart, March 2012

Extended Abstract

Groundwater Recharges in Dryland Alluvial Megafans: Geomorphology versus Lithofacies Controls, Examples from Central Iran

Nasser Arzani

University of Payme Noor, Geology Department, P.O. Box 19395-4697 Tehran, IR Iran, email: arzan2@yahoo.com

Key words: alluvial megafans, climate, groundwater recharge, geomorphology, central Iran

Abstract

Subsurface hydraulic response to surface flood-water recharges in arid environments is controlled by discrete geologic structures and their sedimentary lithofacies in a variety of aquifer settings. Alluvial megafans (>30 km in their length) are the most important underground water resources in arid lands and in an attempt to recharge these aquifers, flood-water harvesting projects have been widely applied during the last two decades in Iran. This study compares two different types of alluvial megafans around the Kohrud Mountain range, for variability of flood-water recharges related to their geomorphology, lithofacies distribution and their evolution under the general trend of Quaternary climatic changes towards an arid to semiarid setting of central Iran. Fan surface mapping (based on 1/50000 topographic maps, satellite images, and fieldwork), revels that flood-water harvesting and groundwater recharge is controlled by the fan-catchment characteristics, particularly the type of source-area geology and flood power (sediment/water ratio), as well as, the locations of fan-intersection point and active depositional lobe. Geomorphic evolution of Quaternary dryland alluvial negafans controls the heterogeneity of sedimentary lithofacies and the general environmental conditions necessary for groundwater recharges with aggradations versus incision corresponding to high magnitude, low frequency flood discharges during times of increased aridity and a differential sediment/water discharge ratios in different source rock geology. Understanding surface and subsurface geological heterogeneity is critical for characterization groundwater recharges in arid environments.

Introduction

The heterogeneity and complex hydrofacies distribution in alluvial fans are the major concerns in predicting and modelling groundwater recharges in these sedimentary and geomorphic systems (e,g. Ritzi et al. 1995, Eschard et al. 1998, Weissmann et al. 1999, 2000). Alluvial-fan morphology and sedimentary style are controlled by tectonics, climate, base level and source lithology and are expressed as aggradations versus incisions/dissections in fan sequences (Blair & McPherson 1994, Blair 1999, Hartley et al. 2005, Clarke et al. 2010). These evolutionary trends are much more pronounced in large alluvial fans (megafans), used for flood-water harvesting projects in drylands. Drylands cover about one third of the Earth's land areas, where desertification, drought and water shortage are the main threat to the environment and its ecosystem (UNEP 1997, Goudie 2002, Reynolds & Smith 2002). Alluvial megafans, which are one of the most important underground water resources in such a harsh environmental setting, have been defined as large fans with >30 km in length (apex to toe) and an extreme low slope gradient of <1° (cf. Arzani 2005). Other terms that have been used to describe these geomorphic features include fluvial fans (e.g. Collinson 1996, Horton & DeCelles 2001); stream-dominated (losimean) alluvial fans (e.g. Weissmann et al. 2002), and distributive fluvial systems (e.g. Weissmann et al. 2010, Hartley et al. 2010). The subsurface hydrofacies potential (water reservoirs) in these large alluvial systems is governed by the gross geometries of the deposited sedimentary bodies and their lithofacies characteristics as they

www.borntraeger-cramer.de
2012/0025/$ 1.25

grade laterally from the coarse gravelly/sandy deposits of the proximal to muddy fine-grained deposits of the distal areas of the megafans (Weissmann et al. 1999, Chakarabatry & Ghosh 2010, Hartley et al. 2010).

Flood-water harvesting projects have been widely applied during the last two decades in Iran. In this paper, two Quaternary alluvial megafans (Soh and Zefreh) from western and southern flanks of Kohrud Mountain range (KMR) in central Iran were selected for a comparative study (Fig. 1). Locations selected for flood-water harvesting projects in these alluvial megafans were related to their geomorphology and fan characteristics under arid to semiarid climatic setting of central Iran.

General setting and methodology

The study area is located along the Gavkhoni-Abarkoh-Sirjan Basin of central Iran, a fault-bounded depression, which is up to 100 km wide and more than 600 km long (Fig. 1A). Alluvial fans source from the side valleys and extend towards the centre of this depression, where it is desert and playa. The present climate of the area is arid to semi-arid and the average rainfall between 30 and 200 mm in the basin, and up to 300 mm in the headwaters of many ephemeral rivers (Arzani 2005, 2007, 2012). Fieldwork and mapping of fan surfaces were based on the 1/50000 topographic maps (Iranian Cartographic Centre) and satellite images (TM, Landsat 4 and ETM Landsat 7) and catchment bedrock geology was cross checked using published 1/100000 geological maps (Zahedi 1973, Radfar et al. 2002).

Flood-water harvesting in megafans

This study compares two types of alluvial megafans around the Kohrud Mountain range, for variability of flood-water recharges related to their geomorpholo-

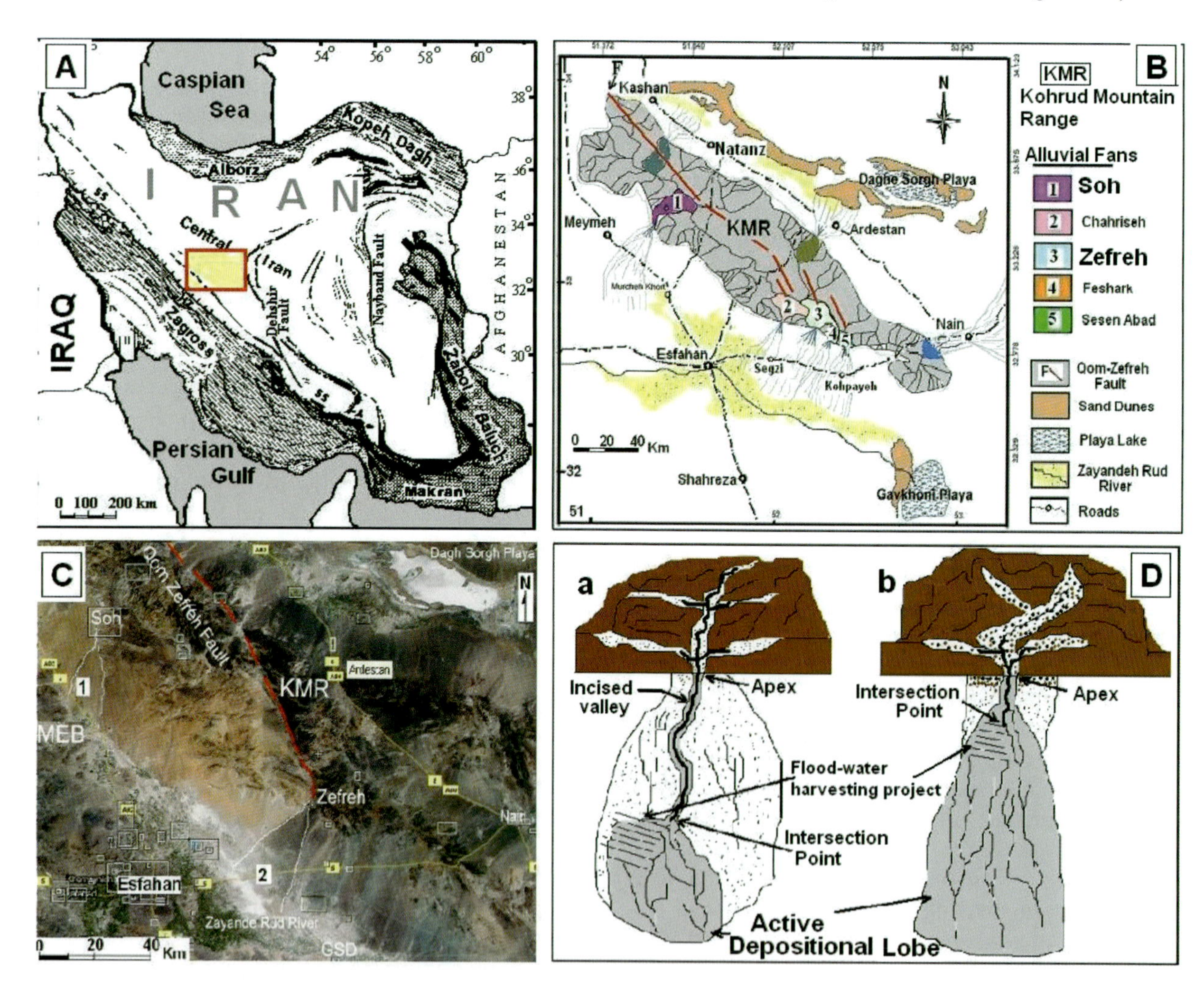

Figure 1: Location, general structural elements (A & B), satellite image (C) and generalized model (D) with megafans undergoing incision/dissection (a) or aggradational fan evolutionary trends with the location of flood-water harvesting projects in the drylands of the central Iran.

gy, source-rock geology, lithofacies distribution and their evolution under the general trend of Quaternary climatic changes towards an arid to semiarid setting of central Iran. These fans exhibit different gross geometries of the deposited sedimentary bodies and lithofacies/hydrofacies characteristics.

Soh fan

The Soh-fan surface is incised by a deep channel and presently its active depositional lobe is far downfan. The location of flood harvesting project of this fan is at its intersection point and 13 km downfan from the apex and toward the medial part, where the incised channel gradually terminates into a recent active lobe (Fig. 1). The distal fan area of this fan extends towards the playa fringe and the sediments of the axial basin river.

The Soh fan, which its catchment geology is in the sedimentary terrains of the KMR covers an area of 437 km^2 and has a radial length of 32.4 km. It originates from a catchment with an area of 178 km^2 and a basin relief of 900 m. The catchment geology is sourced largely in Devonian sandstones/dolomite, Triassic dolomites, Jurassic sandstones/marls, Cretaceous limestones and Eocene conglomerates. Tributary valleys and catchment channels that feed the fan body are mantled with little coarse sediments and at the present arid to semi-arid climate of the area, the harder lithologies provide limited gravels for the bed loads for the periodic flash floods.

The 13 km long and up to 15 m thick exposures along the side walls of the incised Soh-fan channel exhibit an inter-stratification of two main facies, including thick-massive-bedded, coarse gravelly deposits and stratified finer-grained conglomerate to pebbly sandstones. The coarse gravelly facies consists of massive, up to 150 cm thick, beds and comprises ~30% of the exposures and is a clast-rich, unsorted, slightly bouldery, cobble-pebble deposit with subrounded to subangular clasts. This facies, with irregular, crude to planar bedding styles, is exposed mainly at the basal parts of the cut walls of the incised channel.

The location of water harvesting projects in Soh fan coincide with the recent active depositional lobe of this fan, located 13 km downfan, where the recent thin veneer of coarse permeable sediments mantle the fine-grained marly impermeable sediments of the medial to distal fan areas of the main fan body. The remarkable downfan shift of the intersection point, as the result of the incision on fan surface, resulted in this inverse grading and development of drastic permeability barriers for the flood-water recharges in this type of megafans.

Zefreh fan

The Zefreh fan has a simple surface and its intersection point is about 2 km from the apex, where its major shallow channel grades down slope into the flat areas of the flood-water harvesting project. The fan has an area of 394 km^2; a fan length of 35.7 km and a distal-fan expansion angle of about 85°. The proximal terraces and side walls of the main fan channel exhibit alternate massive, coarse-grained, polymictic, clast- or matrix-supported gravels with stratified finer-grained gravels to conglomerates and pebbly sandstones. Zefreh fan, which its catchment geology rests mainly in igneous terrain of Urumieh Doghtar Belt of the KMR, has a catchment area of 184.5 km^2 and its basin relief is 1250 m. The hard lithologies, such as basalt and siliceous-cemented sandstones, andesite and carbonates, cover only about 30% of the catchment area and about 70% of the catchment area is covered with the highly fractured and physically weathered Eocene volcaniclastics of the western side of the Qom Zefreh Fault (Fig. 1B, C). The low angle slope, the short length of the main fan channel and the typical coarse, permeable, gravelly sediments of this fan provided the proximal fan areas as an ideal location for the flood water recharges on this type of fans.

Discussion

Alluvial fans are the major hydrological resources in arid lands (e.g. Cooke et al. 1993, Houston 2002, Parsons & Abrahams 2009). This has been well recognized since more than twenty five centuries ago in drylands of central Iran for position of farms and cities on the fan surface and construction of ancient Qanats ("karize") as underground tunnels conducting water from proximal to distal fan areas (Wulff 1986, Papoli & Labaf 2000, Bouayad 2002, Arzani 2003, 2005).

Groundwater recharges in alluvial megafans is controlled by both their geomorphology as well as their lithofacies distributions. Source rock lithology and dynamic processes, controlled by climate and tectonics, are the major controls for fan evolutions. Bedrock physical and chemical weathering properties directly control sediment flux from the catchment to the basin and thus influence alluvial fan aggradation and or progradation. Weathering is controlled by both tectonics and climate affecting the source-area geology and transport capacity (e.g. Blair 1999, Harvey 2005). Physical weathering in drylands, as an important impact of climate on alluvial fan development, may results in the rapid break-down of clasts and provides the gravels as the sediment source. Although shale and very soft clayey substrate produce large volume of sediment, this is usually very fine-grained and can be easily moved to distal parts of the basin by suspension in episodic, high-magnitude flood events (Blair 1999).

Although the studied alluvial fans derived from the western and southern of KMR are both low-angle, alluvial megafans that developed under arid to semiarid climate, there are key differences that generate important insights into controls on development of different hydrofacies in alluvial megafan. The Zefreh fan has a relatively wider drainage basin and its relief is higher than the Soh fan. The Soh fan is more elongate in shape and its catchment length is larger than the Zefreh fan. The Zefreh fan source's region is mainly physically weathered volcaniclastics; whereas the Soh's source is sedimentary. The Zefreh fan catchment yields much coarse sediments than the Soh fan, as gravel pavements are common in their tributary valleys. The Soh fan is deeply dissected by the main channel with head-cut erosion into a back-filled feeder valley and its intersection point has shifted downfan, whereas the Zefreh fan displays minimal incision and just hosts shallow channels that facilitate flood-water harvesting projects in the proximal part of the fan (Fig. 1D).

The relative erodability of the bedrock and source-area lithology of the KMR alluvial fans appears to be a key factor in controlling the difference in the morphologies and hydrofacies of studied fans. The catchments of the Zefreh and Soh fans are located in the igneous terrain of UDMC and sedimentary terrain in east and west of QZF respectively. The Zefreh fan catchment yields much coarse sediments than the Soh fan. Weathering of the igneous (andesite and basalt) and particularly the volcaniclastic rocks of the Zefreh fan catchment produces significant volume of coarse gravels. The general bedrock lithology and geomorphology of the Zefreh-fan catchment is ideal promoters of high sediment discharges to water ratios and capable of funnelling a high-velocity, sediment-discharged flash flood to the fan apex. As the result of the abundance of coarse-grained sediment derived from mainly volcaniclastic rocks, Zefreh fan remained in aggradational modes with little in the way of substantial fan incision and its intersection point remained near the apex. This is different for the Soh fan with sandstone, limestone, dolomite and marly source regions and a catchment mantled with little colluviums as coarse sediments and consequently a low sediment/water discharge ratio during periods of incision corresponding to times of increased aridity. The incised channel of the Soh fan is relatively deep, long and sinuous, while for the Zefreh fan it is shallow, short and braided. In drylands, sediment supply determines the shape of megafan channel networks and while sinuous channels develop in areas with low rate of sediment supply, braided type channels are characteristics of high sediment supply (Hartley et al. 2010). Moreover, channel aggradation or incision controls the location of intersection point. If the aggradation is sufficient to fill the alluvial-fan incised valley, the intersection point shifts toward the apex and develops positive accumulation space for fan aggradation. Conversely, decrease in sediment supply or increase in flood power may lead to channel degradation and fan incision, shifting the intersection point and its associated deposition downfan (e.g. Weissmann et al. 2002). Sediment supply, which is mainly a function of source area lithology and climate, is a dominant driver behind the development and evolution of alluvial megafans, particularly, in drylands.

Conclusions

Alluvial megafans are important water resources in drylands and their evolutionary trend, from aggradational to incision/dissection modes, could determine their hydrofacies potential flood-water harvesting projects and ground-water recharges in drylands. The large fans undergoing aggradational modes with a simple surface and a short main flood channel and an intersection point located in the proximal areas of the fan, such as Zefreh fan, are more ideal for flood harvesting relative to the dissected/incised megafans with a long intersection point draining into the distal fan areas, such as Soh fan. An understanding of the controlling factors, such as bedrock lithology in this study, on fan development and its geomorphic evolution could provide insight into the characteristic of hydrofacies and their nature for ground-water recharges in drylands.

References

Arzani, N., 2003: The tragedy of ancient qanats in Kavir borders and arid lands, a case study from Abarkoh Plain, Central Iran. – Iranian International Journal of Science **4** (1): 73–86.

Arzani, N., 2005: The fluvial megafan of Abarkoh basin (Central Iran): an example of flash-flood sedimentation in arid lands. – In: Harvey, A. M., Mather, A. E. & Stocks, M. (eds.): Alluvial Fans: Geomorphology, Sedimentology, Dynamics. Geological Society, London, Special Publications, **251**: 41–59.

Arzani, N., 2007: Playa lake level fluctuation and its influence on recent sediments of a terminal fan-playa fringe environment, Abarkoh Basin, Central Iran. – Journal of Sciences, IRI **18** (1): 19–33.

Arzani, N., 2012: Catchment lithology as a major control on alluvial megafan development, Kohrud Mountain range, central Iran. – Journal of Earth Surface processes and Landforms. DOI: 10.1002/esp.3194.

Blair, T.C. & McPherson, J. G., 1994: Alluvial fans and their natural distinction from rivers based on morphology, hydraulic processes, sedimentary processes and facies assemblages. – Journal of Sedimentary Research **64**: 450–589.

Blair, T.C., 1999: Sedimentary processes and facies of the waterlaid Anvil Spring Canyon alluvial fan, Death Valley, California. – Sedimentology **46**: 913–940.

Bouayad, M., 2002: The qanat project, Waterway 14, South–Central Asia, Unesco Publication.

Chakarabatry, T. & Ghosh, P., 2010: The geomorphology and sedimentology of the Tista megafan, Darjeeling Himalaya: Implication for megafan building processes. – Geomorpholog **115**: 252–266.

Clarke, L., Quine, T.A., Nicholas, A., 2010: An experimental investigation of autogenic behaviour during alluvial fan evolution. – Geomorphology **115**: 278–285.

Collinson, J. D., 1996: Alluvial sediments. – In: Reading, H. G. (ed.): Sedimentary Environments: Processes, Facies and Stratigraphy, Blackwell Science: 37–82.

Cooke, R. U., Warren, A. & Goudie, A., 1993: Desert geomorphology. – University College of London. 526 p.

Eschard, R., Lemouzy, P., Bacchiana, C., Desaubliaux, G., Parpant, J. & Smart, B., 1998: Combining sequence stratigraphy, geostatistical simulations, and production data for modeling a fluvial reservoir in the Chaunoy Field (Triassic, France). – AAPG Bulletin **82**: 545–568.

Goudie, A. S., 2002: *Great Warm Deserts of the World.* Oxford and New York. – Oxford University Press.

Hartley, A. J., Mather, A. E., Jolley, E. &Turner, P., 2005: Climatic controls on alluvial fan activity, Coastal Cordillera, northern Chile. – In: Harvey A. M., Mather, A. E. & Stokes, M. (eds.): Alluvial Fans: Geomorphology, Sedimentology, Dynamic. Geological Society London, Special Publication, **251**: 95–115.

Hartley, A. J., Weissmann, G. S., Nichols, G. J. & Warwick, G. L., 2010: Large distributive fluvial systems: characteristic, distribution, and controls on development. – Journal of Sedimentary Research **80** (2): 167–183.

Harvey, A. M., 2005: Differential effect of base-level, tectonic setting and climatic change on Quaternary alluvial fans in the northern Great Basin, Nevada, USA. – In: Harvey A. M., Mather, A. E. & Stokes, M. (eds.): Alluvial Fans: Geomorphology, Sedimentology, Dynamic. Geological Society London, Special Publication **251**: 117–133.

Horton, B.K. & DeCelles, P. G., 2001: Modern and ancient fluvial megafans in the foreland basin systems of the central Andes, southern Bolivia: implications for drainage network evolution in fold-thrust belts. – Basin Research **132**: 43–63.

Houston, J., 2002: Groundwater recharge through an alluvial fan in the Atacama Desert, northern Chili: mechanism, magnitude and causes. – Hydrological processes **16**: 3019–3035.

Papoli Yazdi, M. H. & Labaf Khaniki, M., 2000: The role of qanats in the formation of cultures. – Proceeding of the international symposum on qanats, 1, pp. 1–25.

Parsons, A. J. & Abrahams, A. D., 2009: Geomorphology of desert environments. – 2nd ed., Springer 831 p.

Radfar, J. Kohansal, R. K. & Zolfaghari, R., 2002: Kuhpayeh geological map (1/100000). – Geological Survey of Iran.

Reynolds, J. F. & Smith, D. M. S., 2002: Global Desertification: Do Humans Cause Deserts? – Dahlem Workshop Report 88, Dahlem Univ. Press, Berlin.

Ritzi, R. W., Dominic, D. F., Brown, N. R., Kausch, K. W., McAlenney, P. J. & Basial, M. J., 1995: Hydrofacies distribution and correlation in the Miami valley aquifer system. – Water Resources Research **31**: 3271–3281.

Shukla, U. K., Singh, I. B., Sharma, M. & Sharma, S., 2001: A model of alluvial megafan sedimentation: Ganga Megafan. – Sedimentary Geology **144** (3–4): 243–262.

UNEP. 1997: United Nations Environment Programme. World Atlas of Desertification. – 2nd ed., Edited by N. Middleton and D. Thomas, London, UNEP, 182pp.

Weissmann, G. S., Labolle, E. M. & Fogg, G. E., 1999: Three-dimensional hydrofacies modeling based on soil surveys and transition probability geostatistics. – Water Resources Research **35**: 1761–1770.

Weissmann, G. S., Labolle, E. M. & Fogg, G. E., 2000: Modeling environmental tracer-based groundwater ages in heterogeneous aquifers. – In: Bentley, L. R., Sykes, J. F., Brebbia, C. A., Gray, W. G. & Pinder, G. F. (eds.): Computational Methods in Water Resources, XIII International Conference on Computational Methods in Water Resources, Calgary, Alberta, Canada, June 25–29, 2000, Proceedings: 805–811.

Weissman, G. S., Mount, J. F. & Fogg, G. E., 2002: Glacially driven cycles in accumulation space and sequence stratigraphy of a stream-dominated alluvial fan, San Joaquin Valley, California, U.S.A. – Journal of Sedimentary Research **72**: 240–251.

Weissmann, G. S., Hartley, A. J., Nichols, G. J., Scuderi, L. A., Olson, M., Buehler, H., Banteah, R., 2010: Fluvial form in modern continental sedimentary basins: Distributive Fluvial Systems (DFS). – Geology **38** (1): 39–42.

Wulff, H. E., 1986: The qanats of Iran. – Scientific American: 94–106.

Zahedi, M., 1973: Etude géologique de la région de Soh (W de l' Iran central). – Geological Survey of Iran. Report No. 27: 1–197.

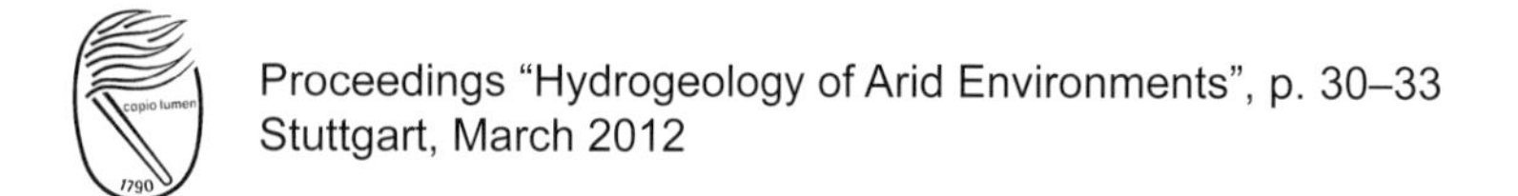
Proceedings "Hydrogeology of Arid Environments", p. 30–33
Stutgart, March 2012

Extended Abstract

Integrated Examinations of Hydraulic Conductivity, Apparent Resistivity and Hydrochemical Characteristics of Aquifers in a View to Minimize Failure Rates of Wells in Weybo River Catchment, Southern Ethiopia, East Africa

Aychluhim D. Damtew

Department of Applied Geology, Ruhr University of Bochum, Universitaetsstr.150, D-44801, Bochum, Germany, email: Aychluhim.Damtew@ruhr-uni-bochum.de

Key words: failure rate, integrated examination, hydraulic conductivity, apparent resistivity, hydrochemical characteristics, Ethiopia

Introduction

Weybo River Catchment, the investigated area, is located adjacent to the South-Western boundary of the Main Ethiopian Rift System. It is confined within the Omo-Ghibe Basin, one of the twelve major river basins of the country. The area is dominantly covered by acid volcanics, both pyroclastic flows and falls, of Miocene to Pliocene age. Climate of the study area ranges from semi-arid in the lowest Weybo valley to humid in the highest mount Damota.

Groundwater is a major source of water supply in the study area. The absence of comparable substitutes for groundwater in the area forces the major actors that are engaged in water supply and sanitation programs to highly depend on the development of groundwater sources. According to the 2006 report of the Southern Region Water Resources Development Bureau, the drinking water coverage of the study area is only 35%. Over 70% of the water schemes are shallow boreholes fitted with hand pumps. Hand dug wells, even though affected by seasonality and poor sanitation due to poor wellhead protection and poor withdrawal techniques, are also considered as good sources.

Even though groundwater sources are major water supply sources in the area, the failure rate of well drilling is high. 1 in every 3.5 to 4 wells gets abandoned and those that are 'productive' yield low to very low (0.75–5 l/s) due either to improper well siting or difficulty of drilling, well caving and loss of circulation being the main difficulties. This high failure rate coupled with the limited financial resources, and the high well drilling and construction costs aggravates the water supply and sanitation situation in the area. As reviewed from hydrogeological investigation reports available at the regional and zonal offices, most well failures are due to poor understanding of the hydro(geo)logic system of the area and high dependence on results from a single investigation method. This study is, therefore, made in a view to minimize the failure rate of well drilling associated with improper well siting through adopting integrated examinations of hydraulic conductivity, apparent resistivity and hydrochemical characteristics of aquifers.

Objectives of the work

The major objective of the work is to map potential groundwater zones in the study area thereby reducing failure rate of wells and exploring high yielding aquifers.

Methodology

A three-step approach that begins with review of existing reports and analyses of inventories of wells, examinations of pumping test and lithological log

www.borntraeger-cramer.de
2012/0030/$ 1.00

data has been followed. The spatial distribution of both productive and abandoned wells is overlain on existing geological map. Additional surveys were done to identify unrecorded abandoned wells, and systematic analysis of the spatial distributions of high and low yielding wells was also made. Unrecorded hand dug wells were also inventoried and a little information on the type of formations drilled were collected. Following these activities, further geological and hydrogeological investigations around the abandoned, low and high yielding wells were done. Vertical electrical surveys along 14 lines using ABEM Terameter-SAS 4000 and profiling surveys across 4 lines around those areas and the unexplored regions of the study area were carried out. A total of 68 water samples from deep and shallow wells, hand-dug wells and springs were collected for major ion analysis. Measurements of insitu physical properties of waters (pH, EC, T, Eh and TDS) from different sources were also done. Data analysis and interpretation is made using available softwares including ArcGIS, Resist, Aquachem, WatBal, XSTAT and other geostatistical softwares.

Results and discussion

It is seen that high yielding wells with higher hydraulic conductivity values, as computed from constant rate pumping test data, are clustered around fractured regions of the catchment.

Two distinct zones are observed after making an integrated analysis of the apparent resistivity values of aquifer layers and their corresponding hydraulic conductivity values. Zone-1 shows a lower apparent resistivity value for aquifers of lower hydraulic conductivity. These aquifer formations, as seen from

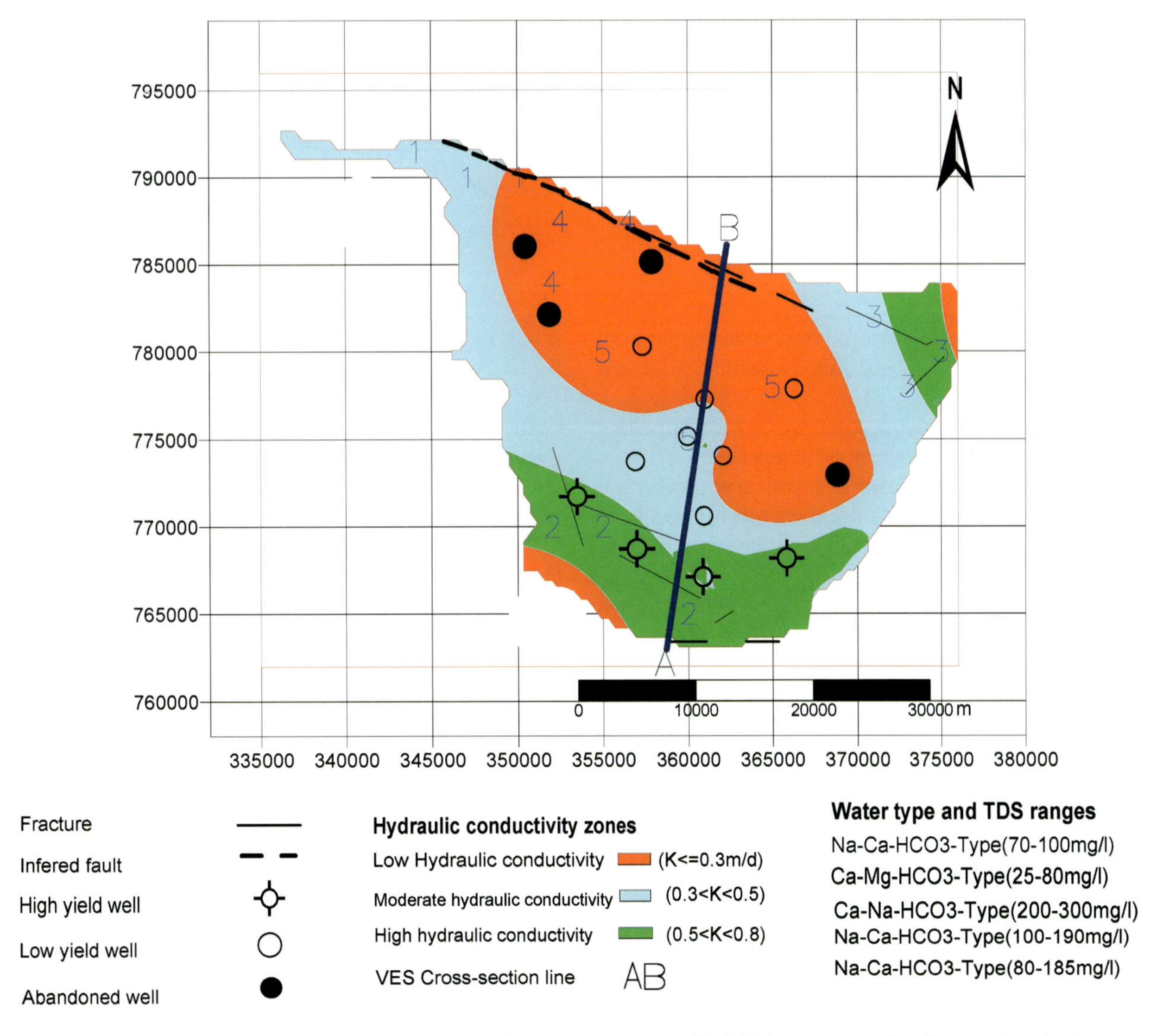

Figure 1: Map showing hydraulic conductivity, fracture zones, wells, VES cross-section line and water types.

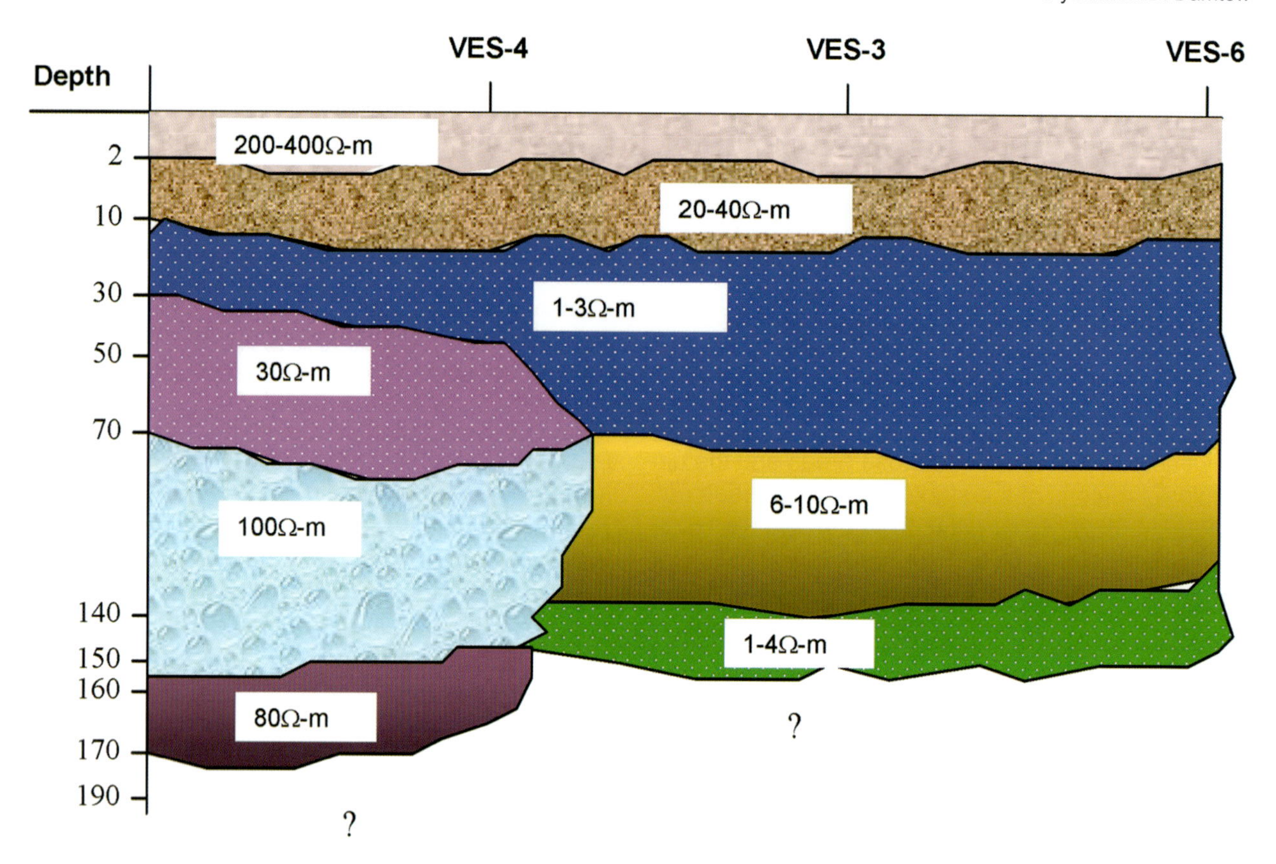

Figure 2: VES Cross-section along line AB.

the borehole logs, are deeply weathered ignimbrite and rhyolite. These boreholes generally yield low (0.75–5 l/s).

Zone-2 clearly reveal layers of lower apparent resistivity but higher hydraulic conductivity. This is still in a very good agreement with the aforementioned observation where aquifers are formed from both fractured and weathered ignimbrites and rhyolites.

Most of the productive and high yielding wells are located around the fractured zone and those that are abandoned are located far away from those fractured areas.

Delineations of recharging, transitional and discharging zones using hydrochemical method best overlaps with those delineated by geomorphological method.

Conclusion

The work reveals that groundwater potential of the area is seen to be controlled by fracturing and degree of weathering. Higher hydraulic conductivity zones are mapped at or near fractured/faulted regions and decreasing away from these structures.

It has also been seen that aquifers that are formed from the deeply weathered ignimbrites, rhyolites and trachytes yield low and possess a lower hydraulic conductivity generally in the order of 0.1–0.5 m/d. The apparent electrical resistivity values do not show a clear contrast between the deeply weathered aquifer formations and the fractured and weathered formations. Lower and nearly similar values (1–10 Ω-m) are obtained in both regions. This indicates that reliance only on interpreted apparent resistivity values of geologic layers may lead to wrong conclusions and eventually to well failures.

Further exploration of hidden fractures and discontinuities are vital in siting productive and high yielding wells. Application of lateral heterogeneity mapping may help locate buried fractures which, indeed, should be supported by field investigations and interpretations of areal and satellite imageries.

Chances of well failures could also be minimized by identifying the different hydrogeologic zones in the process of the investigation. Graphical interpretations of the field and laboratory measured chemical parameters in combination with the geomorphological analysis greatly help classify the catchment in to these zones.

Acknowledgement

Southern Region Water Resources Development Bureau, Ministry of water resources of Ethiopia, Wolyta and Hadia Zones Water Development Offices, and Addis Ababa University are duly acknowledged for their all rounded support.

References

Davidson, A., Moore, J. M., Davies, J.C., Shiferaw, Al. & Tefera, M. 1973: A Preliminary Report on the Geology and Geochemistry of Parts of Sidamo, Gemu Gofa and Kefa Provinces. – Rep. No. 1, Ethiopian Institute of Geological Survey.

Dobrin, M. B., 1976: Introduction to Geophysical Prospecting. – Mc Graw Hill Book Company, New York, 582 pp.

Hem, J. D., 1985: Study and interpretation of the chemical characteristics of natural water. – U.S. Geological Survey Water Supply Paper 2254, 263 pp.

Kazmin, V., 1979: Stratigraphy and correlation of volcanic rocks in Ethiopia. – Note No. 106, Ethiopian Institute of Geological Survey.

Kruseman, G. P. & Ridder, N. A., 1990: Analysis and Evaluation of Pumping Test Data. – Intern. Inst. for Land Reclamation and Improvement, Wageningen, The Netherlands, 377 pp.

Ministry of water resources of Ethiopia, Japanese International Cooperation Agency, 2003: Geophysical Survey, Groundwater Development and Water Supply Training Project, 9th Groundwater Investigation Training Course Manual. – Addis Ababa, Ethiopia.

Ministry of Water Resources of Ethiopia, 1996: Omo-Gibe Basin Integrated Development Master Plan. – Vol. 6,7,8,14, Library of Ministry of Water Resources, Addis Ababa, Ethiopia.

Mohr, P. A., 1960: Report on a Geological Excursion through Southern Ethiopia. – Bull. Geophys. Obs. Institute of Geological Survey of Ethiopia, Addis Ababa, Vol. 3, 9–20 pp.

Southern Region Water and Mines Resources Development Bureau, 2004: Borehole site selection report around Wolyita and Hadia Zones. – Unpublished report, Awassa, Ethiopia.

Southern Region Water and Mines Resources Development Bureau, 2003: Water points inventory report. – Unpublished report, Awassa, Ethiopia.

Proceedings "Hydrogeology of Arid Environments", p. 34–35
Stuttgart, March 2012

Extended Abstract

How Much Water is Left? – An Economic Approach to the Quantification of Non-Renewable Groundwater Resources

Heiko Dirks[1], Siegfried Holtkemper[2], Mohamed Al-Saud[3], Randolf Rausch[2]

[1] Dornier Consulting, Riyadh, Saudi Arabia, email: heiko.dirks@gizdco.com
[2] GIZ IS, Riyadh, Saudi Arabia
[3] Ministry of Water & Electricity, Riyadh, Saudi Arabia

Key words: non-renewable groundwater resources, groundwater management, groundwater economy

Introduction

In arid regions, the fossil groundwater resources of large sedimentary basins are often used to satisfy municipal, agricultural, and industrial water demands. In contrast to limited available renewable water resources, these non-renewable water resources bear a huge potential to satisfy any present water demands, while neglecting the future ones. Lack of water management during the last decades led to overexploitation of aquifers, large cone of depressions, and water quality deterioration.

A key question in management of non-renewable groundwater resources is: how much water is left? In the classical approach, the available volumes are calculated considering geological, hydrogeological and technical aspects like aquifer dimensions, aquifer parameters, water quality, pumping height, and drilling depth (GDC 1980).This is done in a static approach or, more elaborated, with the help of a groundwater flow model.

Methodology

An economic approach that aims to answer the question first determines the cost for groundwater abstraction. Second, it assigns a value to the groundwater resource. Following the economic approach, groundwater is considered a resource, if costs of the groundwater abstraction and supply are less than the value of the groundwater. If costs are higher than the value, the groundwater is not a resource, and demand is better satisfied by an alternative supply (e.g. desalinated sea-water). Figure 1 shows the flowpath of an economic approach to non-renewable groundwater resources quantification.

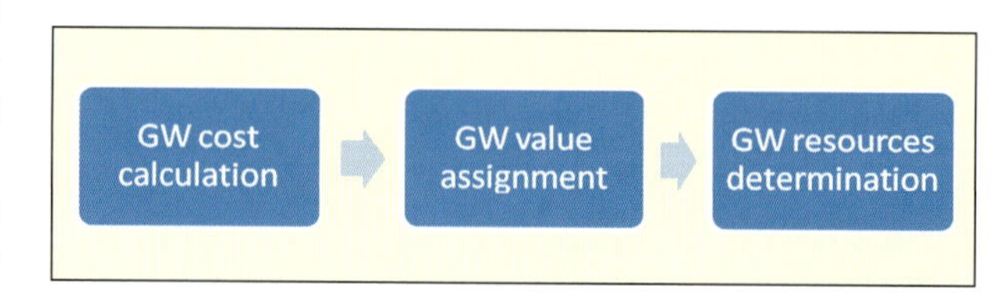

Figure 1: Flowpath of an economic approach to non-renewable groundwater resources quantification.

Groundwater costs: The thresholds defining technical limits of groundwater abstraction (e.g. a pumping height of 300 m, or a TDS limit of 2,000 mg/l) try to determine, how much effort is feasible to access the groundwater. Effort with regard to pumping capacities, water treatment, or drilling works can easily be expressed as costs. Besides the parameters considered in the classical quantification approach, the transportation costs of the groundwater to the location of demand are also included in the cost calculation. Figure 2 shows the major parameters determining the cost of groundwater abstraction.

Groundwater value: A prerequisite for assigning any value to groundwater is the assumption that groundwater is a scarce good. This is in contrast to often practiced present abstraction schemes, which are solely demand driven. In the approach

www.borntraeger-cramer.de
2012/0034/$ 0.50

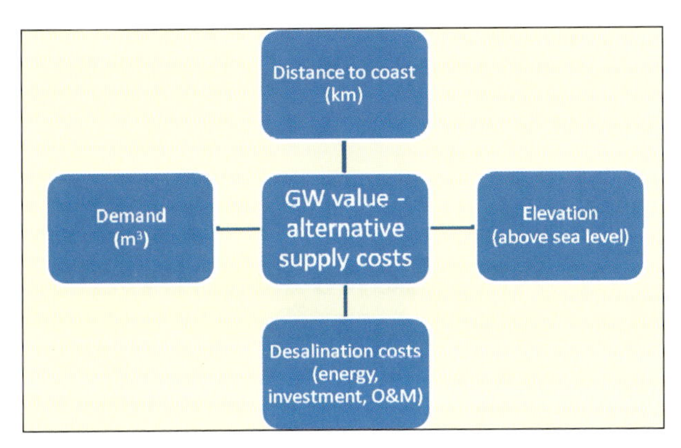

Figure 2: Parameters determining the value of non-renewable groundwater.

proposed here, the value of non-renewable groundwater is determined by the opportunity costs of any alternative water source, which is commonly desalinated sea-water. The value of the groundwater resource is equal to the cost for any alternative renewable water supply, and depends on parameters like location of demand (its distance to coast, elevation), energy costs, and investment and O&M costs for the alternative supply. Figure 3 shows the major parameters determining the value of non-renewable groundwater.

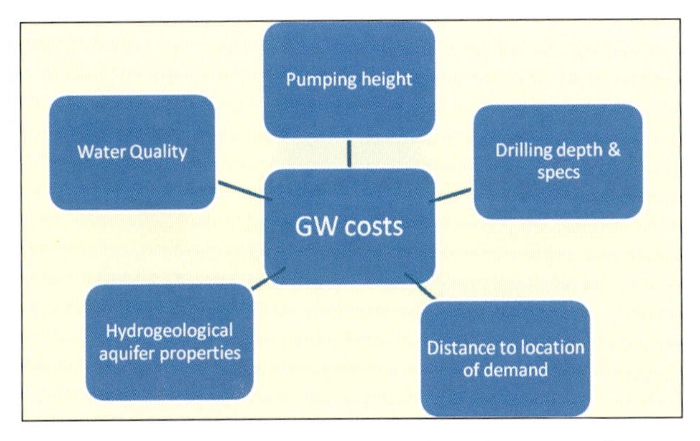

Figure 3: Parameters determining the cost of non-renewable groundwater.

Conclusions

An economic approach to the quantification of non-renewable groundwater resources puts an end to classic resources calculations, which often vary in the result by orders of magnitude for the same aquifer, depending on which parameters found reasonable. It considers both supply and demand for the resources quantification, and allows for a reliable future prognosis of these.

An economic approach takes into account the spatial and temporal variability of cost and value of the resource. It integrates the single technical limits, as a result the "nearest water is the cheapest", and hence it is feasible to treat it with more effort and pump it from higher depth than groundwater from far away from the demand center. Temporal developments e.g. of pump height and water quality can be considered and quantified in terms of costs, providing the basis for multi-generation water policies and tariffs. As cost and value of the groundwater are highly variable in space and time, a coupling of reliable regional groundwater flow models and economic models is a must to deliver useful results.

An economic approach of non-renewable groundwater resources quantification clearly prioritizes the use for drinking water, as it delivers the highest economic benefit. Any agricultural use of non-renewable groundwater is economically inefficient, as the value of the water is much less than in domestic use. The breakdown to cost and value of the groundwater resource enables easy and comprehensive groundwater management decisions.

References

GDC – Groundwater Development Consultants (International) Ltd., 1980: Umm Er Radhuma Study; Riyadh (Ministry of Agriculture and Water).

Proceedings "Hydrogeology of Arid Environments", p. 36–39
Stuttgart, March 2012

Extended Abstract

Investigations of Recharge to an Arid Zone Freshwater Lenses

A. Fadlelmawla[1], K. Hadi[1], K. Zouari[2]

[1] Kuwait Institute for Scientific Research, Water Resources Division, email: afadl@safat.kisr.edu.kw
[2] International Atomic Energy Agency, Hydrology Program

Key words: groundwater recharge, salinization processes, Kuwait, subsurface runoff, freshwater lenses, chloride mass balance

Introduction

Despite the extreme aridity of the area, the infiltration of the rainwater and runoff from the occasional rainstorms of Kuwait is known to produce freshwater lenses in the natural depressions of the northern part of the country (Parsons Corporation 1964). Al-Raudhatain depression (Fig. 1) has the largest freshwater accumulation in Kuwait, which occurs as lenses floating on top of the regional brackish groundwater. The importance of these lenses as the country's only natural freshwater reserve and its fragile stability call for thorough management schemes for its utilization. In this context, several studies were dedicated to these depressions (Parsons Corporation 1964, Bergstrom & Aten 1964, Omar et al. 1981, Al-Sulaimi et al. 1988, and others). These studies have provided a wealth of in-

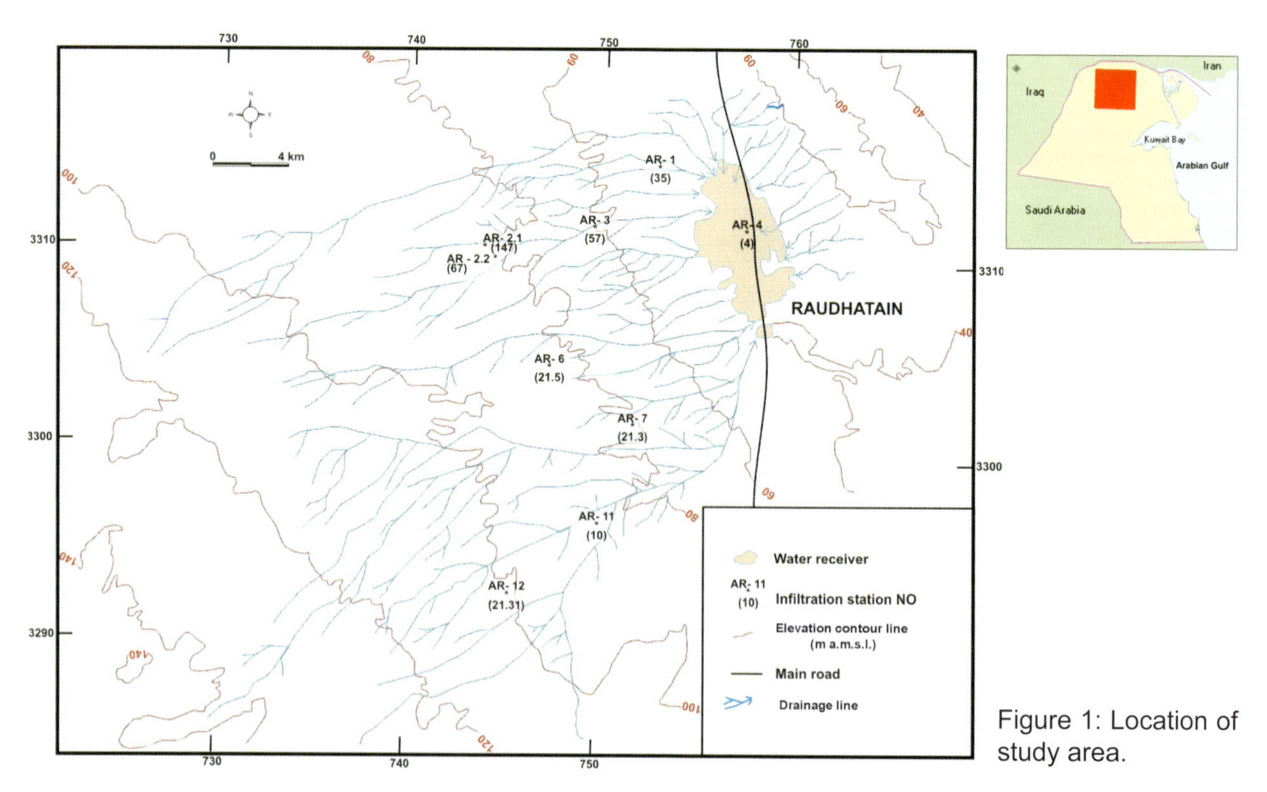

Figure 1: Location of study area.

www.borntraeger-cramer.de
2012/0036/$ 1.00

formation that characterizes, in details, many of the hydrogeological aspects of the area. In this paper, the aim is to present a coherent description of the recharge and salinity evolution processes as well as estimate the recharge rates at and in the vicinity of the depression.

According to Al-Sulaimi et al. (1988) the drainage system that flows into Al-Raudhatain depressions consists of 12 wadis (total area of 670 km^2 that drain into the main depression from all directions (Fig. 1). The drainage basins of these streams have a total surface area of about 670 km^2.

The unsaturated zone at Al-Raudhatain is about 30 m depth of poorly sorted gravelly sand with thin layers of medium grained sandstone. The aquifer is mostly porous sandstone of about 50 m thickness, followed of undetermined thickness of siltstone (Parsons Corporation 1964).

Materials and Methods

The X-ray diffraction (XRD) technique for semi-quantitative measurement was used to determine the various minerals abundance in the 92 collected soil samples. A total of 161 groundwater samples were collected from 43 wells. All samples were collected from the depression and its immediate vicinity with an average sampling depth of 11 m below groundwater table. The distribution of samples on the above groups is as follows; 43 samples for major anions and cations, 43 samples for trace elements, 31 samples for ^{13}C and ^{14}C, and 44 samples for ^{18}O, ^{2}H and ^{3}H. For ^{14}C analysis, and to avoid complications of transporting large volumes of groundwater, field precipitation of bicarbonate was conducted to satisfy the IAEA laboratory requirements of 2.5 g of carbon. Two geochemical models were utilized to reconstruct the geochemical conditions/reac-

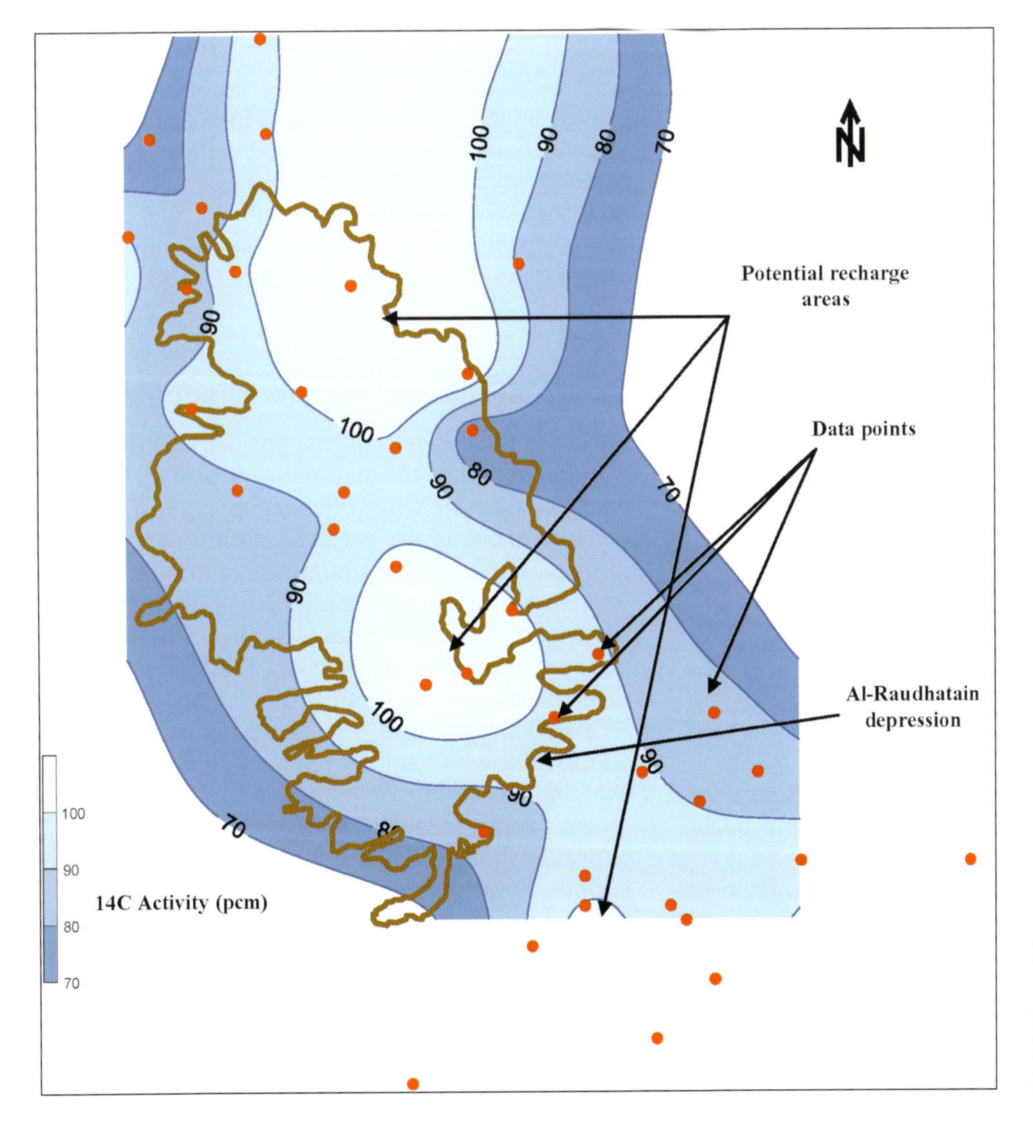

Figure 2: Carbon-14 activities contour map highlighting potential recharge areas.

tions that controlled the evolution of the observed groundwater chemistry. Those models are: a speciation model (WATEQ4F; Ball & Nordstrom 1992) to calculate the equilibrium status, species activities and minerals saturation indices, and a mass balance model (NETPATH; Plummer et al. 1991) to interpret net geochemical mass-balance reactions of plausible phase masses. Recharge to the groundwater was estimated using chloride balance method as described in (Bazuhair & Wood 1996).

Results and Discussions

Recharge characterization and estimation. The ^{14}C activities were high, ranging between 80 and 109% of the atmospheric concentrations, with an average of 91%, while $\delta^{13}C$ varied from –13.39 to –6.1‰ v-PDB. Overall, 55% of the samples show apparent ages less than 500 years. These results are in agreement with those obtained by Bergstrom & Aten (1964). Only 40% of the analysed samples contained detectable ^{3}H in low concentrations (<1 TU). Nonetheless, mere detection of ^{3}H implies that the groundwater sample is younger than 45 years old. Combining the implications from ^{14}C and ^{3}H results one can conclude the following; the sampled groundwater at the depression and its vicinity contain significant portions of recharges from recent (mostly less than 50 years old) rainfall events.

Figure 2 shows three areas where ^{14}C activities are highest, since these areas also have the freshest groundwater, they were considered locations of recharge. More importantly, these observations indicated the locality of recharge.

Figure 3 is the $^{18}O/^{2}H$ relationship; on the same Figure, the Global Meteoric Water Line ($\delta^2H = 8\ \delta^{18}O + 10$; Craig 1961), North Oman Meteoric Water Line, and the United Arab Emirates (UAE) local meteoric water line ($\delta^2H = 8\ \delta^{18}O + 17$; Zouari 2004). Examining this Figure, it can be seen that the samples with TDS<1000 mg/L, i.e. least mixed with regional groundwater, are in good agreement with the UAE meteoric water line (EMWL) confirming: (1) recharge by infiltration from local rainfall events and (2) minor evaporation during recharge.

Geochemistry of recharge. The equilibrium state of the groundwater samples is dominated by sulphate complexes ($CaSO_4$, $MgSO_4$ and $NaSO_4$). The groundwater is under-saturated with respect to halite, which might be due to the absence or insignificant presence of this mineral in the top soil and solid phase matrix, as was evident in the results of the XRD analysis. Gypsum, dolomite and calcite are all close to equilibrium. Results of solid-liquid phases mass transfer calculations by NETPATH show that

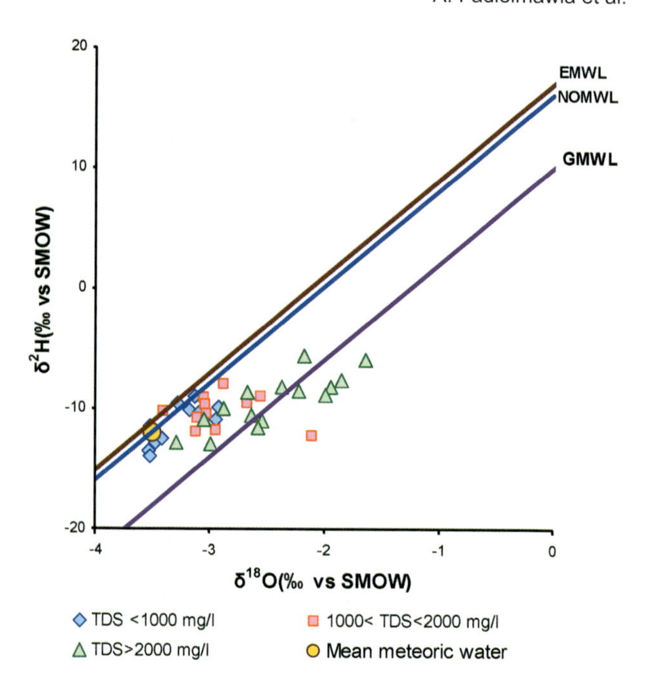

Figure 3: Oxygen-18/^{2}H diagram of Al-Raudhatain groundwater.

mass transfer between illite and the groundwater is minimal confirming the implications given above on illite equilibrium. Calcite is mostly precipitating except for few localities. Both dolomite and gypsum are dissolving in all samples; however, the later is contributing more to the liquid phase ionic composition.

Conclusions and Further Discussions

Local rainfall events are responsible for freshwater accumulations at the depressions of northern Kuwait as suggested from the comparison of $^{18}O/^{2}H$ concentrations in the groundwater to the local meteoric line. Such recharging rainstorms have contributed to the lenses during the last 30 to 40 years as evident from the ^{3}H results. Given the low infiltration rates at the depression, the lack of evaporation signs on the fresh groundwater evident by $^{18}O/^{2}H$ activities, and the localization of recharge supported by ^{14}C activities at the depressions: the likely sequence for recharge is through infiltration at the wadis followed by subsurface runoff (i.e. rainwater that infiltrates the top soil and flows in a surface layer to its discharge location) towards the depression. At the depression the subsurface runoff percolates to the groundwater. Other contributions from the observed ponds at the depression are also likely; however in insignificant volumes as discussed above. It remains to indicate that portion of the infiltrated waters at the wadis percolates to the regional groundwater, which is evident in low salinity pockets outside the depression.

The salinization of the recharge water is controlled by three processes, evaporation, dissolution of the saturated and unsaturated zones minerals, and mixing with the regional brackish to saline groundwater system. The evaporation process is taking place mostly during the early stages of infiltration, subsurface runoff and percolation to the water table; nonetheless, its magnitude is minimal as evident from the relation of $\delta^{18}O/\delta^{2}H$ and the local meteoric water line. During these early stages and during the presence of the water in the upper parts of the aquifer, the geochemical interactions between the solid and liquid phases are the main process for increasing the salinity from nearly nil to around 1000 mg/L. The main geochemical processes controlling the salt content of the water at this stage are dissolution of albite and gypsum and precipitation of calcite. As the water percolates deeper or in localities adjacent to brackish water at shallow depths, mixing with the brackish/saline water contributes significantly to the salinity of the lens. The effect of the mixing process is evident from the deviation of the $\delta^{18}O/\delta^{2}H$ of the groundwater samples with salinities exceeding 1000 mg/L from the meteoric water line. It was estimated using stable isotopic results that the mixing process is contributing about 20% of the salts at this stage. During the final stage and closer to the brackish/saline regional groundwater, the impact of mixing grows to more than 60% of the salts; nonetheless, the dissolution remains a significant contributor. Recharge, based on chloride balance method estimates, averaged around 25mm. Though this estimate is significantly high for an arid, it agrees with the described recharge process that suggests subsurface runoff from substantial areas to the depression.

References

Al-Sulaimi, J., Amer, A., Saleh, N., Salman, A. S., Ayyash, J. & El-Sayed, M. I., 1988: Delineation of watersheds in Kuwait and their geomorphological characteristics. – Report no. KISR2856, Kuwait Institute for Scientific Research, Kuwait.

Bazuhair, A. S. & Wood, W. W., 1996: Chloride mass-balance method for estimating ground water recharge in arid areas: example from western Saudi Arabia. – Journal of Hydrology **186**: 153–159.

Ball, J. W. & Nordstrom, D. K., 1992: WATEQ4F: a program for the calculating speciation of major, trace and redox elements in natural waters. – International Ground Water Modeling Center (IGWMC), USA.

Bergstrom, R. & Aten, E., 1964: Natural recharge and localisation of fresh groundwater in Kuwait. – J. Hydrol. **2**: 213–231.

Craig, H., 1961: Isotopic variations in meteoric waters. – Science **133**: 1702–1703.

Omar, S. A., Al-Yacoubi, A. & Senay, Y., 1981: Geology and groundwater hydrology of the State of Kuwait. – J. Arabian Gulf and Arabian Peninsula Studies **1**: 5–67.

Parsons Corporation, 1964: Groundwater Resources of Kuwait, vols I, II and III. – Ministry of Electricity and Water, Kuwait.

Plummer, L. N., Prestemon, E. C. & Parkhurst, D. L., 1991: An interactive code (NETPATH) for modelling NET geochemical reactions along a flow path. – US Geol. Survey Water Resources Investigations Report 91–4078. USGS, Reston, Virginia, USA.

Zouari, K., 2004: Isotopes hydrology techniques in water resources management. – Assessment of Artificial Groundwater Recharge:Tawiyaen and Wurrayah Dams (UAE). Unpublished IAEA Report.

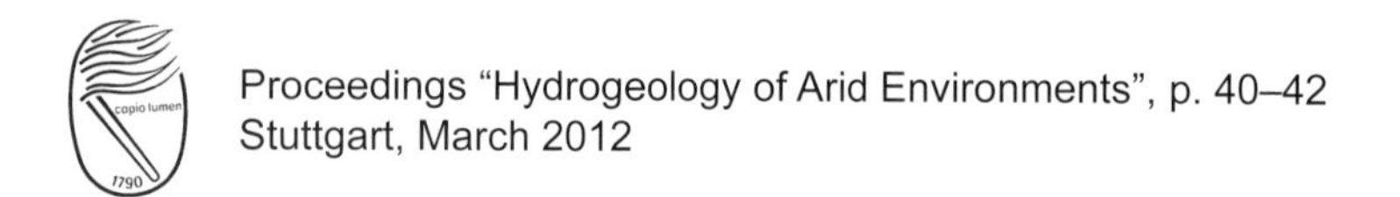
Proceedings "Hydrogeology of Arid Environments", p. 40–42
Stutgart, March 2012

Extended Abstract

Historical Background of Water Resources and Current Management Initiatives in the Semi-Arid Plain of Souss-Chtouka (Morocco)

Y. Fakir[1], M. Le Page[2], A. Aslikh[3], A. Fanzi[3], M. Huber[4]

[1] UCAM, University Cadi Ayyad of Marrakech, Morocco, email: fakir@ucam.ac.ma
[2] CESBIO, Joint Research Unit (CNRS,UPS,CNES,IRD), France, email: michel.lepage@cesbio.cnes.fr
[3] Agency of the Hydraulic Basin of Souss-Massa-Draa, Morocco, email: haaslikh@gmail.com, abdfanzi@gmail.com
[4] geo:tools, Germany, email: markus.huber@geo-tools.de

Key words: groundwater, hydrology, drought, DSS

Introduction

In arid and semi-arid regions, naturally limited water resources are often concomitantly affected by water shortage and increasing water demands. This is the current case of Morocco, located in the extreme north-west of Africa. Moreover, several recent studies dealing with the projected global changes in water ressources (Alcomo et al. 2007, kundzewicz et al. 2008) place the country in the regions that will be affected in the future by more severe water stress. Therefore, reconstructing of water resources management policies is a major key factor for resolving current and future water scarcity problems.

The study area and Historical Background of its Water Resources

The Souss-Chtouka plain (5500 km^2) is the most southern atlantic plain of Morocco. The region is a dynamic socio-economic pole. It is the largest exporter of vegetable and citrus products in Morocco, and the Agadir, the main city of the region, is the first seaside resort. The climate is of semi-arid type. The surface water resources are composed of the Souss river and its tributaries (Fig. 1). The rivers have intermittent flow regime, because the dry season is typically very long (6–8 months/year). Groundwater is provided by the alluvial unconfined aquifer of Souss-Chtouka (Dijon 1969). It flows generally from east to west. Preferential recharge areas are located mainly along the rivers (Bouchaou et al. 2008).

The surface water mobilization has been performed by the construction of 6 dams, the first one at 1972 and the last one at 2002. Artificial recharge by water release from Aoulouz dam has been applied since 1991 along the upstream part of Souss river. The Aoulouz dam feds also the Guerdane chanel, constructed in 2009 and destined to irrigate 10000 ha of citrus in the El Guerdane perimeter, previously irrigated only by groundwater only.

The use of private groundwater pumping for irrigation began in 1940. At this time, the agriculture was traditional, using derivation of surface water, springs and khettaras. In 1940, irrigation pumping was estimated to 8 hm^3. Consumption of groundwater for agricultural purposes has rapidly increased to 22.6 hm^3 in 1956, 85.5 hm^3 in 1963 and 124.3 hm^3 in 1969 (Combe & El Hebil 1977). Recently, the groundwater abstraction is estimated around 500 hm^3 (Fig. 2).

Issues and Impacts of Water Exploitation

Since the early 1970s, the frequency of droughts has increased. The latter have caused decrease in available surface water, reduction in aquifer's recharge and increase in water abstraction. The aquifer of Souss-Chtouka which initially had the best offer in water resources, has suffered from heavy exploitation. The exploited groundwater is devoted to agricultural development and drinking water supply

www.borntraeger-cramer.de
2012/0040/$ 0.75

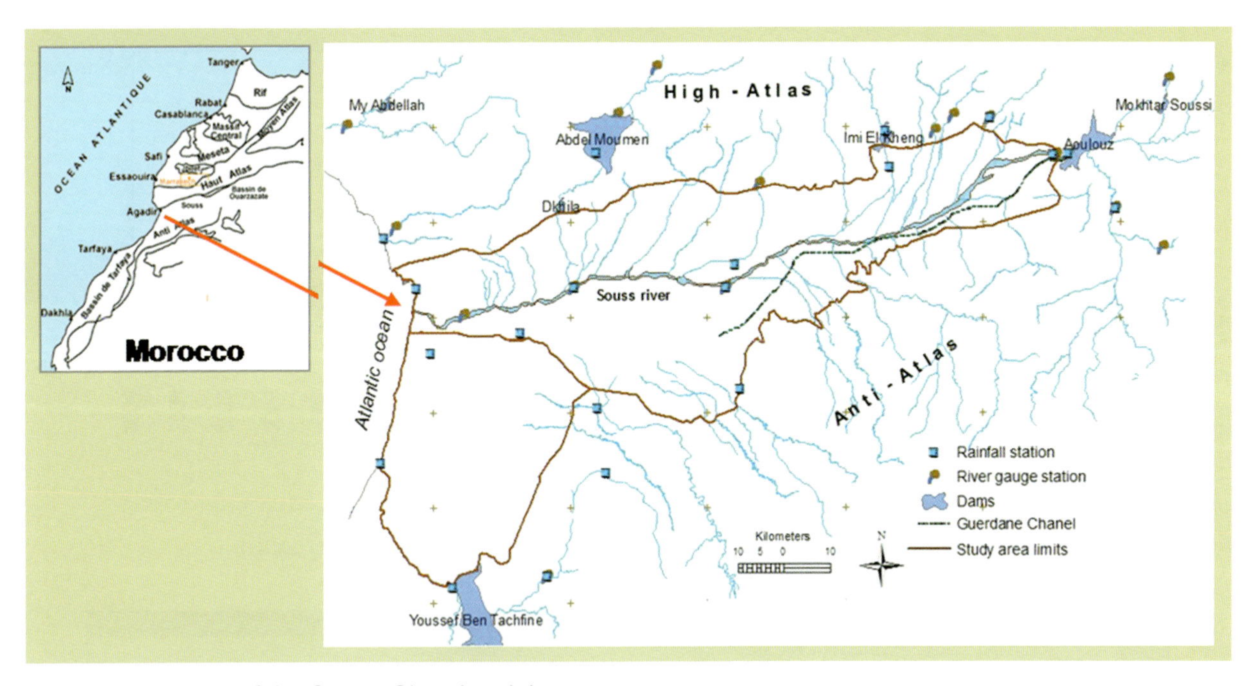

Figure 1: Location of the Souss-Chtouka plain.

both for urban and rural population. It also has filled deficits related to surface water availability as well as deficits induced by delays in the implementation of water projects, in applying saving water practices and in managing the increasing water demands.

Despite efforts at surface water mobilization, the degradation of groundwater was important, as assessed by various indicators: long-term declines in groundwater levels, drying of springs and Khettaras, drops of well fields productivity and relocation of pumping wells.

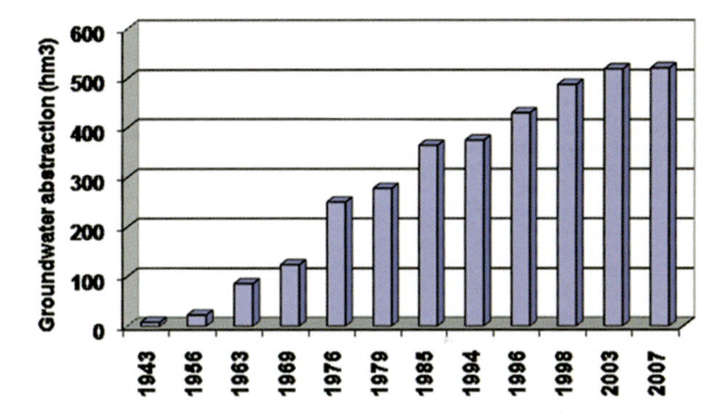

Figure 2: Historical Evolution of groundwater abstraction for irrigation.

Current Management Initiatives

During the last years, the water managers implemented new polices for a better integrated water management, through both technical and regulatory measures. In this context, a groundwater agreement was recently established between the Hydraulic Agency of Souss-Massa-Draa (ABHSMD) and its partners aiming to elaborate a participatory policy of groundwater protection. On another hand, a Decision Support System (DSS) for integrated management of water resources is under development in the framework of AGIRE program (GIZ, ABHSMD). The DSS is similar to the one applied for the Haouz basin (Fakir et al. 2010, Le Page et al. accepted) and composed of a GIS which gathers spatial and temporal data, SAMIR (Satellite Monitoring for Irrigation) is a tool devoted to the estimation of agricultural water demand by remote sensing imagery, WEAP21 (Water Evaluation and Planning system) used for integrated water resources planning (Fig. 4) and MODFLOW which is the linked code for groundwater flow modeling. From a technical side the DSS aims to compare, spatially and temporally, monthly sectoral water demands with regards to available surface and groundwater resources within the period since 2003. The impact of the present management policy on the whole water system will then be assessed for various hydrological conditions.

References

Alcamo, J., Florke, M. & Marker M., 2007: Future long-term changes in global water resources driven by socio-economic and climatic changes. – Hydrologi-

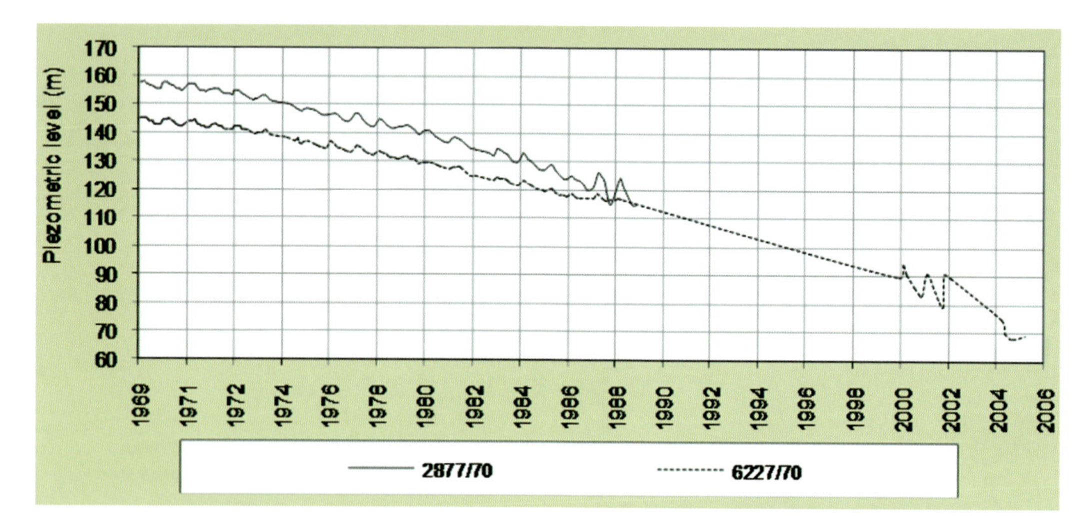

Figure 3: Decline of piezometric levels in one of the most affected sector (El Guerdane).

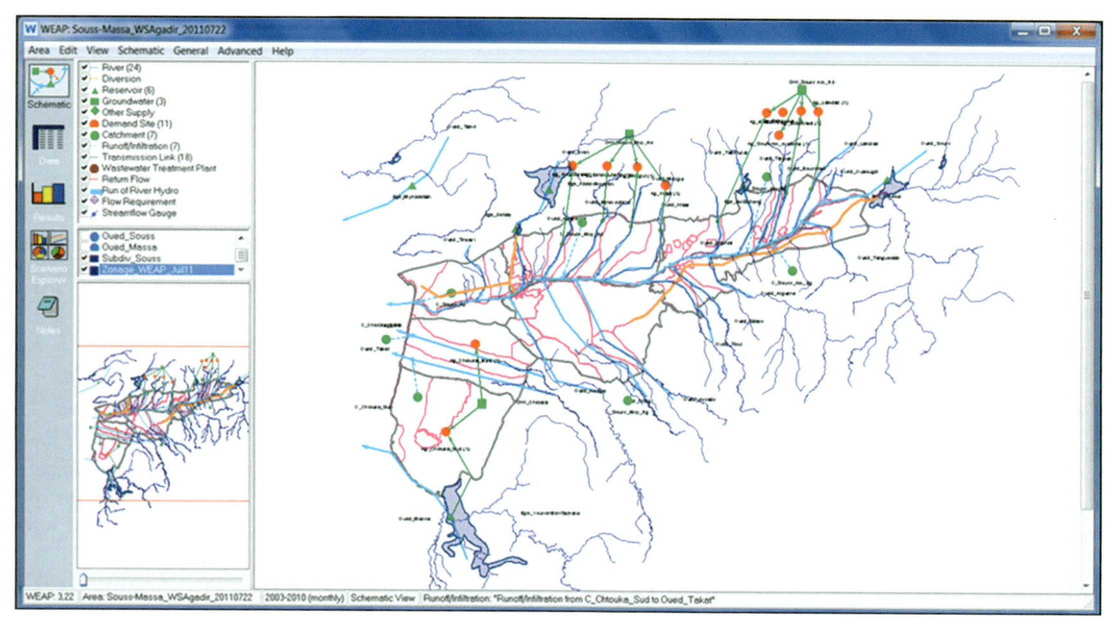

Figure 4: Schematic presentation of WEAP model of the Souss-Chtouka basin.

cal Sciences, Journal des Sciences Hydrologiques, **52**(2): 247–275.

Bouchaou, L., Michelot, J. L., Vengosh, A., Hsissou, Y., Qurtobi, M., Gaye, C. B., Bullen, T. D. & Zuppi G. M., 2008: Application of multiple isotopic and geochemical tracers for investigation of recharge, salinization, and residence time of water in the Souss–Massa aquifer, southwest of Morocco. – Journal of Hydrology (2008) **352**: 267–287.

Combe, M. & El Hebil, A., 1977: Vallée du Souss, in Ressources en eau du Maroc, t3. Domaines atlasique et sud-atlasique. – Notes et Mémoires du Service Géologique Maroc **231**: 169–201.

Dijon, R., 1969: Etude hydrogéologique et inventaire des ressources en eau de la vallée du Souss. Notes et Mémoires du Service Géologique Maroc 214, 291.

Fakir, Y., Berjamy, B., Tilborg, H., Huber, M., Wolfer, J., Le Page, M. & Abourida A., 2010: Development of a decision support system for water management in the Haouz-Mejjate plain (Tensift basin, Morocco). – XXXVIII IAH Congress, Krakow, 12–17 September 2010.

Kundzewicz, Z. W., Mata, L. J., Arnell, N. W., Doll, P., Jimenez, B., Miller, K., Oki, T., Şen, Z. & Shiklomanov, I., 2008: The implications of projected climate change for freshwater resources and their management. – Hydrological Sciences Journal **53** (1): 3–10

Le Page, M., Fakir, Y., Berjamy, B., Bourgin, F., Jarlan, L., Abourida, A., Benrhanem, M., Jacob, G., Huber, M., Sghrer, F., Chehbouni, G., accepted: An integrated DSS for groundwater management based on Remote Sensing. The case of a semi-arid aquifer in Morocco. – Journal of Water Resources Management.

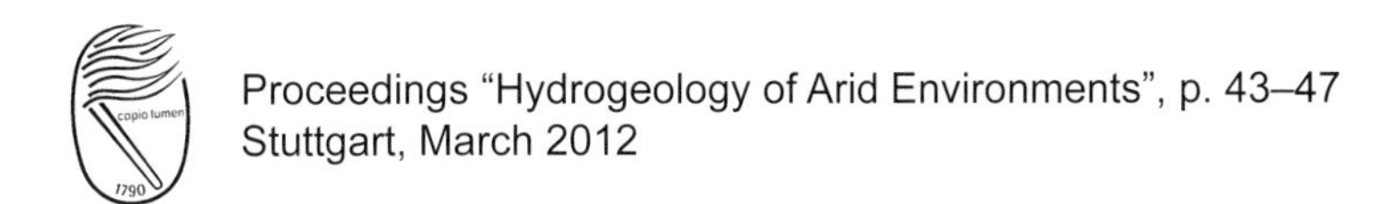

Proceedings “Hydrogeology of Arid Environments”, p. 43–47
Stuttgart, March 2012

Extended Abstract

Saltwater Intrusion Balances in the Nubian Aquifer System

W. Gossel[1], A. Sefelnasr[2], P. Wycisk[3]

[1,3] Martin Luther University Halle, Inst. for Geosciences and Geography, Dept. Hydrogeology and Environmental Geology, V.-Seckendorff Platz 3, D-06120 Halle, email: wolfgang.gossel@geo.uni-halle.de
[2] Geology Department, Assiut University, Assiut, Egypt, email: ahmed.sefelnasr@daad-alumni.de

Key words: Nubian Aquifer System, saltwater intrusion, regional groundwater modelling, groundwater balances, long-term modelling

Introduction

The Nubian Aquifer System is one of the biggest groundwater systems in the world with an area of about 3 million km^2 and a maximum thickness of about 4000 m. The position in the Northeast of the Sahara, a desert with a precipitation in this part of less than 10 mm/year and a potential evapotranspiration of 6000 mm/year makes the aquifer the most important water resource in the biggest part of Egypt, the eastern part of Libya and the northern part of Sudan. Even a small part of Chad is tangented by this aquifer system. The management of this water ressource is highly ranked in the different development scenarios of the adjacent countries. Diverse publications focussed this topic in the last decades (Thorweihe & Heinl 1999, Ebraheem et al. 2002, 2003, 2004, Gossel et al. 2004, 2008, Sefelnasr 2007). The area of the Nubian Aquifer System is shown in Figure 1.

In the Northern part of the aquifer system an interface between saltwater and freshwater exists that was assumed to be stable. The oases Siwa and Jaghbub in the Northern part of Egypt, several small oases in Libya and the Qattara depression suffer from this saltwater ingression. Even in the Nile Delta the saltwater-freshwater interface reaches several tens of kilometres to the South of the coastline and affects even the North of Cairo. The origin of the saltwater intrusion and the fluctuation dynamics during the Pleistocene was only recently clarified by numerical groundwater modelling approaches. Vertically high resoluted and density driven groundwater flow and transport modelling shows the high fluctuation of the interface in the last 140000 years. The influences regarded in these models range from time dependent groundwater recharge to the seawater level changes. The model results show, that the most important impact is the seawater level. An interesting question that leads also to the influence of the recent development scenarios is that of water balances and water exchange between the Mediterranean Sea and the Nubian Aquifer System. Model results in this question may help to answer the question if and how the interface can be repelled more to North or how groundwater discharge for higher consumption in the Oases will influence the dynamics of the interface.

Hydrogeological overview

The Nubian Aquifer System is built in its main parts by Paleozoic to lower Cretaceous sandstones. During the Cretaceous the sedimentation changes in the Northern part of the area to claystones and in the Tertiary limestones cover the Mesozoic rocks. An overview of the surface hydrogeology gives the map in Figure 1. The hydrogeological structures in the North are shown in a cross-section in Figure 2. Big basin structures of the Paleozoic lead to a differentiation of aquifer thickness, groundwater flow patterns and groundwater potential. In the big Egyptian Oases depressions of the surface lead to reduced depths to groundwater and even – in former times – to springs. Even nowadays some of the wells in the oases are free flowing. Hydrogeological evaluation of the geological formations makes it possible to

www.borntraeger-cramer.de
2012/0043/$ 1.25

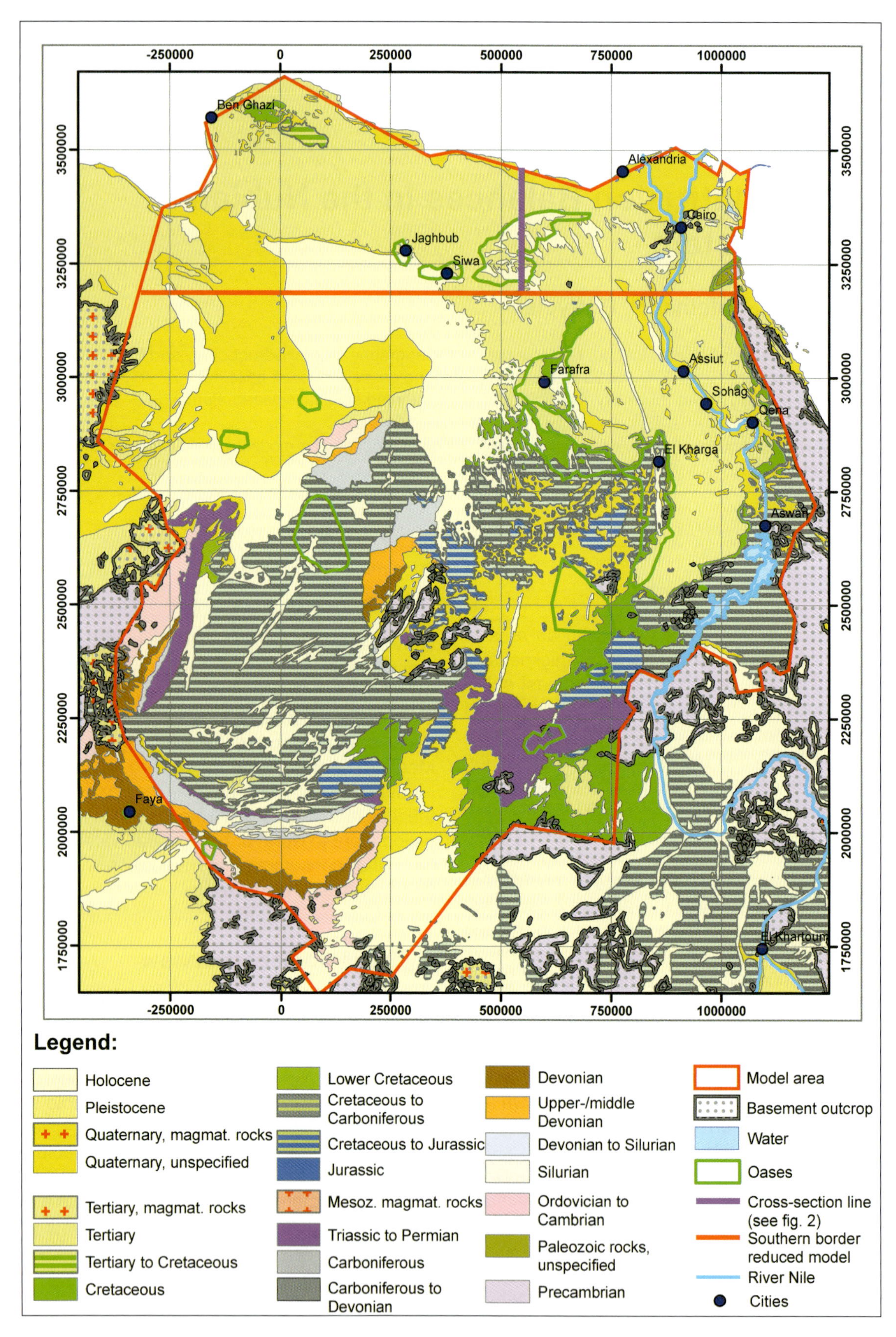

Figure 1: Hydrogeological map of the classified surface geology. The reduced model area and the cross-section line for Fig. 2 are outlined.

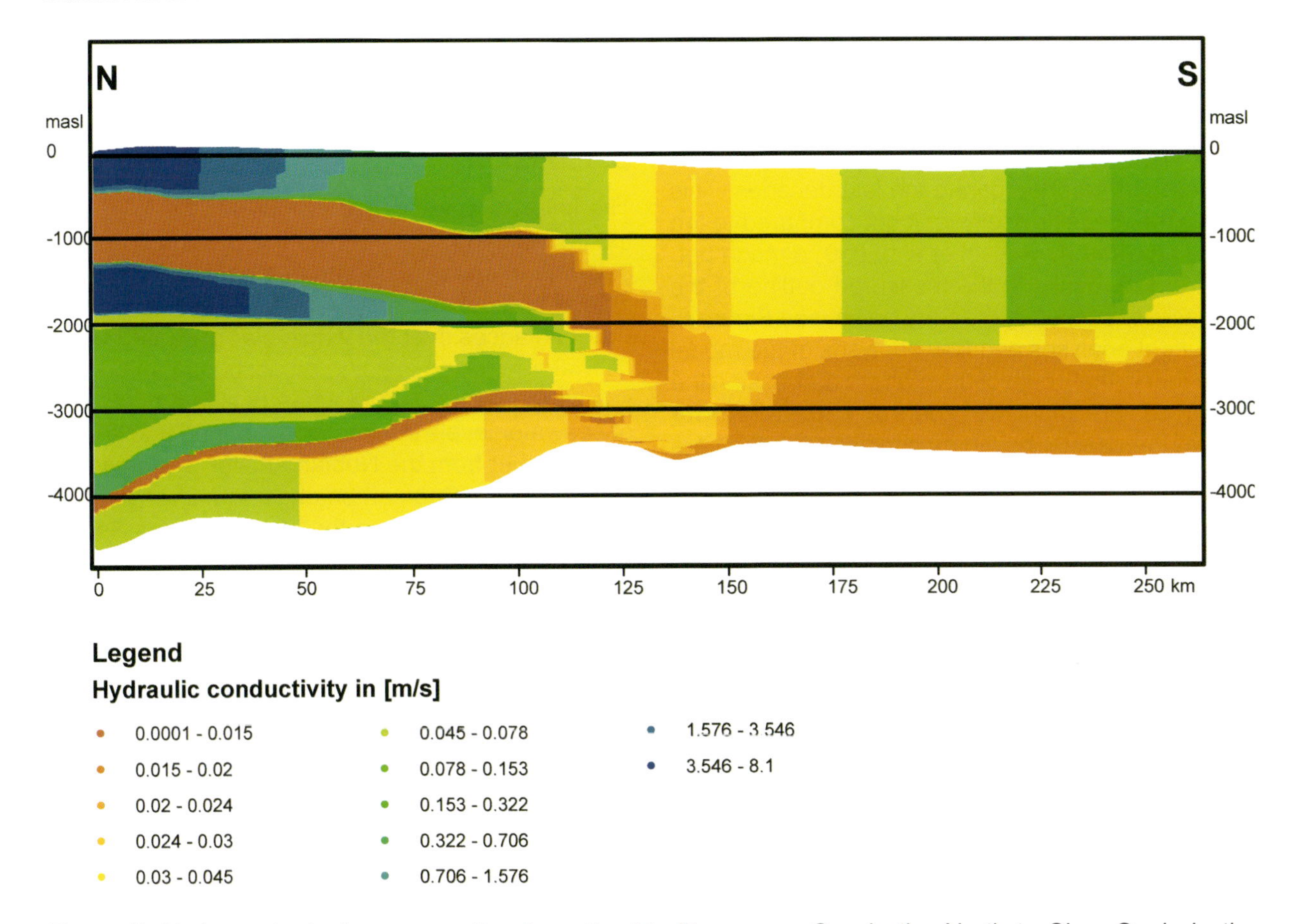

Figure 2: Hydrogeological cross section from the Mediterranean Sea in the North to Siwa Oasis in the South. The cross-section line is shown in Figure 1.

represent the diverse geology by 8 hydrogeological layers as shown and described in detail in Gossel et al. (2004).

Modelling approach

To get a profound knowledge of the amount and the dynamics of the groundwater balances also for distinct layers, the numerical groundwater model was set up with a high vertical discretization. Each hydrogeological layer was divided into 5 numerical layers. Additionally the horizontal resolution was enhanced at the coastline. This increased resolution made a reduction to the Northern part of the model area necessary, otherwise the calculation times would have been too long (approximately 7 weeks with the reduced model). The reduced model area is outlined in Figure 1. For the density driven approach the saltwater with a general salt concentration of 35000 mg/L was assumed to have a density of 1.025 kg/L whereas the freshwater was modelled with a density of 1.000 kg/L.

The model was set up based on a GIS database for the hydrogeological structures and parameter distributions. Compared to the model of the total area of the Nubian Aquifer System a higher vertical discretization was necessary to get better results for the groundwater balances. The boundary conditions were identified quite simple as shown in Gossel et al. (2010) with mainly no flow boundaries in the East and West. At the Northern boundary the coastline of the Mediterranean Sea was implemented as a Dirichlet boundary condition with time varying seawater levels. Additionally the River Nile was implemented with Dirichlet boundary conditions that changed the levels forced by the seawater levels. The boundary condition in the South was derived from the results of the big model area. The goundwater recharge was also implemented as reported in Gossel et al. (2010).

The finite element modelling tool Feflow was used to calculate the model on a high end PC with 12 cores, 12 GB RAM and SSD.

Results

The only possibility to calibrate the model according to the objectives is the recent outline of the saltwater-freshwater interface. Changes of the interface are not reported during the last decades. This makes the model in itself weak but compared to the model reported by Gossel et al. (2010) a much better fit of the results to the measured outline of the interface is reached in the Sirte basin, see Figure 3.

Budget analyses and flow analyses of the model show that in the South of the Qattara depression the differences are about two orders of magnitude lower than at the coastal boundary condition but they are changing widely over the model time. The differences between inflow and outflow are again two orders of magnitude lower than the exchange fluxes and indicate therefore a high stability over time. In general the model shows that the influence of groundwater recharge is quite low compared to the influxes from the Mediterranean Sea.

Going into more detail, the lower layers show higher infiltration rates than the upper layers. Additionally layers with higher hydraulic conductivity lead to a faster progress and deeper intrusions. Therefore the intrusion starts faster in the upper layers than in the lower layers.

Discussion

The influx of seawater of several Million m^3/d over the last 140000 years shows the problems to push back the interface. The intruded saltwater sums up in total to a volume of several km^3. Additionally the water balances over the whole glacial show that the saltwater intrusion is going on according to the high seawater levels. Only in times of a fast drawdown of seawater levels the intrusion stopped and an outflow of water from the Nubian Aquifer System to the Mediterranean Sea can be observed. Even in times of lowest water levels in the Mediterranean Sea an influx of a few Million m^3/d is registered. This influx is driven by the higher density of the saltwater compared to the freshwater in the aquifer. The analysis of the balances of the numerical layers shows this in more detail.

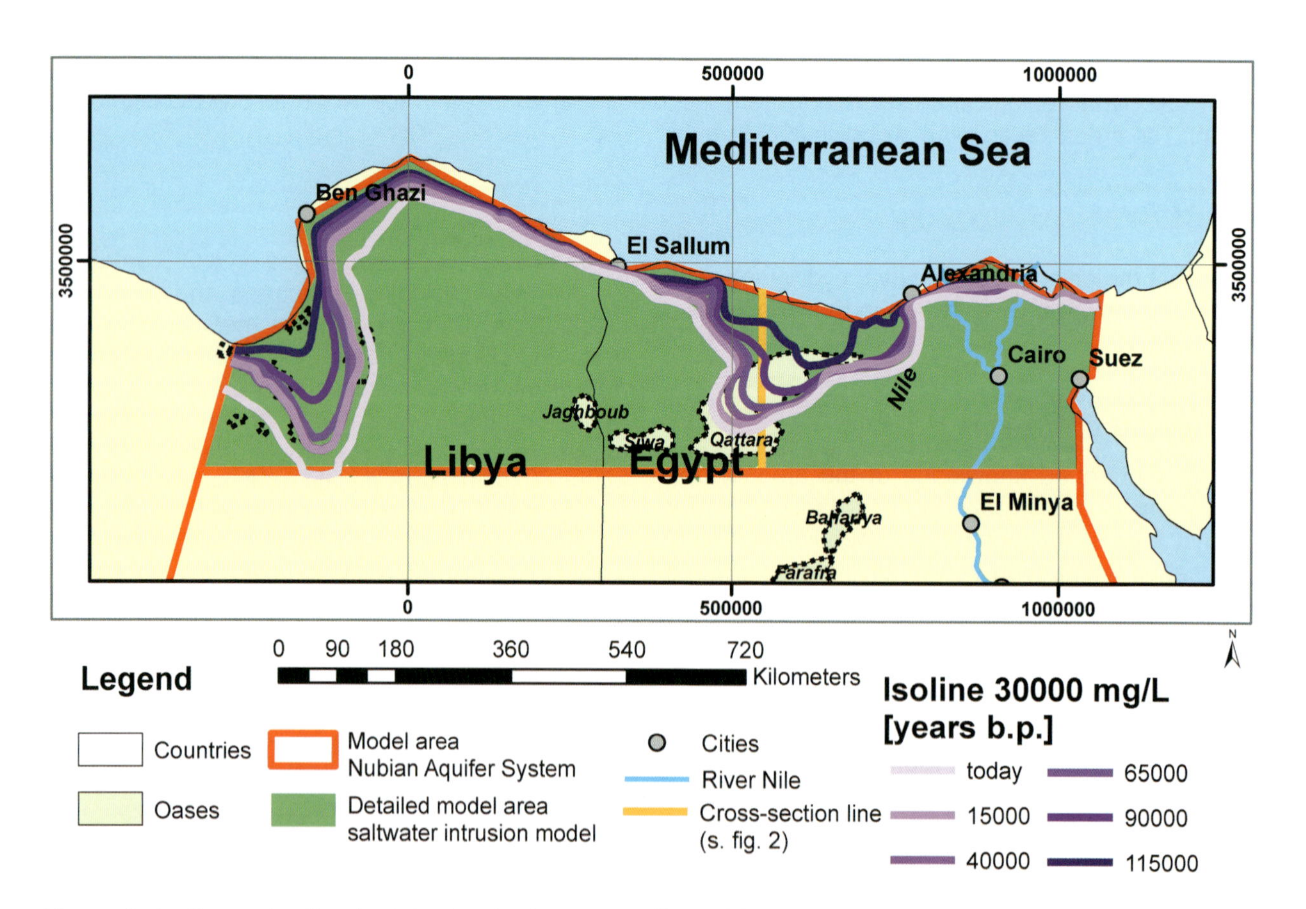

Figure 3: Isolines of saltwater intrusion in the top aquifer: 30000 mg/L isoline in the years 115000, 90000, 65000, 40000, 15000 before present and the recent distribution from the model.

References

Ebraheem, A. M., Riad, S., Wycisk, P. & Seif El Nasr, A. M., 2002: Simulation of impact of present and future groundwater extraction from the non-replenished Nubian Sandstone Aquifer in SW Egypt. – Environmental Geology **43**: 188–196.

Ebraheem, A. M., Garamoon, H. K., Riad, S., Wycisk, P., Seif El Nasr, A. M., 2003: Numerical modeling of groundwater resource management options in the East Oweinat area, SW Egypt. – Environmental Geology **44**(4): 433–447.

Ebraheem, A. M., Riad, S., Wycisk, P.& Seif El Nasr, A. M., 2004: A local scale groundwater flow model for groundwater resources management in Dakhla Oasis, SW Egypt. – Hydrogeology Journal **12**(6): 714–722; Berlin.

Gossel, W., Ebraheem A. A. & Wycisk, P., 2004: A very large scale GIS-based groundwater flow model for the Nubian Sandstone Aquifer in Eastern Sahara (Egypt, northern Sudan and Eastern Libya). – Hydrogeology Journal **12**(6): 698–713; Berlin.

Gossel, W., Ebraheem, A. A., Sefelnasr, A. M. & Wycisk, P., 2008: A GIS-based flow model for groundwater resources management in the development areas in the eastern Sahara, Africa. – In: Adelana, S. M. A. & MacDonald, A. M. (eds.): Applied groundwater studies in Africa. Balkema, Rotterdam.

Gossel, W., Sefelnasr, A. M. & Wycisk, P., 2010: Modelling of paleo-saltwater intrusion in the northern part of the Nubian Aquifer System, Northeast Africa. – Hydrogeol. Journal **18**/6: 1447–1463.

Sefelnasr, A. M., 2007: Development of groundwater flow model for water resources management in the development areas of the Western Desert, Egypt. – Diss., Martin Luther University Halle-Wittenberg, 188 pp., ULB Sachsen-Anhalt. http://sundoc.bibliothek.uni-halle.de/diss-online/07/07H178/prom.pdf.

Thorweihe, U. & Heinl, M., 1999: Grundwasserressourcen im Nubischen Aquifersystem (Groundwater ressources in the Nubian Aquifer System). – In: Klitzsch, E. & Thorweihe, U. (eds.): Nordost-Afrika: Strukturen und Ressourcen (Northeast Africa: Structures and ressources). Deutsche Forschungemeinschaft. Wiley-VCH.

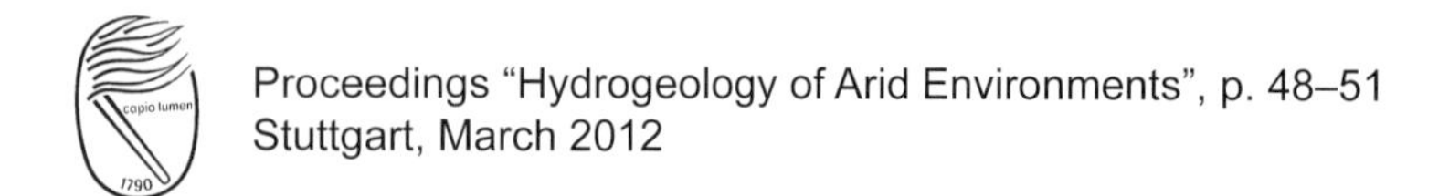

Proceedings "Hydrogeology of Arid Environments", p. 48–51
Stuttgart, March 2012

Extended Abstract

Development of a 3D Groundwater Flow Model in Semi-Arid to Arid Region: The Western Drainage Basin of the Dead Sea (Israel and West Bank)

Agnes Gräbe [1,3], Tino Rödiger[1], Olaf Kolditz[2,3], Karsten Rink[2], Thomas Fischer[2], Feng Sun[2], Wenqing Wang[2]

[1] Helmholtz Centre for Environmental Research - UFZ, Department Catchment Hydrology, 06120 Halle/Saale, Germany, email: agnes.graebe@ufz.de
[2] Helmholtz Centre for Environmental Research – UFZ, Department of Environmental Informatics, 04318 Leipzig, Germany
[3] TU Dresden, Applied Environmental System Analysis, 01062 Dresden, Germany

Key words: numerical groundwater flow modelling, Dead Sea, groundwater recharge, semi-arid to arid conditions

Abstract

The SUMAR-Project is concerned with the sustainable management of the western Dead Sea basin. Significant for the analysis of the water balance is although the drop about 20 m of the level of the Dead Sea since the 1960s and thereby enhancing the drainage of the adjacent aquifers (Möller et al. 2007). The study area comprises the subsurface catchment of the western Dead Sea basin and is affected by the cretaceous formation of the Judea Group. Recent studies (Guttman & Zuckerman 1995, Guttman 2000) have constructed a two-dimensional (one horizontal layered) numerical simulation of groundwater flow in this region (Laronne Ben-Itzhak & Gvirtzman 2005). These previous groundwater flow models focused either in the modelling the eastern and western Judea Group aquifer (area between Judean Mountain and Dead Sea shoreline) or modelling the northern part of the Negev desert in the south of the research area. In this study we gives an overview about the developed regional groundwater flow model of the eastern and southern Judea group aquifer which was achieved by the scientific software OpenGeoSys (OGS).

Introduction

Water is a scarce resource in the semi-arid to arid regions around the Dead Sea and groundwater is the only fresh water source in this area. For a sustainable management of the western Dead Sea basin a characterisation and modelling of the available water resource and the water balance are important. In this study we gives an overview about the developed regional groundwater flow model of the eastern and southern Judea group aquifer which was achieved by the scientific software OpenGeoSys (OGS). OGS is specialized in coupled hydro systems processes and calculates the groundwater flow in porous and fractured media of the aquifer system based on the finite element method (OGS 2011). Because of scarce hydrogeological data we established different model scenarios to represent the steady state of the hydraulic character of the study area. By using the model independent nonlinear parameter estimation code PEST we calibrated the numerical model on the base of hydraulic heads.

Conception Model

The study area is located at the western side of the Dead Sea, covering a total area of ~4000 km^2. The hydrological features of the study area are fractured bedrock and variable karstified rocks of aquifers of the Lower (Lower Cenomanian) and the Upper Judea Group (Upper Cenomanian - Turonian). Both aquifer together has a thickness of 800 to 850 m. Both aquifers are separated by a marly aquiclud. The highest elevated region in the west and the en-

www.borntraeger-cramer.de
2012/0048/$ 1.00

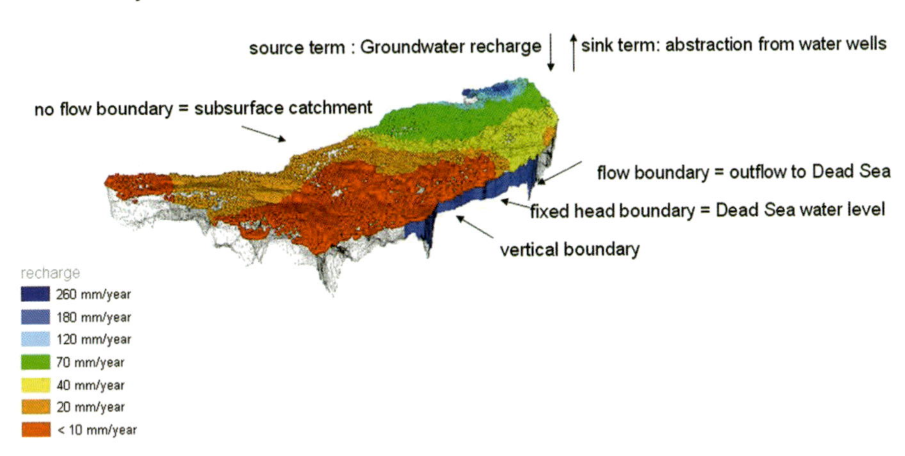

Figure 1: Spatial distribution of boundary conditions.

tire western water divide is dominated by a large N-S directed anticline. From there, the strata dip generally towards the east, the reason why the deposits of the Lower Judea Group aquifer crop out in the western part of the catchment area. Towards the Jordan Valley, the Lower Judea Group Aquifer is covered by the layers of the Upper Judea Group Aquifer. Close to the Jordan Valley, the base of the Senonian Mount Scopus Group is preserved on the top of the Upper Judea Group Aquifer. In the outcrop area both aquifers are characterized as unconfined. Local confined conditions occur where clayey or marly depositions intervene. However, towards the Lower Jordan Valley, the hydrological conditions change from unconfined to confined, forced by the covering low permeable layers. The groundwater level decreases in both aquifers as the strata dip to the east. Significant for the analysis of the water balance is although the drop about 20m of the level of the Dead Sea since the 1960s and thereby enhancing the drainage of the upper and lower aquifer (Möller et al. 2007). The irregular distribution of annual precipitation ranges from more than 700 mm on the top of the Judea Mountains to 50 mm and less in the Jordan Valley. Hence, the recharge area is strong delimited to the high elevated outcrop areas of the aquifers in the western parts of the study area.

Structural Model

We developed a structural model on the base of 579 boreholes stratigraphies, digitalized geological maps, faults, geological cross sections and geological data tables in ArcGIS. The discretization of the model domain is based on a large set of geological input data and little information about hydrological data. It takes into account surface information derived from a DEM of the region with a resolution of 30 meters as well as the course of a number of wadis and wells located in that area. The resulting finite element mesh consists of 114 327 nodes and 184 481 tetrahedral- and prism-elements. These elements have been assigned properties of 38 different materials (Rink et al. 2011). The result was a very precise 3D-Layer structural model with good mesh quality (Rink et. al. 2011b).

Boundary Conditions

The numerical groundwater flow model is based on the structural model. As starting conditions physical and chemical properties of fluid and solid phase were set as well as the material properties representing the geological formations.

The following boundary condition has been assigned (Fig. 1):

- no flow boundary conditions at the northern, western and southern border of the study area
- flow boundary condition at the eastern border of the study area which is at the same time the groundwater outflow to the Dead Sea
- constant head boundary along the Dead Sea (= water level of the Dead Sea)
- source terms = groundwater recharge, steady state model is based on the estimation of groundwater recharge by Guttman (2000).

Model Settings

For the multi-complex study area a 4-layer model was created, which contains the main geological layers of the Judea Group Aquifer system and replicates in the finite-circuit model approach the highest spatial complexity due to the distribution of hydraulic conductivity. The upper and lower aquifers are characterized by limestone and dolomite, and thus a similar stratigraphy. Therefore, the two aquifers were parameterized with the similar kf values of $3x10^{-6}$ m / s. For the aquiclude that separates the two aquifers with an average thickness of approximately 10m, a constant hydraulic conductivity of $5x10^{-9}$ m / s is assumed. Because of the poorly hydraulic conductivity of rock and sediment formation of Quaternary to Senon a constant kf-value of $5x10^{-8}$ m / s was adopted.

The 4-layer model was spatial changed in areas with important hydrogeological feature such as anticline, synclinal or fault structures (Fig. 2). (1) Along the Dead Sea transform faults that characterize this region significantly, a more permeable kf-value was assumed. These fault zones found along the Dead Sea were also reflected in parallel fault zones at the east of Hebron and Herodion: $1.4x10^{-4}$ m / s to $8x10^{-5}$ m / s. (2) Next to the shoreline of the Dead Sea there are partly large-scale zone of Quaternary sediments, which are an alternation of gravels, sands, clays and salt layers. These layers impede the groundwater inflow to the Dead Sea and were therefore given with impermeable kf-values: $5x10^{-9}$ m / s. (3) A great influence on the flow patterns of groundwater from the upper and lower aquifer has the synclinal fold near Beer Sheva and Arad. Because of this trough structure, the groundwater is channeled to the north basin of the Dead Sea. For this reason more conductive geological structure kf values were assumed: $4.5x10^{-6}$ m / s. (4) In the areas around Jerusalem to Bethlehem an accumulation of faults can be observed. With the overall high availability of water from humid conditions in the mountain area karstic conditions occurs in the near surface aquifers with a hydraulic conductivity of $1x10^{-5}$ m / s. (5) Next to the intensively used well field Bani Naim inhibitory kf values were assigned because there is a short distance in anticline and syncline structures to act as a hydraulic barrier: $1x10^{-7}$ m / s.

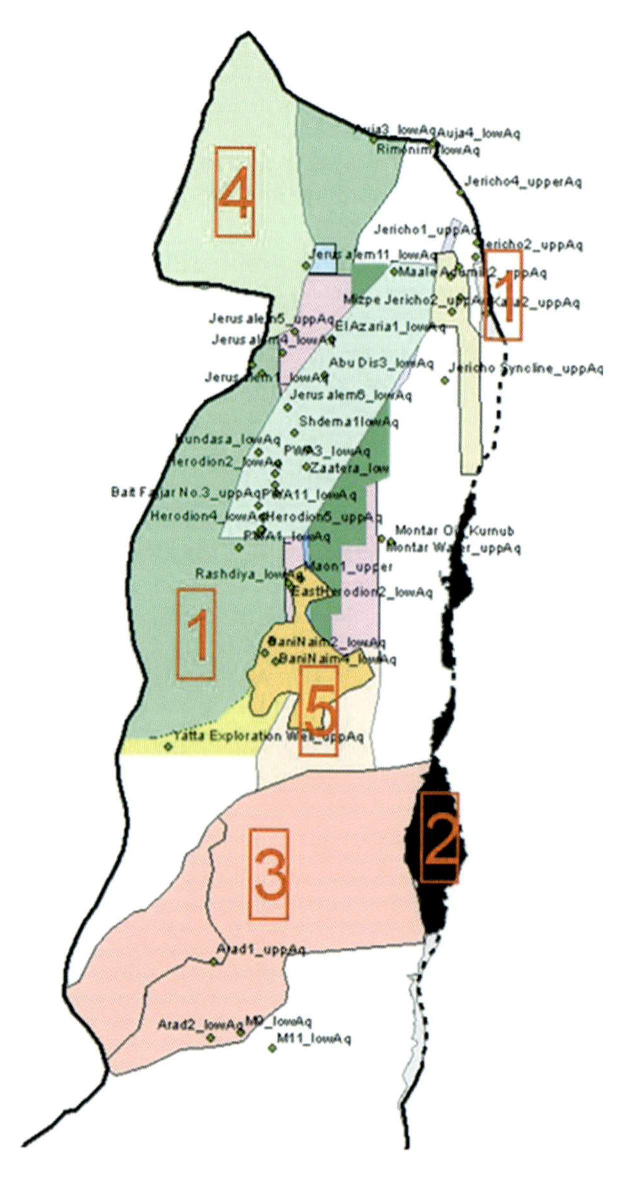

Figure 2: Overview of spatial distributed hydraulic conductivities of the upper and lower Judea Group Aquifer.

Model results

The steady state model applies the groundwater heads of appropriate wells of the 13 wells of the upper and 24 wells of the lower Judea Group Aquifer in 2006 (Fig. 3). The deviation of hydraulic heads over all wells ranges 1–89 m. The correlation coefficient, as a measure of goodness of fit, is 0.917. Generally correlation coefficients above 0.9 can be considered as acceptable (Hill 1998). The discharge along the eastern model boundary to the Dead Sea is in total 116 MCM/years.

Conclusion

In this work for the first time a 3D groundwater model was developed for the subsurface catchment of the Western Dead Sea Escarpment. The model was calibrated with all available and consistent data and the groundwater flow regime was investigated under steady state conditions. As a result of the parameter identification we found the hydraulic conductivity for the upper and lower Judea Group Aquifer ranges from 2.0×10^{-3} to 1.0×10^{-7} m/s. For more details on the study, the interested reader is referred to presentation on the conference and to Gräbe et al (2012).

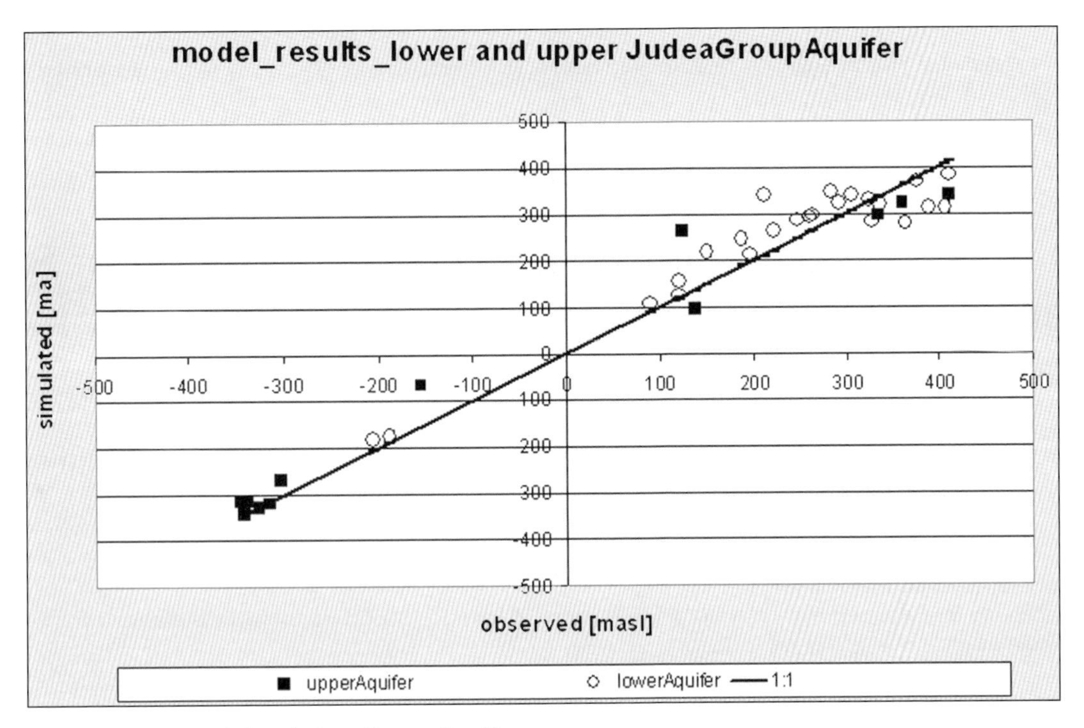

Figure 3: Model results of the Judea Group Aquifer.

Acknowledgements

The authors thank the German Federal Ministry of Education and Research for funding the SUMAR-Project (grant code: 02WM0848). Moreover, Helmholtz Impulse and Networking Fund through Helmholtz Interdisciplinary Graduate School for Environmental Research (HIGRADE) supported this work.

References

Baida, U., Goldschtoff, Y. & Qidron, I., 1978: Numerical model of the Cenomanian aquifer in the southern Yarkon-Taninim basin (Beer Sheva area). – Tahal, Water Planning for Israel, Rep. 01/78/78, p. 45

Baida, U., 1986: The basin of the Yarkon-Taninim aquifer: aspects of quality and quantity. – Proc. Int. Assoc. Hydrol. **86(6):** 51–57.

Goldschtoff, Y. & Schachnai, E., 1980: The Yarkon-Taninim aquifer in the Beer Sheva area: outline and calibration of the flow model. – Tahal, Water Planning for Israel, Rep. 01/80/58, p. 20.

Gräbe et al., 2012: Development of a 3D groundwater flow model in semi-arid to arid region: the western drainage basin of the Dead Sea (Israel and West Bank). in prep.

Guttman, Y., 1987: Salinization along the western border of the Jordan Valley and Dead Sea. – In: Terra Nostra, The 13th Meeting on the Dead Sea Rift as a unique global site. Dead Sea, Israel: The German Israeli Foundation for Scientific Research and Development.

Guttman, Y., 1988: A two layers model of flow regime and salinity in the Yarkon-Taninim aquifer. – Tahal, Water Planning for Israel, Rep. 01/88/23, p. 17.

Guttman, Y. & Rosenthal, A., 1991: Mizpe Jericho region – Study of the salinity sources and water potential. – Tahal Consulting Engineers Ltd. 01/91/46.

Guttman, Y. & Zuckerman, H., 1995: Flow Model in the Eastern Basin of the Judea and samaria Hills. – Tahal Consulting Engineers Ltd. 01/95/66.

Laronne Ben-Itzhak, L. & Gvirtzman, H., 2005: Groundwater flow along and across structural folding: An example from the Judean Desert, Israel. – J. Hydrol. **312**: 51–69.

Möller, P., Rosenthal, E., Geyer, S., Guttman, J., Dulski, P., Rybakov, M., Zilberbrand, M., Jahnke, C. & Flexer, A., 2007: Hydrochemical processes in the lower Jordan valley and in the Dead Sea area. – Chem. Geology **239**(1-2): 27–49.

OGS, 2011: OpenGeoSys Project, www.opengeosys.net

Rink, K., Fischer, T., Gräbe, A. & Kolditz, O., 2011a: Visual Preparation of Hydrological Models. Proc of MODELCARE 2011.

Rink, K., Kalbacher, T. & Kolditz, O., 2011b: Visual data management for hydrological analysis. – Environ. Earth Sci., DOI: 10.1007/s12665-011-1230-6.

Rosenthal, A. & Kronfeld, J., 1982: $^{234}U/^{238}U$ disequilibrium as an aid to the hydrological study of the Judea Group aquifer in eastern Judea and Samaria, Israel. – J. Hydrol. **58:** 149–158.

Sun, F., Shao, H., Kalbacher, T., Wang, W., Yang, Z., Huang, Z. & Kolditz, O., 2011: Groundwater drawdown at Nankou site of Beijing Plain: model development and calibration. – Environ Earth Sci, DOI: 10.1007/s12665-011-0957-4.

Weiss, M. & Gvirtzman, H., 2007: Estimating Ground Water Recharge using Flow Models of Perched Karstic Aquifers. – doi: 10.1111/j.1745-6584.2007.00360.x

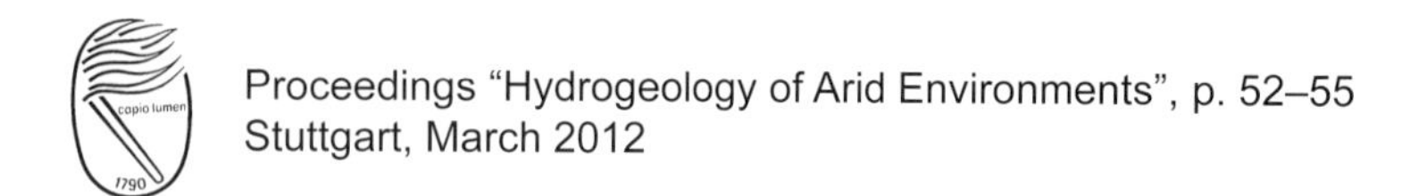
Proceedings "Hydrogeology of Arid Environments", p. 52–55
Stuttgart, March 2012

Extended Abstract

Combining Field Measurements and Modeling of Soil Water Dynamics to Quantify Groundwater Recharge in Dryland Savanna, Namibia

Alexander Gröngröft[1], Lars Landschreiber[1], Nikolaus Classen[1], Wim Duijnisveld[2], Annette Eschenbach[1]

[1] Institute for Soil Science, University of Hamburg, Allende-Platz 2, 20146 Hamburg, Germany, email: a.groengroeft@ifb.uni-hamburg.de
[2] Federal Institute for Geosciences and Natural Resources, Stilleweg 2, 30655 Hannover, Germany, email: Wilhelmus.Duijnisveld@bgr.de

Key words: soil water balance, rain-use efficiency, modeling, Savanna

Semi-arid savannas are defined by characteristic vegetation pattern consisting of a mixture of grasses with interspersed trees or shrubs and with a significant influence of grazing mammals (Scholes & Walker 1993). The ecology of these systems is under study for long term; however the role of the soil-vegetation interaction within the water cycle is not fully understood. The competition for soil water is regarded as a significant and suppressive driver of woody vegetation on grasses and other herbaceous plants (Smit & Rethman 2000). Bush encroachment which is an enormous problem for many countries in the African savanna zone (Klerk 2004) thus is seen as a result of changing grazing influence with implications on the local to regional water balance (Huxman et al. 2005).

The central Namibian thornbush savannas do exhibit strong signs of bush encroachment. Here, from 2007 onwards, the soil water contents and hydraulic heads are measured automatically at four positions. Aim of the study is to analyse the interaction of soil properties, vegetation structure and soil water balance in a water-restricted environment. Special emphasis is given to the role of *Acacia* shrubs and trees on modifying local water balance budgets.

To analyse and extrapolate the measured data modeling runs with the SWAP software (Kroes et al. 2008) have been conducted, resulting also in daily groundwater recharge values at the bottom boundary of the soil domain, here 1 m below soil surface. This approach based on studies of the unsaturated zone is mostly applied in semi-arid environments if the unsaturated zone is generally thick (Scanlon et al. 2002).

Figure 1 gives an overview on the relevant input data for the water balance modeling for both systems: Vertical layering, the corresponding soil physical properties, the proportional vertical root distribution and the leaf area indices in relation to vegetation development stages. Whereas the soils properties of both positions are quite similar the root distribution of grass and tree-dominated positions differ significantly.

Within Figure 2 a comparison of measured and modeled soil water content data are given for the period 15.09.2007–25.05.2010 for the *Acacia*-site. For all three seasons, for which rainfall amounts were substantial above average, elevated soil moisture contents reached the lower profile boundary (= saprolite layer) in February and subsequent water fluxes to deeper layers occurred. The modeled fluxes (see Table 1) at the bottom boundary where 204, 188 and 49 mm for the three seasons, respectively.

To overcome the disadvantages of untypically wet years in the monitoring period 30 years of synthetic weather data were generated based on weather data of Windhoek with a statistical approach. With these weather data soil water balance simulation runs were conducted with the optimized soil-physical and vegetation input data for the both

www.borntraeger-cramer.de
2012/0052/$ 1.00

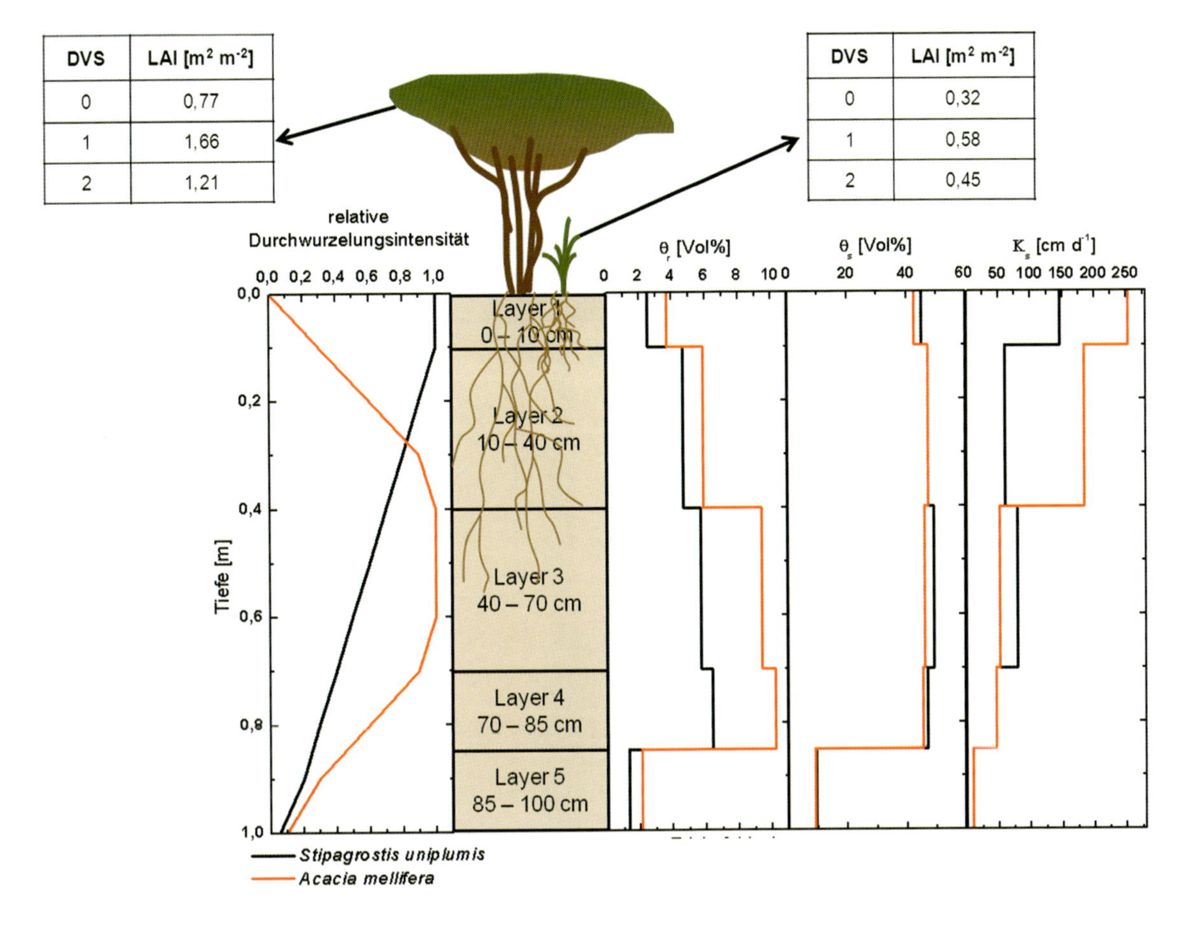

Figure 1: Overview on input data for water balance modeling.

sites compared. The long-term averages of input and output data are given in Table 2.

With respect to bottom boundary flux, the modeling results in only slight differences between both vegetation types. As suspected from other regions with semi-arid conditions (overview see Scanlon et al. 2006), the mean boundary flux is small (1.4–2.1 mm a^{-1}). Within the 30 year period water percolation in 1 m depth occurred only in 3 seasons with exceptional rainfall amounts.

The presented results are still preliminary, as pedo-transfer-functions and parameters regarding vegetation properties (leaf area indices, root distribution) have to further fine-tuned based on measured data. In our case study the condition of a thick unsaturated zone is fulfilled although the occurrence of a small saprolite layer above a fissured bedrock restricts the application of soil water balance models to the upper 1 m but roots of the woody vegetation do grow deeper and are most likely able to extract water from zones below the simulated depth. Thus we name the simulated flow bottom flux instead of groundwater recharge, as further reductions of the flows by transpiration are generally possible.

Additionally the spatial significance of the results has to be questioned. The studied soils (Haplic Luvisols, chromic) represent a significant proportion of the landscape (Haarmeyer et al. 2010) however there are patches of Calcisols as well as Vertisols which differ in water balance and might be responsible for higher groundwater recharge values.

References

Haarmeyer, D. H., Luther-Mosebach, J., Dengler, J., Schmiedel, U., Finckh, M., Berger, K., Deckert, J., Domptail, S.E., Dreber, N., Gibreel, T., Grohmann, C., Gröngröft, A. Haensler, A., Hanke, W., Hoffmann, A., Husted, L. B., Kangombe, F. N., Keil, M., Krug, C. B., Labitzky, T., Linke, T., Mager, D., Mey, W., Muche, G., Naumann, C., Pellowski, M., Powrie, L. W., Pröpper, M., Rutherford, M. C., Schneiderat, U., Strohbach, B.J., Vohland, K., Weber, B., Wesuls, D., Wisch, U., Zedda, L., Büdel, B., Darienko, T., Deutschewitz, K., Dojani, S., Erb, E., Falk, T., Friedl, T., Kanzler, S.-E., Limpricht, C., Linsenmair, K.E., Mohr, K., Oliver, T., Petersen, A., Rambold, G., Zeller, U., Austermühle, R., Bausch, J., Bösing, B. M., Classen, N., Dorendorf, J., Dorigo, W., Esler, K.J., Etzold, S., Graiff, A., Grote-

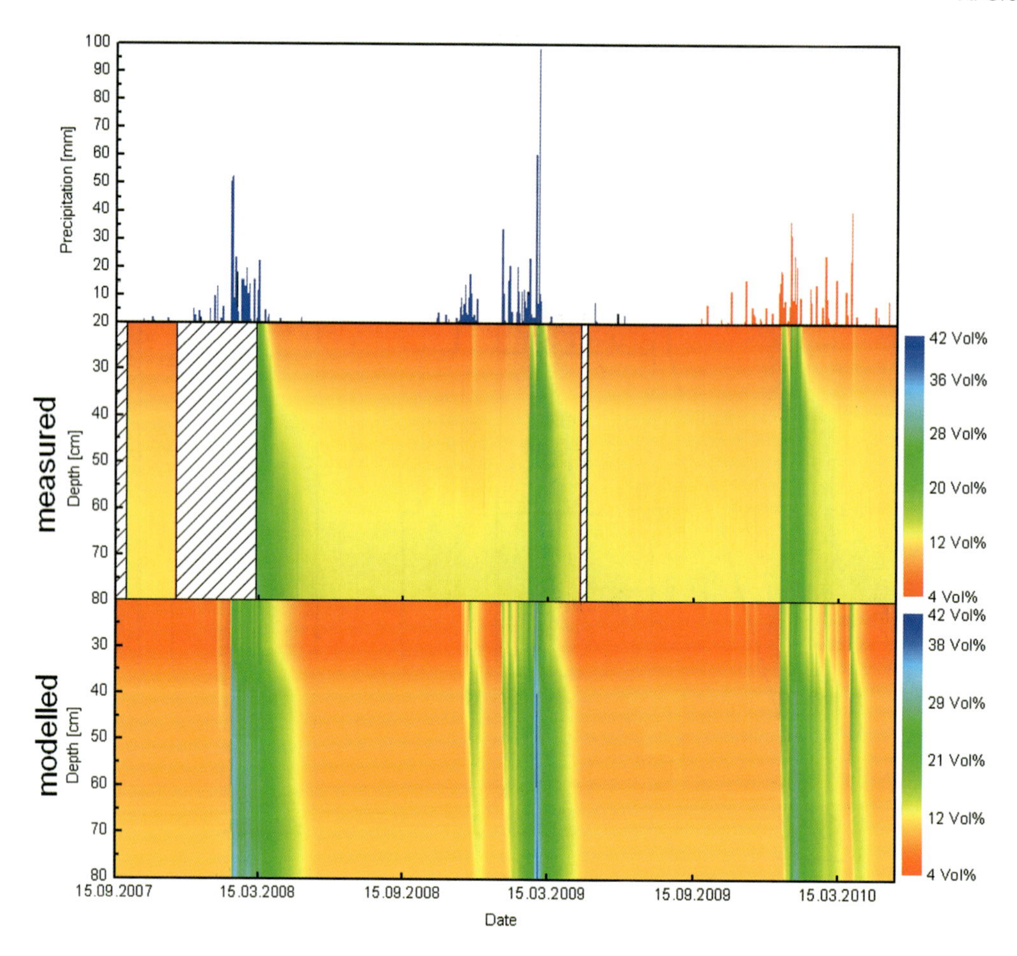

Figure 2: Rainfall (top) and depth distribution of soil water content for the *Acacia* site (measured data middle, modeled data bottom).

Table 1. Annual water balance budgets for two sites and three season (2007/08 to 2009/10).

	Grass-dominated site			**Acacia-dominated site**		
	07/08	**08/09**	**09/10**	**07/08**	**08/09**	**09/10**
Precipitation [mm]	+612.2	+529.3	+363.8	+465.4	+597.4	+505.4
Interception [mm]	–6.4	–7.7	–7.0	–15.8	–21.1	–19.1
Runoff [mm]	–7.9	0.0	0.0	0.0	–2.7	0.0
Transpiration [mm]	–143.9	–203.2	–128.0	–141.7	–220.9	–240.6
Evaporation [mm]	–217.4	–255.1	–231.5	–103.7	–157.6	–197.2
Bottom boundary flow [mm]	–237.0	–55.8	0.0	–204.3	–187.5	–49.4

Table 2. Mean annual water balance budgets for two sites and 30 years.

	Precipitation [mm]	**Interception [mm]**	**Transpiration [mm]**	**Evaporation [mm]**	**Bottom boundary flow [mm]**
Acacia-dominated	+375.7	–15.3	–143.8	–214.9	–2.1
Grass-dominated	+375.7	–5.7	–92.1	–276.8	–1.4

husmann, L., Hecht, J., Hoyer, P., Kongor, R.Y., Lang, H., Lieckfeld, L. A. B., Oldeland, J., Peters, J., Röwer, I. U., September, Z. M., Sop, T. K., van Rooyen, M. W., Weber, J. Willer, J. & Jürgens, N., 2010: The BIOTA Observatories. ed. Klaus Hess Publishers, Göttingen & Windhoek.

Huxman, T. E., Wilcox, B. P., Breshears, D. D., Scott, R. L., Snyder, K. A., Small, E. E., Hultline, K., Pockman, W. T. & Jackson, R.B., 2005: Ecohydrological Implications of Woody Plant Encroachment. Ecology **86**: 308–319.

Klerk, N. D., 2004: Bush encroachment in Namibia. Report on Phase 1 of the bush encroachment research, monitoring and management project. – Ministry of Environment and Tourism, Windhoek.

Kroes, J. G., Van Dam, J.C., Groenendijk, P., Hendriks, R. F. A. & Jacobs, C. M. J., 2008: SWAP version 3.2. Theory and user manual Alterra, Wageningen.

Scanlon, B. R., Healy, R. W. & Cook, P.G., 2002: Choosing appropriate techniques for quantifying groundwater recharge. – Hydrogeology Journal **10**: 18–39.

Scanlon, B. R., Keese, K. E., Flint, A. L., Flint, L. E., Gaye, C. B., Edmunds, W. M. & Simmers. I., 2006: Global synthesis of groundwater recharge in semiarid and arid regions. – Hydrological Processes **20**: 3335–3370.

Scholes, R. J. & Walker, B. H., 1993: An African savanna: synthesis of the Nylsvley study Cambridge University Press. – Cambridge.

Smit, G. N. & Rethman, N. F. G., 2000: The influence of tree thinning on the soil water of a semi-arid savanna of southern Africa. – Journal of Arid Environments **44**: 41–59.

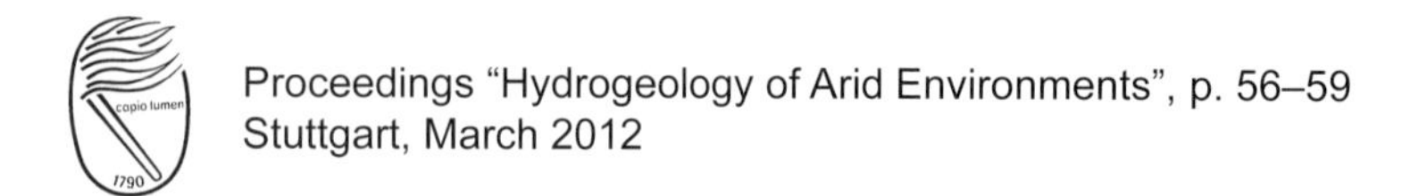
Proceedings "Hydrogeology of Arid Environments", p. 56–59
Stuttgart, March 2012

Extended Abstract

Evaluating Uncertainty Introduced to MABIA-WEAP-FAO56 Soil Water Balance Simulation Model by Using Limited Meteorological Data

M. Jabloun[1], A. Sahli[1,*], V. Hennings[2], W. Muller[2], J. Sieber[3], D. Purkey[3]

[1] Institut National Agronomique de Tunisie. 43 Av. Charles Nicolle, 1082 Tunis Mahrajène-Tunisia
[2] Bundesanstalt für Geowissenschaften und Rohstoffe, Stilleweg 2 D-30655 Hannover-Germany
[3] Stockholm Environment Institute-US Center. 11 Curtis Avenue, Somerville, Massachusetts 02144-USA.
* Email corresponding author: sahli_inat_tn@yahoo.fr

Key words: FAO 56 Soil water balance model, FAO Penman–Monteith equation, limited data, MABIA, WEAP

1. Introduction

Irrigated agriculture is the primary user of diverted water globally, reaching a proportion that exceeds 70–80% of the total in the arid and semi-arid zones (Fereres & Soriano 2007). The rapid increase of the world population and the corresponding demand for extra water by sectors such as industries and municipals, forces the agricultural sector to use irrigation water more efficiently. Particularly in Mediterranean areas where water resources are limited and irrigation is a necessary part of agricultural practices, accurate estimates of the crop water requirement (ET) are critical in order to make informed decisions regarding water management. Since the ET varies over the growing season, farmers will adjust the irrigation frequency and/or application depth during the growing season. Making regular direct infield measurements of plant and/or soil water status to schedule irrigation is usually too laborious, time consuming, difficult, or expensive for individual farmers (Jones 2007). As a result it has been suggested that a good precision in the application of irrigation can potentially be obtained by the use of 'soil water balance calculations', where the soil moisture status change is estimated by the difference between the inputs (irrigation plus precipitation) and the losses (runoff plus drainage plus crop evapotranspiration). FAO 56 Paper is a standard reference for crop evapotranspiration (ET_c) and irrigation water requirement (Allen et al. 1998). The proposed methodology for computing actual crop evapotranspiration (ET_a) was based on the application of a dual soil water balance (DSWB) at the top soil and the root zone layers and the use of the reference evapotranspiration (ET_o) and the dual crop coefficient method that separates evaporation from transpiration. Subsequent papers have demonstrated the accuracy of the FAO-56 method for several crops and weather conditions (Allen 2000 for cotton, Liu & Pereira 2000 for wheat and maize, Jabloun & Sahli 2006 for potato and green pepper, Goodwing et al. 2006 for peach). Actually, most organisations are moving towards wider use of this approach (ASCE in USA, DWLBC in Austria, ect...) and many reliable tools that allow an easy adoption of the dual Kc methodology were developed for field scale (MABIA-ETc by Jabloun & Sahli 2005, SimDualKc by Rolim et al. 2007) and for regional scale (MABIA-Region by Jabloun et al. 2011, WEAP by Seiber 2011). However, for daily ET_a calculation, the FAO-56 DSWB requires daily data on maximum and minimum air temperature, relative humidity, solar radiation and wind speed (u) to compute the daily ET_o and to adjust the basal crop coefficient to the local climatic conditions. Unfortunately, for many locations, as is the case for Tunisia, such meteorological variables are often incomplete and/or not available. A number of studies on accuracy of FAO-Penman-Monteith method for estimating ET_o using limited climatic data were published (Popova et al. 2006, Jabloun & Sahli 2008, Kwon & Choi 2011). However, the studies on applicability and performance of using ET_o es-

www.borntraeger-cramer.de
2012/0056/$ 1.00

timated values with missing meteorological parameters in the FAO 56 - DSWB were limited. Based on these considerations, the objective of this study is to quantify the uncertainty that arises in the water balance terms i.e. crop evapotranspiration, irrigation requirement and drainage when site-specific climate data are missing.

2. Materials and Methods

2.1. Climatic data

To evaluate the performance of the FAO-56 DSWB when using ETo estimations from limited climatic data according to the FAO-56 PM method, daily data recorded at five meteorological stations managed by the Tunisian National Meteorological Institute (INM) are used. The respective locations are described in Table 1.

Table 1: Geographical coordinates and characteristic coefficients of the studied sites.

Stations	North latitude	Altitude (m)	k_{rs} ($°C^{-0.5}$)
Beja	36'44"	158	0.1596
Kairouan	35'40"	60	0.1714
Sidi-Bouzid	35'00"	354	0.1699
Tunis	36'50"	3	0.1817
Zaghouan	36'26"	156	0.1670

All selected weather stations have good quality daily data records from 1994 to 2000 including daily maximum and minimum temperatures and relative humidity, sunshine duration and wind speed. In this study four scenarios of climatic data availability and five methods to estimate ET_o were considered (Table 2). Except for Hargreaves equation, the FAO-Penman-Monteith method was used to estimate ETo (Allen et al. 1998).

2.2. Crops, soil and field parameters

In order to understand the impacts the climatic missing data might have on the use of the "FAO 56 dual soil water balance model" four irrigated crops i.e. wheat, potato, tomato and olive were considered in the five different locations and for the seven growing seasons. Field data used for simulations was summarized in Table 3.

2.3. Calculation model and tool

Calculations were made using the MABIA-Region software (Jabloun et al. 2011). This tool integrates the FAO 56 - DSWB model proposed by Allen et al. (1998) and uses the dual crop coefficient method that separates evaporation from transpiration. Regarding the Climate, Soil, Crop and Irrigation inputs which could be hard to define by farmers, two databases with a list of crops and soils that include all the required parameters to use in the FAO 56 - DSWB model are added. Furthermore, to minimize the impact of the lack of agro-climatic data at field scale, the tool includes in addition options for choosing (i) different methods to estimate reference evapotranspiration (ETo) depending on climatic data availability and (ii) different hydraulic pedotransfer functions to estimate soil water characteristics depending on the availability of data about physical soil components. Also, the software offers a choice of one or more irrigation scheduling criteria (fixed depth, fixed interval, variable depth and variable interval) according to water availability at field level. Same procedures and functionalities were also implemented in the new version of WEAP software (Seiber 2011).

2.4. Statistical analysis

The results of ETc, irrigation (I) and drainage (D) estimations obtained when using ETo calculation scenarios mentioned above were compared with results of the computation when ETo was estimated by the full data sets, which were taken as reference (ref). Following the suggestion of Jacovides & Kontoyiannis (1995), the performance of the estimated

Table 2: Scenarios of climatic data availability and ETo estimating methods.

ETo estimation Methods	Temperatures (°C)	Relative Humidity (%)	Sunshine Duration (hours)	Wind Speed (m/s)
	T_{max} & T_{min}	HR_{max} & HR_{min}	n	u_2
ETo-Ref	+	+	+	+
ETo-K_{rs}	+	+	-	+
ETo-T_{min}	+	-	+	+
ETo-T_{max}-T_{min}	+	-	-	+
ETo-Hargreaves	+	-	-	-

Table 3: Trials base information referring to crop, soil and irrigation system.

Crop Parameter	Wheat	Potato	Tomato	Olive
Plantation date	**15th November**	**15th February**	**01st March**	**01st March**
Initial Stage (d)[a]	55	20	30	30
Dev. Stage (d)	70	25	40	90
Mid. Stage (d)	50	35	50	60
Late Stage (d)	35	25	30	90
Root length (m)	1.0	0.6	0.8	1.5
TAW (mm/m)	145	145	145	145
Initial WC	FC	FC	FC	FC
Irrigation f_w	1.0	0.6	0.6	0.2

Table 4: Comparison between Irrigation (I) and Drainage (D) computed by the FAO 56-DSWB from ETo estimated with full data set and when ETo is estimated by considering different scenarios of climatic data availability for the different crops.

		ETo estimation Methods							
		ETo-K_{rs}		ETo-T_{min}		ETo- T_{max}-T_{min}		ETo-Harg.	
		MBE	RMSE	MBE	RMSE	MBE	RMSE	MBE	RMSE
Wheat	ETo (mm)	17	19	-36	44	25	68	38	78
	I (mm)	10	26	-27	46	21	72	31	76
	D (mm)	3	15	7	14	6	16	3	13
Potato	ETo (mm)	12	14	-21	26	10	38	23	47
	I (mm)	8	17	-25	34	4	38	16	47
	D (mm)	-3	8	-2	10	-2	10	-3	12
Tomato	ETo (mm)	23	26	-42	48	12	67	43	87
	I (mm)	24	37	-37	49	15	69	44	88
	D (mm)	-2	9	1	7	0	9	-1	12
Olive	ETo (mm)	25	31	-72	82	16	115	53	139
	I (mm)	15	60	-42	72	16	105	39	120
	D (mm)	-4	10	3	24	4	21	4	22

water balance terms, obtained using a given data, was evaluated using the statistical parameters: Mean Bias Error (MBE) and Root Mean Square Error (RMSE).

3. Results and Discussion

The findings suggest that when there is only limited data, it is better to estimate ET_0 using FAO Penman–Monteith method and procedures than using Hargreaves method. The higher deviations on ET_0 occur when only available information is minimum and maximum air temperature and when using the Hargreaves equation (Table 4).

The comparison of the irrigation requirements and deep percolation estimates for the studied crops using ET_0 computed from limited data to those computed with full data set revealed that the differences are acceptable considering the 5 locations and the 7 years studied. Both the Mean Bias Error (MBE) and the Root Mean Square Error (RMSE) of the ET_0 component were small (Table 4), leading to small errors in the water balance estimates. As for ETo, the higher deviations of crop irrigation requirement and deep percolation amounts occur when only minimum and maximum air temperature are the available information. These deviations on the water balance components were significantly higher when ETo calculated using the Hargreaves equation is considered in the soil water balance. For example, for irrigated wheat, the MBE values ranged from –27 to 30 mm and RMSE from 25 to 75 mm for the irrigation requirements, for an overall mean irrigation requirement of about 400 mm. Concerning deep percolation under the wheat rooting depth, the MBE and RMSE values ranged from 3 to 7 mm and from 13 to 16 mm, respectively, for an overall mean of about 90 mm. Same trends were also observed with potato, tomato and olive crops where an average irrigation requirement of about 322, 635 and 715 mm were obtained, respectively. The average drainage rates of the four crops, as estimated by the FAO 56-DSWB and using the different ETo calculation procedures, were 28, 13 and 18 mm over their respective growing season.

Conclusion

The FAO-PM ETo estimation method gave a better estimation of the soil water balance term of the winter wheat, potato, tomato and olive compared to the one derived from the Hargreaves equation.

References

Allen, R. G., Pereira, L. S., Raes, D. & Smith, M., 1998: Crop evapotranspiration: Guidelines for computing crop requirements. – Irrigation and Drainage Paper No. 56, FAO, Rome, Italy.

Allen, R. G., 2000: Using the FAO-56 dual crop coefficient method over an irrigated region as part of an evapotranspiration intercomparison study. – J. Hydro. **229**: 27–41.

Fereres, E. & Soriano, M. A., 2007: Deficit irrigation for reducing agricultural water use. – J. Exp. Bot. **58**: 147–159.

Goodwin, I., Whitfield, D. M. & Connor, D. J., 2006: Effect of tree size on water use of peac (*Prunus persica* L. Batsch). – Irrig. Sci. **24**: 59–68.

Jabloun, M. & Sahli A., 2005: MABIA-ETc: a tool to improve water use in field scale according to the FAO guidelines for computing water crop requirements. – In: CD-Rome The Improving Water Use Efficiency in MEDiterranean Agriculture Workshop, Rome, Italy, 29–30 September 2005.

Jabloun, M. & Sahli, A., 2006: Development and testing of an irrigation scheduling model. – In: CD-Rome proceeding of The First International Conference on the Theory and Practices in Biological Water Saving (ICTPB) Section 3: Agronomic water saving methods and projects, China, Beijing, May 21–25, 2006.

Jabloun, M. & Sahli, A., 2008: Evaluation of FAO-56 methodology for estimating reference evapotranspiration using limited climatic data Application to Tunisia. – Agric Water Manag. **95**: 707–715.

Jabloun, M., Sahli, A. & Mougou, A., 2011: MABIA-REGION: A Regional irrigation evaluation and scheduling tool based on the Dual Crop Coefficient Method with Extensions to the original FAO-56 procedure. – In: CD-Rome Abstract of the Second WEAP Regional Conference "Applying a Decision Support System as a Tool for Integrated Water Resources Management and Climate Change Adaptation"-ACSAD-BGR, Jordan, May 03-04, 2011.

Jones, H. E., 2007: Monitoring plant and soil water status: established and novel methods revisited and their relevance to studies of drought tolerance. – J. Exp. Bot. **58**: 119–130.

Kwon, H. & Choi, M., 2011: Error assessment of climate variables for FAO-56 reference evapotranspiration. – Meteorol. Atmos. Phys. **112**: 81–90.

Liu, Y. & Peirera, L. S., 2000: Validation of FAO methods for estimating crop coefficients. – Trans. of the CSAE **16**: 26–30.

Popova, Z., Kercheva, M. & Pereira, L. S., 2006: Validation of the FAO methodology for computing ETo with missing climatic data application to South Bulgaria. – Irrig. Drain. **55**: 201–215.

Rolim, J., Godinho, P., Sequeira, B., Paredes, P. & Pereira, L. S., 2007: Assessing the SIMDualKc model for irrigation scheduling simulation in Mediterranean environments. – Opt. Med. **B56**: 49–61.

Seiber, J., 2011: Highlights of the New Version of WEAP. – In CD-Rome Abstract of the Second WEAP Regional Conference "Applying a Decision Support System as a Tool for Integrated Water Resources Management and Climate Change Adaptation"-ACSAD-BGR, Jordan, May 03-04, 2011.

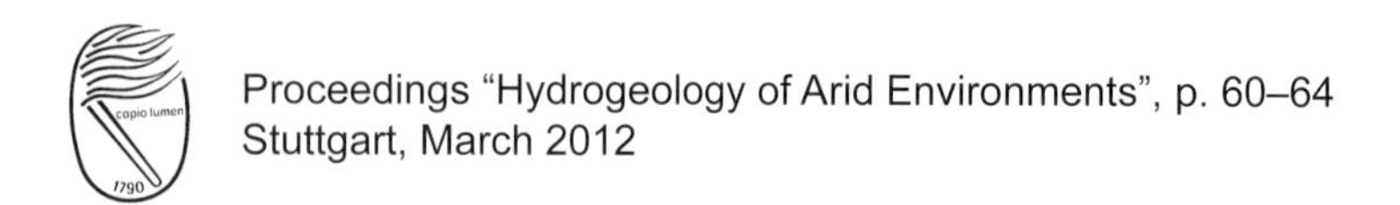

Proceedings "Hydrogeology of Arid Environments", p. 60–64
Stuttgart, March 2012

Extended Abstract

Large Scale Groundwater Recharge Estimation with Hydrological Models in Arid Environments – Case Study Arabian Peninsula –

B. Keim[1], R. Rausch[2], M. Al-Saud[3], H. Pfäfflin[4], A. Bárdossy[5], D. Bendel[6], M. Lorenz[7]

[1] Ingenieurgesellschaft Prof. Kobus und Partner GmbH , Hessbrühlstr.21, 70565 Stuttgart,
email: keim@kobus-partner.com
[2] GIZ IS, P.O. Box 2730, Riyadh 11461, Kingdom of Saudi Arabia, email: Randof.Rausch@gizdco.com
[3] Ministry of Water & Electricity, Saud Mall Center, Riyadh 11233, Kingdom of Saudi Arabia,
email: malsaud@mowe.gov.sa.
[4] Ingenieurgesellschaft Prof. Kobus und Partner GmbH , Hessbrühlstr.21, 70565 Stuttgart,
email: pfaefflin@kobus-partner.com
[5] Institut für Wasser- und Umweltsystemmodellierung - Universität Stuttgart, Pfaffenwaldring 61, 70569 Stuttgart,
email: Andras.Bardossy@iws.uni-stuttgart.de
[6] Institut für Siedlungswasserbau, Wassergüte- und Abfallwirtschaft - Universität Stuttgart, 70569 Stuttgart,
email: david.bendel@iswa.uni-stuttgart.de
[7] Schreiberstr. 36, 70199 Stuttgart, email: manuellorenz@gmx.de

Key words: recharge, arid, external drift kriging, stochastic precipitation simulation

Introduction

The estimation of groundwater recharge is a crucial factor for groundwater projects. Currently, large scale (100,000 up to 1,000,000 km²) projects have no single, reliable method that allows recharge estimation to have a high accuracy. In arid environments precipitation and consequently groundwater recharge are characterized by a high temporal and spatial variability. Direct measurements present only local values. Even tracer methods show disadvantages. An example is the assumption of negligible runoff for the Chloride method or the limitation of accurate estimations of the porosity for the Tritium method. To overcome these difficulties, hydrological modeling has been chosen for the case study. In general, the availability of data for the calibration of the hydrological models for surface-runoff is poor. Infiltration measurements have been carried out to reduce uncertainty in the modeling. The main focus of the presentation will be on the approach that combines ground-based point measurements, satellite precipitation patterns, and stochastic simulated precipitation as an input for the modeling.

Hydrological Conditions in the Kingdom of Saudi Arabia

The long-term mean of measured precipitation ranges from approximately 31 mm/a at the station of the airport at As Sulayyil up to 256 mm/a at the station of the airport of Abha in the Asir Mountains. The mean precipitation of 20 airport stations in the Kingdom of Saudi Arabia has been calculated at 109 mm/a.

Potential evapotranspiration has been calculated from the data of meteorological stations using the FAO Penman-Monteith approach. The annual potential evapotranspiration averaged over the 14 stations in Saudi Arabia has been calculated at 6.9 mm/d. This equals an annual total potential evapotranspiration of 2,533 mm/a. In the North potential evapotranspiration varies from 2.5 mm/d in winter to 8–12 mm/d in summer with an annual mean of 6.4 mm/d (2,336 mm/a). The Southern stations Sharurah and As Sulayyil yield 4.5 mm/d in winter and 9 mm/d in summer with an annual mean of 7.2 mm/d (2,628 mm/a).

www.borntraeger-cramer.de
2012/0060/$ 1.25

Study Area and Overview on the Hydrological Modeling

The study area covers the center of the kingdom (see Fig. 1). In order to determine the groundwater recharge within the study area it is important to know, if there occurs inflow of surface water flows into the study area from outside. Therefore, an extended study area has been identified.

In the Arabian Peninsula case study, the excess of precipitation and groundwater recharge has been modeled with the help of the soil moisture accounting routine provided with the hydrological model program HEC-HMS. The soil moisture accounting routine is a physical approach that allows a continuous modeling of all processes involved. The advantage of this procedure for modeling precipitation excess is that the amount of water leading to surface runoff and the amount of water infiltrating into the soil is calculated in one step.

In a quantitative assessment of surface water flow the hydrological model has to consider surface-runoff, channel flow, and bed infiltration. HEC-HMS is used for this purpose. This modeling package provides several options.

The hydrological modeling has to take realistic space-time patterns of rainfall and the main hydrological properties of the soil and the catchments as well as the water losses due to evapotranspiration into consideration.

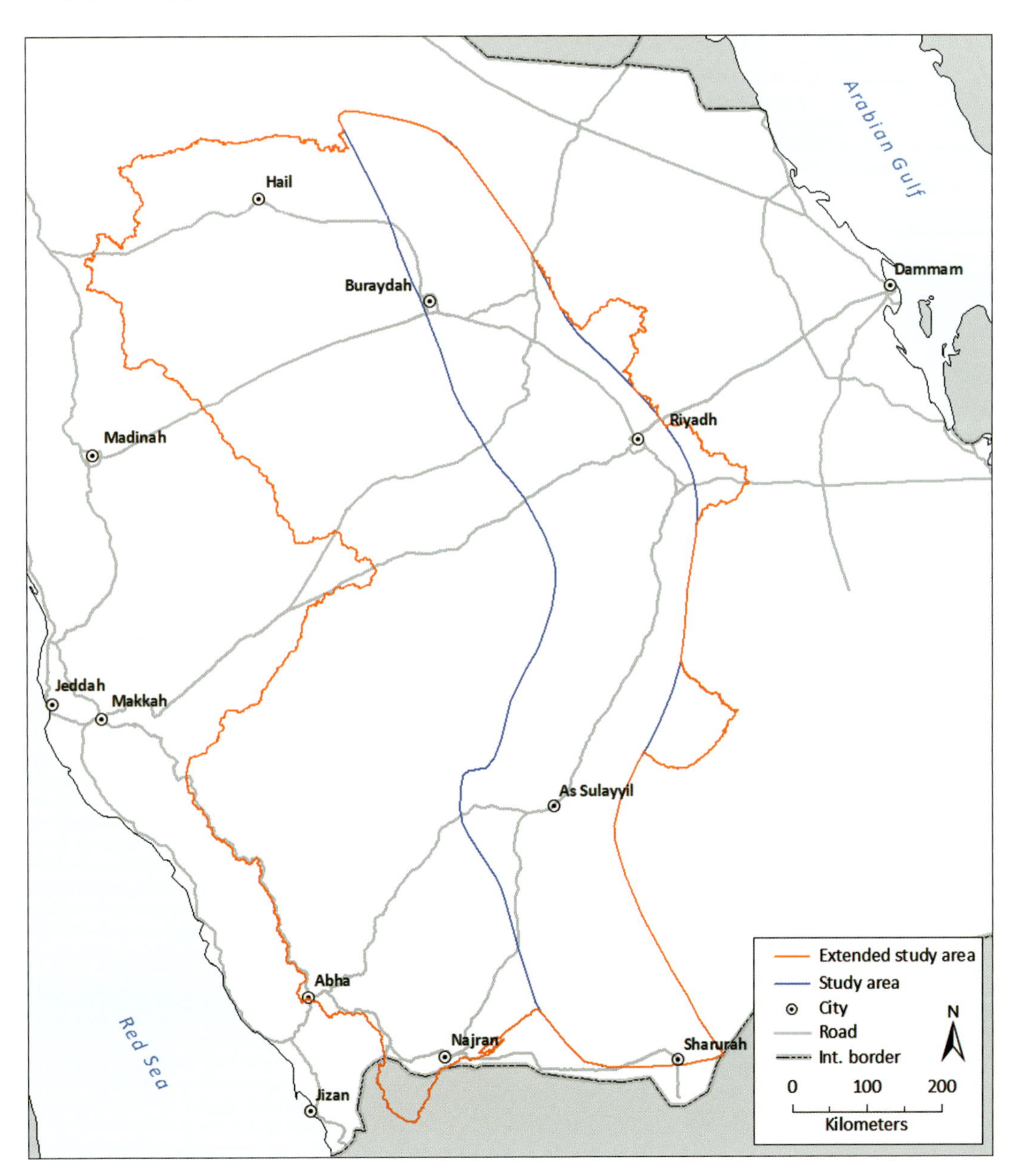

Figure 1: Location map and study area.

Infiltration Measurements

Infiltration measurements were conducted to reduce uncertainty in modeling. They were focused on the infiltration rates of different drainage features. In this drainage features, the main infiltration and groundwater recharge processes are taking place. The infiltration test sites were classified according to the following encountered drainage features:

Wadi: which represents a riverbed with its floodplains,

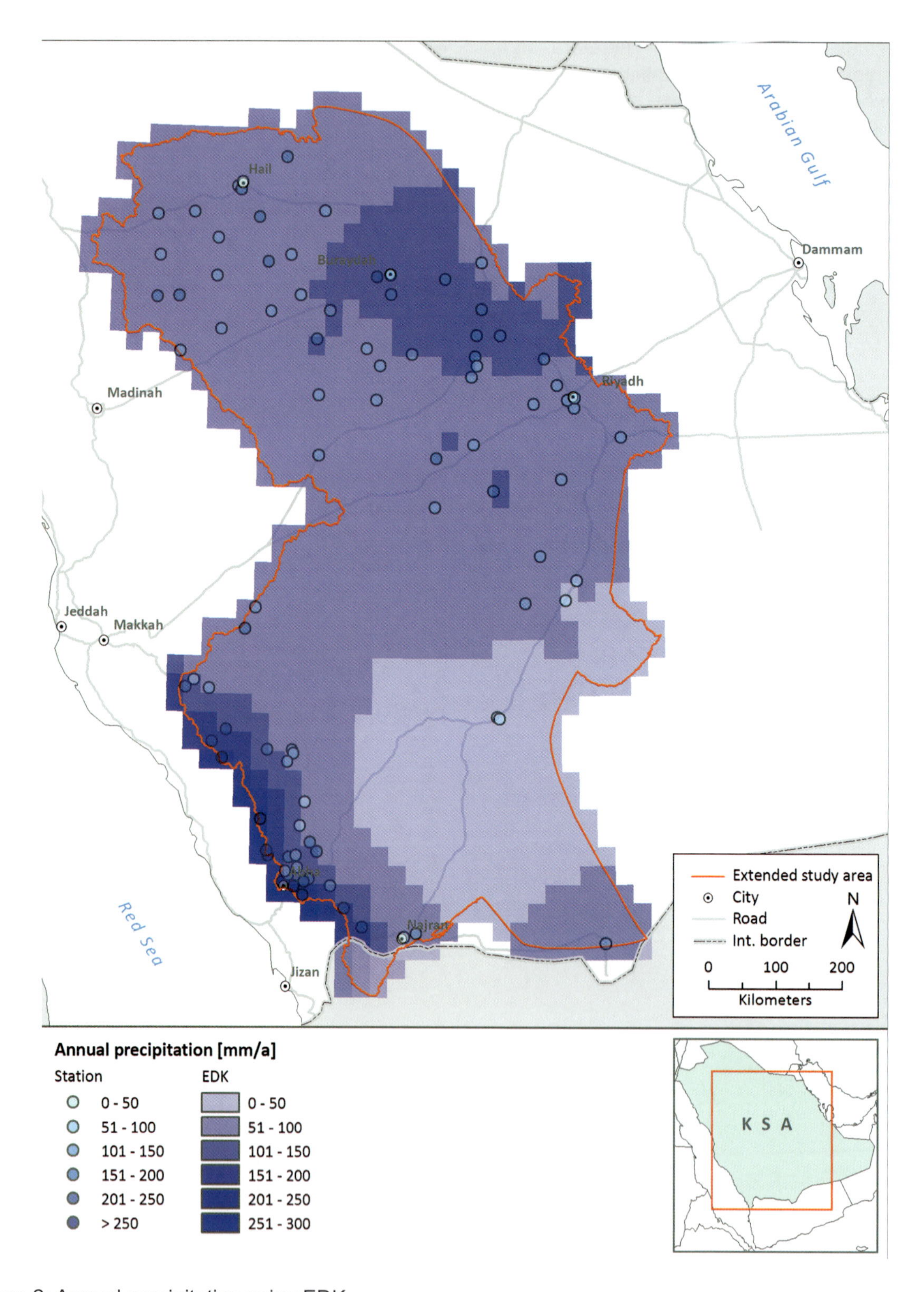

Figure 2: Annual precipitation using EDK

Table 1: Summary of results from infiltration tests sorted by drainage type.

Drainage feature	Minimum infiltration rate [mm/h]	Maximum infiltration rate [mm/h]	Medium infiltration rate [mm/h]	Medium hydraulic conductivity [m/s]
Wadi	8.0	183.6	58.9	$1.6 \cdot 10^{-5}$
Sink	40.5	180.0	81.1	$2.3 \cdot 10^{-5}$
Pond	1.0	34.5	17.4	$4.8 \cdot 10^{-6}$

Sink: an area, which represents the outlet of a wadi,

Pond: an area within a sink or floodplain where water collects after flood events.

A total of 18 soil infiltration tests within the Wadi As Sahba drainage basin were conducted in order to determine the in-situ infiltration rates. A standard set of double ring infiltrometers was used to measure the constant infiltration rate according to DIN 19682-7 1997. The results are presented in Table 1.

External Drift Kriging with Satellite Precipitation Patterns and Ground-Based Point Measurements Used as an Input for a Precipitation Simulation

Large parts of the study area are only sparsely equipped with measuring stations. Yet, there is a need for reliable boundary conditions (i.e. precipitation) for the hydrological model. During the past years Kriging has shown to offer several advantages. It is possible to incorporate additional variables, which are available at each estimation point, into the equation system, to reduce estimation variances while enhancing spatial level of detail.

Regionalization by Kriging requires variograms of the precipitation values. A variogram indicates how much the values of two points separated by a certain distance are expected to differ from one another. The TRMM data set (Remotely sensed rainfall estimates from NASA's Tropical Rainfall Measuring Mission) is available at a resolution of 0.25 by 0.25°, providing more than 3500 precipitation estimates for each month from 1998 to 2010 on a regular grid.

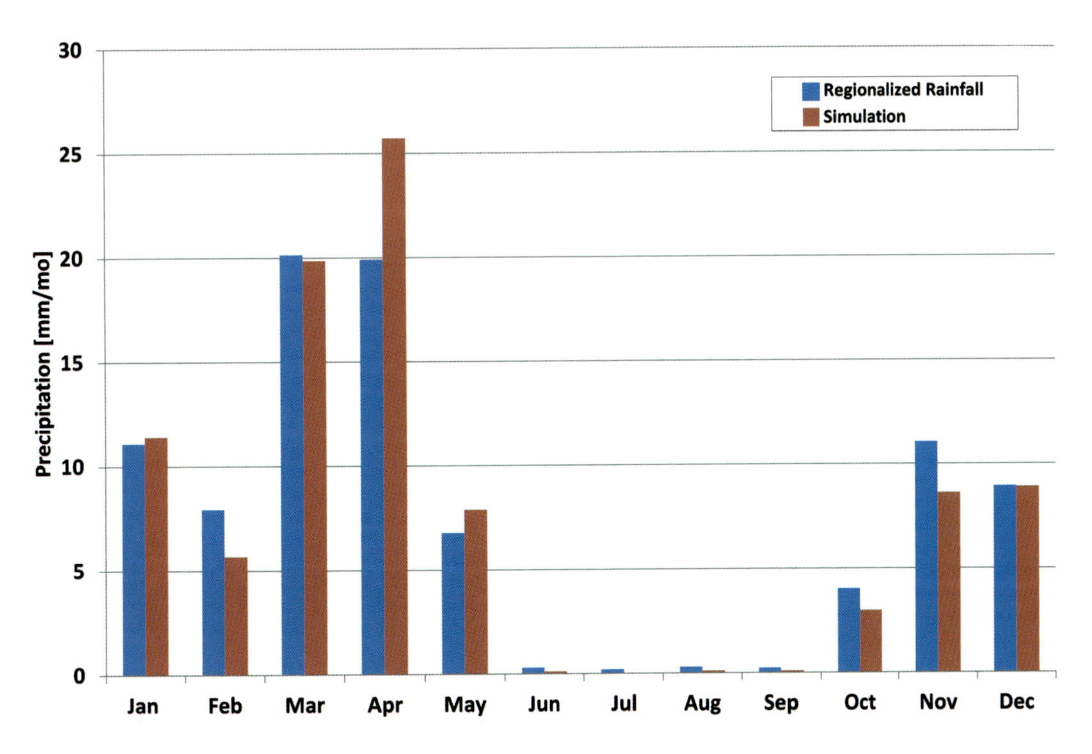

Figure 3: Comparison of simulated and regionalized monthly means of rainfall representing the period 1979 to 2010.

Ordinary Kriging assumes a constant mean within the whole domain, which is not the case here. For example, in the Asir Mountains much higher precipitation means are expected than in the Northern region around Tabouk. External Drift Kriging (EDK) can use additional information that is known at unmeasured points when it is linearly related to the measurement values. The best correlation of known covariates with measured rainfall exists for TRMM rainfall estimates. Other studies often choose the terrain elevation as a covariate, but for the selected measurement stations TRMM yields a much higher correlation. Thus, using TRMM as a covariate can greatly enhance the regionalization when chosen as a drift variable. Figure 2 shows the results of the EDK.

Space Time Precipitation Simulation

In the study area precipitation is characterized by a high temporal and spatial variability. Rainfall is sporadic and, therefore, a statistically random event. In order to determine the groundwater recharge and surface-runoff, long-term time series of precipitation with good spatial coverage are needed. A precipitation simulation approach has been used for the study. The so-called space-time precipitation simulator is a stochastic model.

The simulation takes into account the precipitation properties (such as relative monthly frequency of rainy days, intensities, and amounts) of the ground-based measurements. Due to the stochastic approach, the simulated time series are not conditional to the measured time series at a location.

Figure 3 shows the comparison of the simulated monthly means with the regionalized rainfall. In accordance with regionalized rainfall, the simulation reproduces the annual change. The main rainfall period is from November until May. High rainfall is focused in the period between March and April. The maximum occurs in April. Between June and October there is only little rainfall. The number of simulated rainy days (averaged over all the cells) yields an average of 20.9 days per year. The analysis of the measurements at the rainfall stations in the area yields a figure of 20.0 rainy days per year.

Proceedings "Hydrogeology of Arid Environments", p. 65–70
Stuttgart, March 2012

Extended Abstract

Stable Isotope Studies in Semiarid, Karstic Environments Reveal Information for Sustainable Management of Water Resources in Damascus, Syria

Paul Koeniger[1], Mathias Toll[1], Thomas Himmelsbach[1], Khaled Shalak[2], Ahmed Hadaya[2], Refaat Rajab[1]

[1] BGR Federal Institute for Geosciences and Natural Resources, Groundwater Resources, Stilleweg 2, 30655 Hannover, Germany, email (corresp. author): paul.koeniger@bgr.de
[2] DAWSSA Damascus Water Supply and Sewerage Authority

Key words: stable isotopes, karst groundwater, Figeh spring system, Syria

Introduction

Damascus, the capital of the Syrian Arabic Republic, is most likely facing problems of a future drinking-water supply as well as sustainable use and management of water resources due to rapid population growth within the next decades. The population of Damascus City will rise to 8.3 million inhabitants by 2020 and 13.8 million by 2040 according to the Water Resources Information Center (WRIC) of the Ministry of Irrigation (MoI). Approximately 60% of the water demand of the city is delivered from the Figeh Spring System which is located about 20 km to the west of Damascus in the Anti-Lebanon Mountains. Multiple isotope hydrological investigations have been carried out by various institutions, for instance BGR, Arab Center of Studies of the Arid Zones and Dry Lands (ACSAD), Syrian Atomic Energy Commission (AECS), since the early 1970s to support a sustainable management of the available water resources (see Table 1).

The stable isotopes (deuterium, ^{2}H and oxygen-18, ^{18}O) of water are known to add substantial information for water resources management in arid and semiarid environments (e.g. evaporation loss, groundwater recharge, nutrient transport). Stable

Table 1: Compilation of earlier isotope hydrological studies relevant for the Figeh Spring System with data availability (internal reports or published data), monitored hydrological components precipitation (P), groundwater (GW) and others (e.g. river water, surface water, soil water) and isotopes that were measured.

Isotope studies at Figeh spring	Availability	Hydrological data	Isotope data
Sogreah (1973)	report	P, GW	^{2}H, ^{3}H
Geyh et al (1983)	report	GW	^{2}H, ^{18}O, ^{3}H, ^{14}C
Selkhozpromexport (1986)	report	P, GW, others	^{18}O, ^{3}H
Droubi (1989)	report	P, GW	^{2}H, ^{18}O, ^{3}H, ^{14}C
Kattan (1996a)	publ.	P	^{2}H, ^{18}O, ^{3}H
Kattan (1997b)	publ.	P, GW	^{2}H, ^{18}O, ^{3}H
Kattan (1997a)	publ.	P	^{2}H, ^{18}O, ^{3}H
Al-Charideh (2011)	publ.	P, GW	^{2}H, ^{18}O, ^{3}H

www.borntraeger-cramer.de
2012/0065/$ 1.50

isotopes in precipitation reflect altitudinal, latitudinal, seasonal, amount and continental fractionation effects which can be used to describe recharge altitude of groundwater, recharge amounts through unsaturated zone water movement or short-term residence times. This work aim to compile, re-evaluate and extend stable isotope studies of the Figeh Spring System. With isotope hydrological studies it is expected to better describe i) mean altitudes of groundwater recharge areas of the Figeh Spring System, ii) mean residence times of groundwater and iii) contributions of fast (e.g. events or snowmelt) and slow (karst reservoirs) groundwater in the system.

Study site and methods

The recharge area of the Figeh Spring System includes a northeast trending part of the Anti-Lebanon Range that is bounded on the northwest by the Beeka Valley rift system and by the Barada River to the south (Lamoreaux et al. 1989). Refer to Kattan (1997b) and Lamoreaux et al. (1989) for a detailed description of the local geological and hydrogeological setting. The climate of the study area is of a Mediterranean type characterized by cool winter and dry summer months with two transitional periods in spring and autumn (Selkhozpromexport 1986, Kattan 1997b). Most rainfall occurs during the winter period. Long-term average precipitation of the area varies from 90 mm y^{-1} to the east of Damascus to 1800 mm y^{-1} in the upper parts of the Mount Hermon area. For the Figeh spring catchment weekly precipitation data is available at the stations Serghaya, Zabadani, Madaya, Bloudan (1,550 m asl) and Huraireh beginning from September 2009 to June 2011. Snow samples were collected irregu-

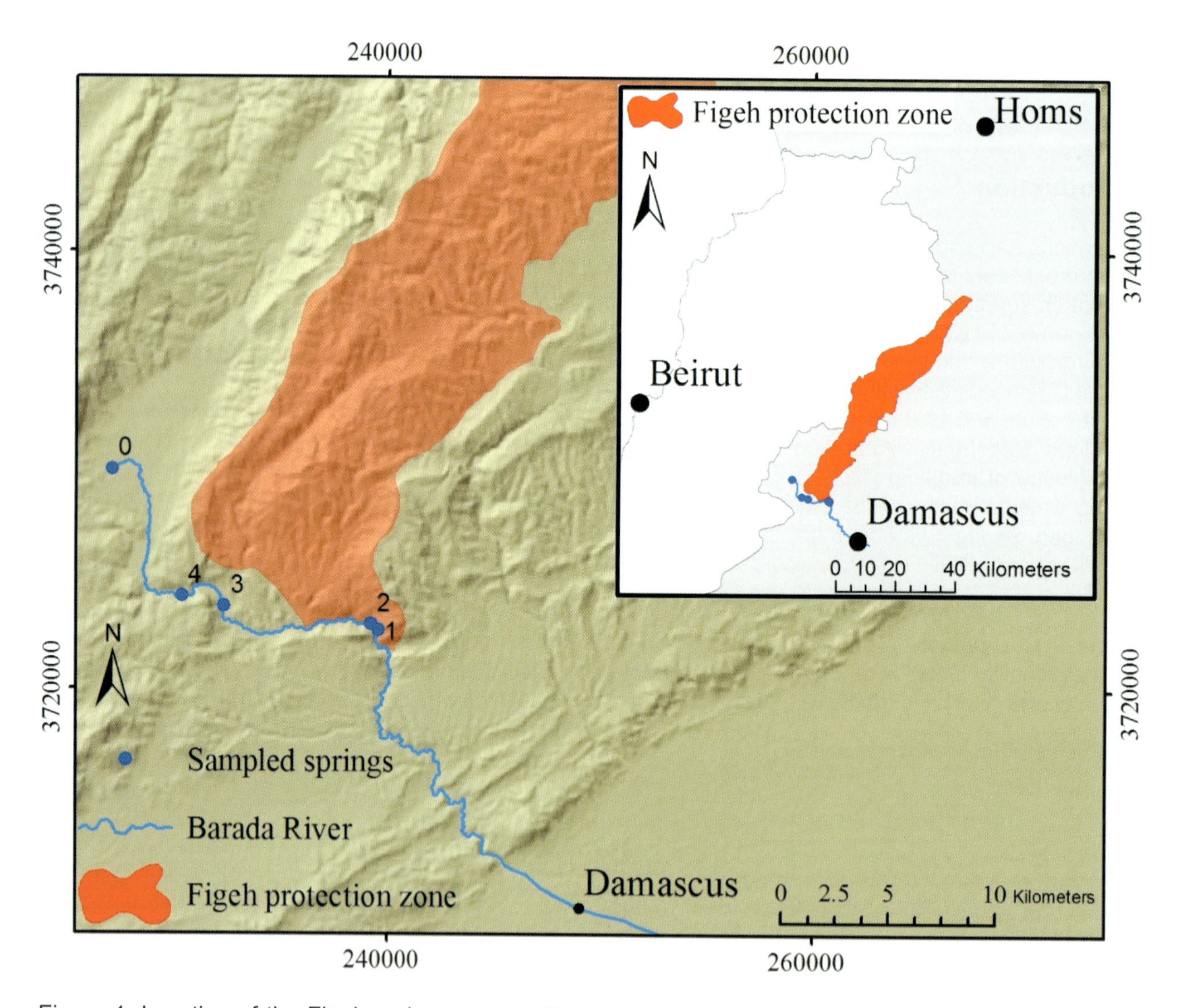

Figure 1: Location of the Figeh spring system with protection zone and sampled springs (1 Figeh Main, 2 Haroush, 3 Kefar Aloumed, and 4 Ein Habeeb), 0 is Barada spring.

Table 2: Elevation, mean discharge, isotope and deuterium excess (DE) values of springs collected during March to September 2011)

Springs	elevation (m asl)	discharge (m^3s^{-1})	n (–)	δ^2H (‰ VSMOW)	$\delta^{18}O$ (‰ VSMOW)	DE (‰ VSMOW)
Figeh Main	830	7.7*	102	–50.4	–8.73	19
Haroush	851		86	–44.5	–7.96	19
Kefar Aloumed	942		60	–41.4	–7.36	17
Ein Habeeb	1025		54	–41.0	–7.44	18

*after Al-Charideh (2011)

larly in the catchment. Discharge observations are available for the Figeh spring from June 2010 to July 2011 as well as temperature and conductivity data in 15 minutes resolution.

Major karst springs of the system are the Figeh Main Spring (830 m a.s.l.), Figeh Side Spring (830 m a.s.l.), Harouch Spring (851 m a.s.l.) and Deir Moukaren well gallery (910 m a.s.l.). For this work we studied stable isotope patterns of four springs in the vicinity of the Figeh spring system that contribute two-thirds of the water supply of Damascus. More than 300 samples were collected from Figeh Main, Haroush, Kefar Aloumed, and Ein Habeeb springs (see Fig. 1 and Table 2) in approximately weekly time resolution between March and September 2011. Other important springs in the wider vicinity of the Figeh Spring System are the Barada Spring (1115 m a.s.l.), and Boukein Spring (1273 m a.s.l) that were not considered during this sampling campaign.

The collected water samples were carried to Germany and analyzed for δ^2H and $\delta^{18}O$ simultaneously using a Picarro L2021-i cavity ring down laser spectrometer after vaporization (VAP 214 vaporizer) at the isotope laboratory of the BGR in Hannover. All samples were measured at least four times and the reported value is the mean value. All values are given in the standard delta notation in per mil (‰) vs. VSMOW. Raw data were corrected for memory effect and excluded if necessary. The data sets were corrected for machine drift during the run and normalized to the VSMOW/SLAP scale. External reproducibility, defined as standard deviation of a control standard during all runs, was better than 1.0‰ and 0.30 ‰ for δ^2H and $\delta^{18}O$, respectively.

Results and discussion

In winter season 2010/2011 precipitation occurred mainly between October and April with highest records of over 100 mm on 12/13 December for the Bloudan station. Snow accumulation and storage is usual between November and March in higher elevations. Spring discharge curves of Figeh Main spring show peak flow values of 14 $m^3 s^{-1}$ in early April 2011 and low flow values of less than 3 $m^3 s^{-1}$ during the summer period.

Stable isotope values in a range between –9.04 ‰ to –7.88 ‰ and –51.8 ‰ to –43.3 ‰ for $\delta^{18}O$ and δ^2H, respectively, were observed for Figeh Main spring during the period May to September 2011. Mean values (see Table 2) agree well with earlier reported stable isotope values (Kattan 1997 a, b, Al-Charideh 2011). For the other springs the stable isotope values are in a range of –8.52 ‰ to –7.37 ‰, and –47.6 ‰ to –41.7 ‰ for Haroush spring, –7.92 ‰ to –6.97 ‰, and –43.5 ‰ to –40.5 ‰ for Aloumed spring and –7.72 ‰ to –7.08 ‰ and –43.0 ‰ to –38.5 ‰ for Ein Habeeb spring, for $\delta^{18}O$ and δ^2H respectively.

Separate clusters were observed for Figeh and Haroush spring stable isotope data on a δ^2H vs. $\delta^{18}O$ plot (see Fig. 2). Kefar Aloumed and Ein Habeeb springs plot approximately on the same level. Earlier published stable isotope values for precipitation collected at Kadmous (750 m asl) and Bloudan (1550 m asl) by AL-Charideh (2011) indicate a clear isotope effect of heavy isotope depletion with increasing elevation. Interestingly the Figeh Main spring with lowest elevation indicate recharge areas with highest mean elevation, whereas Haroush indicate lower mean catchment elevations. According to stable isotope patterns, approximately same mean elevations of recharge catchments can be expected for Kefar Aloumed and Ein Habeeb spring.

The higher resolution sampling that was conducted during May to September 2011 reflects significant variability in stable isotope patterns of Figeh and Haroush spring that indicate other runoff forming processes (e.g. snowmelt influence, mixing of various karst reservoirs) (Fig. 3).

The isotope pattern is similar for δ^2H and $\delta^{18}O$. However, deuterium excess values (DE) show trends of lower peaks during the late summer month, indicating evaporation influence.

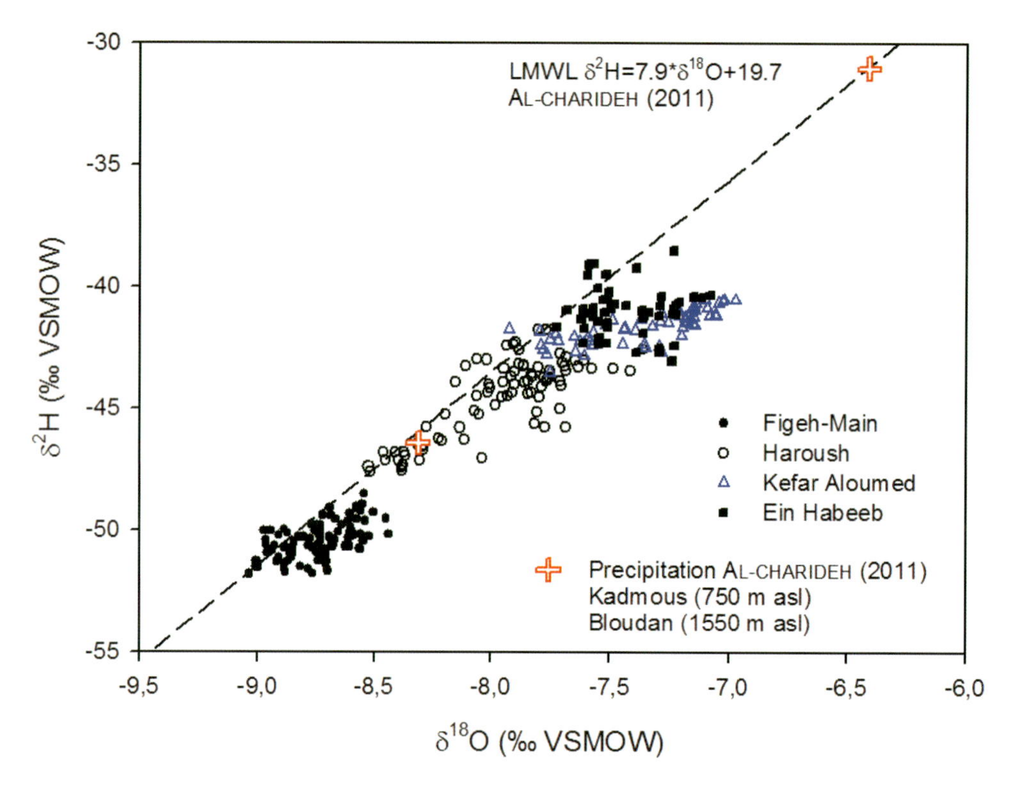

Figure 2: δ^2H vs. $\delta^{18}O$ plot of samples collected at the springs during March to Sept. 2011 in relation to precipitation values reported by Al-Charideh (2011).

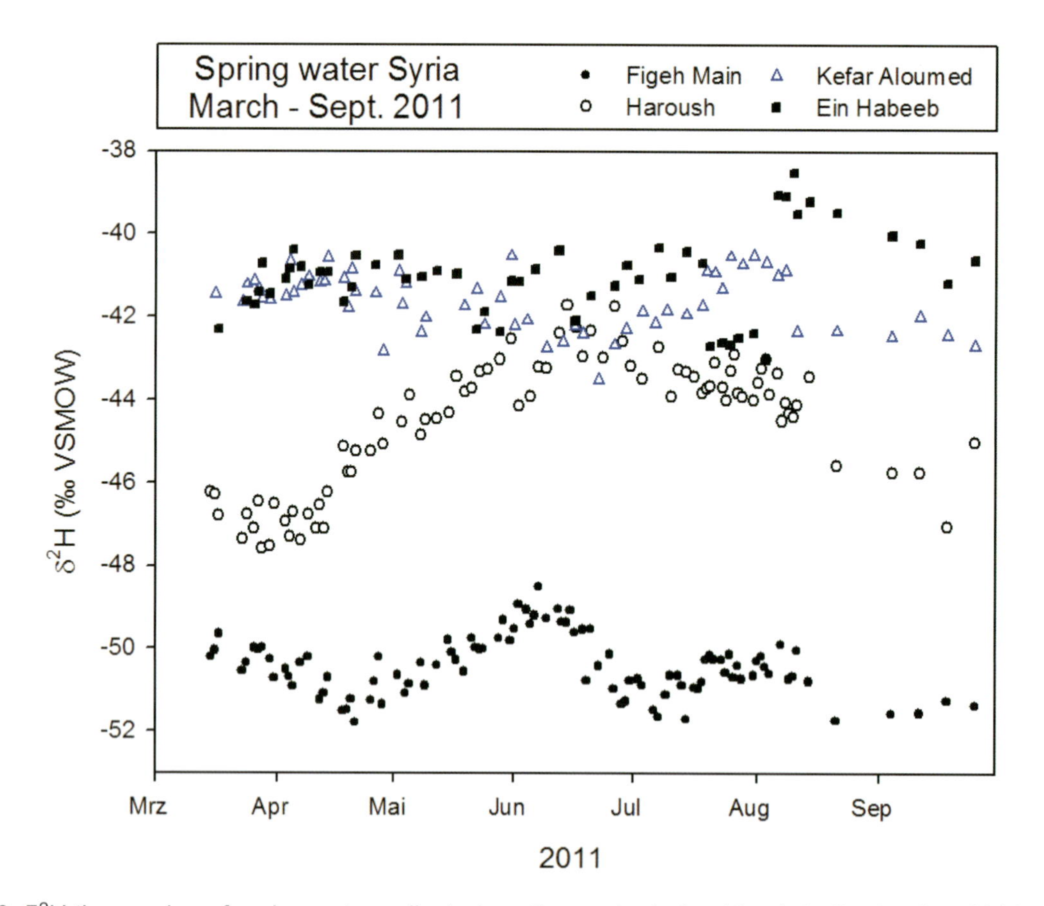

Figure 3: δ^2H time series of spring water collected continuously during March to September 2011.

Earlier isotope hydrological studies were conducted in the 1970s from a French consultants group (Sogreah 1973), within a Syrian-German cooperation (Geyh et al. 1983), a Russian consultants group (Selkhozpromexport 1986) and Droubi (1989) in the 1980s. Studies compiled by Kattan (1996a, b, 1997a, b), working at the AECS, were conducted and published since the early 1990s and in a Syrian-German technical cooperation in the last decade work was conducted from German groups (Geyh 2004) (see Table 1). It has to be considered that those samples were analyzed from different workers and measured in various laboratories so that overall uncertainty might be higher than those given for each laboratory. For the Figeh Main Spring mean water ages ranging from 3720 to 4570 BP were estimated using the carbon-14 method by Geyh et al. (1983), Droubi (1989) and samples collected in 2009, respectively. Tritium values were measured by Sogreah (1973), Selkhozpromexport (1986), Droubi (1989), Kattan (1997a, b) and Al-Charideh (2011). An interpretation using an exponential model and precipitation input data of the GNIP station Beit Dagan (IAEA/WMO 2002) lead to a mean residence time of 40 years for base flow data (Droubi 1989) and up to 50 years by Kattan (1997b).

Stable isotope data were collected by Sogreah (1973), Selkhozpromexport (1986), Droubi (1989), Kattan (1997a, b), and Al-Charideh (2011). Unfortunately early investigators only collected ^{2}H (Sogreah 1973) or ^{18}O (Selkhozpromexport 1986, Droubi 1989). Continuous measurements were made primarily for tritium, and therefore, only a few years of data exist for stable isotopes. During later studies, stable isotope data in precipitation was measured at the locations Bloudan, Damascus, Aleppo, Homs, Izraa, Kouneitra, Palmyra, Suweida and Tartus in Syria during the period of December 1989 to April 1990 by Kattan (1997a). This data is also published in the Global Network of Isotopes in Precipitation database (GNIP) (IAEA/WMO 2002). An interpretation of an altitude effect and groundwater isotope data for an estimation of elevation of recharge areas were given by Geyh et al. (1983), Kattan (1997b) and Al-Charideh (2011).

Continuous time series of stable isotopes from springs were reported in Sogreah (1973), and Kattan (1997b). A short term variability of stable isotopes indicates either contributions from fast groundwater components (snow melt signature or intense rainfall events) or mixing of groundwater of different ages. Sogreah (1973) report data from Figeh Main Spring, Figeh East Spring (which later disappeared after a tunnel was built, according to oral information from May Al-Safadi), and Bissane which refers to a spring in Kefr el Aloumeed (Aïn Bissane). Kattan (1997b) reports data from Figeh Main Spring, Figeh Side Spring, Harouch Spring, Barada Spring and Barada River. The Sogreah (1973) study shows $\delta\,^{2}H$ values that plot close to –50 ‰ with lowest values during spring 1972. Values from Kattan (1997b) are more depleted at about –52 ‰ with lowest values between August and December 1989. Both studies additionally show $\delta\,^{2}H$ values from other springs that are enriched in comparison to Figeh Main Spring. Relatively enriched values found at Bissane (Sogreah 1973) and at Barada and Harouch springs (Kattan 1997b) could be interpreted with a lower mean altitude of the contributing recharge areas. Decreasing values in spring 1972 and in March to April 1990 might be directly caused by spring snowmelt events.

The high resolution and continuous stable isotope data presented here indicate that further information (e.g. on aquifer vulnerability) can be drawn from spring observations. It is expected that such data stable isotope data sets will give further insights into short-term variability of the Figeh Main Spring and mimic the influence of fast groundwater components.

Acknowledgments

We would like to thank DAWSSA for sample collection and cooperation and Dieter Plöthner for advice and comments on an earlier draft of the manuscript. The work is conducted within the framework of the Syrian-german-technical-cooperation project “Protection of the Figeh Spring System” funded by the German Federal Ministry for Economic Cooperation and Development (BMZ).

References

Al-Charideh, 2011: Environmental isotope study of groundwater discharge from large karst springs in West Syria: a case study of Figeh and Al-sin springs. – Environ. Earth Sci. **63**: 1–10.

Droubi, A., 1989: Isotopic and chemical studies for Fijeh spring in SAR. ACSAD DM/t-65, Arab League, Water studies department, (47 pp in Arabic version; 21 pp without figures and tables, English version) (unpublished).

Geyh, M. A., Wagner, W. & Khouri, J., 1983: Altitude effect for Syria and the palaeoclimate. – BGR-Report, Hannover, (unpublished).

Geyh, M. A., 2004: Isotope hydrological study in the Aleppo and Yarmouk regions. – End-of-Mission Report 28 May–4 June 2004, ACSAD Damascus (unpublished).

IAEA/WMO, 2002: Global Network of Isotopes in Precipitation. – The GNIP Database. Accessible at: http://isohis.iaea.org.

Kattan, Z., 1996a: Environmental isotope study of the major karst springs in Damascus limestone aquifer systems: Case of the Figeh and Barada springs. – IAEA TECDOC-890 Isotope field applications for groundwater studies in the Middle East: 127–150.

Kattan, Z., 1996b: Chemical and environmental isotope study of precipitation in Syria. – IAEA TECDOC-890 Isotope field applications for groundwater studies in the Middle East: 185–202.

Kattan, Z., 1997a: Chemical and environmental isotope study of precipitation in Syria. – Journal of Arid Environments **35**: 601–615.

Kattan, Z., 1997b: Environmental isotope study of the major karst springs in Damascus limestone aquifer systems: case of the Figeh and Barada springs. – Journal of Hydrology **193** (1–4): 161–182.

Lamoreaux, P. E., Huges, T. H., Memon, B. A., Lineback, N., 1989: Hydrologic assessment – Figeh spring, Damascus, Syria. – Environmental Geology Water Science, Vol. **13**: 73–127.

Selkhozpromexport, 1986: Water resources use in Barada and Auvage basins for irrigation of crops, Syrian Arab Republic. – USSR Ministry of Land Reclamation and Water Management, Moscow.

Sogreah, 1973: Etude Hydrogeologique de la source Figeh, rapport final, (F. Bazin, R. Huber, J. M. Dujardin, A. Serrano, P. Jardin), Grenoble, France.

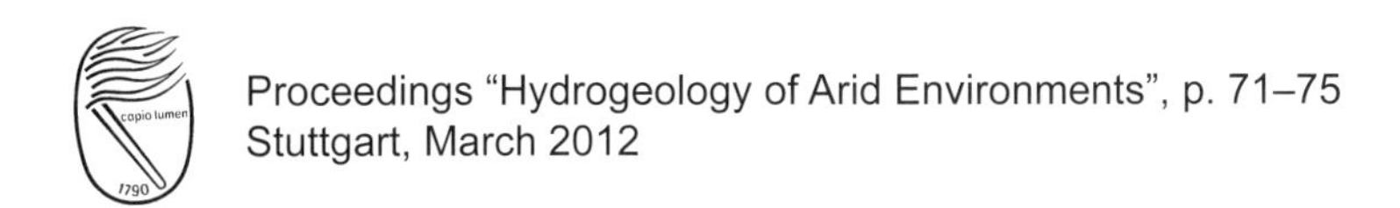
Proceedings "Hydrogeology of Arid Environments", p. 71–75
Stuttgart, March 2012

Extended Abstract

Preparatory Hydrogeological Investigations for the Large-Scale Strategic ASR-Project in the Liwa Desert of the Abu Dhabi Emirate

Georg Koziorowski

Deutsche Gesellschaft für Internationale Zusammenarbeit (GIZ), P.O. Box 44401, Khalifa Street, Liberty Tower, Abu Dhabi, United Arab Emirates, email: georg.koziorowski@giz.de

Key words: artificial groundwater recharge (ASR), Abu Dhabi Emirate, hydrogeology, shallow aquifer system

Introduction

The municipal water supply in the Emirate of Abu Dhabi, in particular of the Capital Abu Dhabi City, depends almost completely on desalinated seawater (DSW), which is provided by state-of-the-art plants, mainly operating in a coupled process for electricity generation and desalinated seawater production. However, over the year, electricity demand and water consumption vary significantly. During peak electricity demand, excess desalinated seawater may be produced, which cannot be used adequately due to insufficient storage capacity and there was even reported discharge of excess desalinated water into the Arabian Gulf. Nevertheless, desalination plants are vulnerable to environmental hazards and unfriendly acts while an appropriate back-up supply system for seasonal and long-term demand or in case of emergency is still missing. At present, the storage capacity for supplying drinking water to the Capital hardly lasts for more than two days.

In 1997, the Consortium of the German International Cooperation (GIZ, former GTZ) and Dornier Consulting (DCo) proposed the use of surplus DSW for replenishing depleted local aquifers to the Government of Abu Dhabi. Later, this initial concept developed into the broader approach of creating a major strategic freshwater reservoir in the subsurface to guarantee the Emirate's drinking water supply also in case of emergency considering that underground storage in suited aquifers offers convincing advantages if compared to any possible engineering structure of meaningful storage capacity.

In the following years, the search for locations suited for artificial groundwater recharge, storage and recovery (ASR) was countrywide intensified as part of the Groundwater Assessment Project Abu Dhabi (GIZ/DCo, 1995–2005) and when advantage of the Western Region became apparent, in 2001and 2002 GIZ/DCo conducted an ASR-feasibility study for its central part, the wider Liwa area. According to the promising results of this study, the ASR-pilot project Liwa was constructed in 2003 and operated in 2004, designed for demonstrating the efficiency of ASR in the Emirate's remote dune sand desert environment on a larger scale. Different recharge and recovery concepts had been tested. All in all, more than two million cubic metres of DSW have been introduced into the subsurface and highly efficiently recovered. The successful outcome of the pilot project formed the basis for designing the main ASR-project. In 2010, construction works commenced in the central western desert of the Abu Dhabi Emirate for implementing the large-scale "Strategic Water Storage and Recovery Project (SWSR) in Liwa", designed and supervised by GIZ/DCo. A minimum of 23 million cubic metres of DSW shall be introduced into the local shallow dune sand aquifer in order to create a major artificial subsurface reservoir, guaranteeing the safe drinking water supply of the Emirate's Capital for up to three months in case of emergency. In addition, the project helps to operate Abu Dhabi's desalination plants highly efficiently. Once implemented, this project will represent a benchmark for water management in arid regions.

www.borntraeger-cramer.de
2012/0071/$ 1.25

Available hydrogeological information

An indispensible prerequisite for the implementation of any ASR-measure is sufficient information on the local hydrogeological settings and hydrodynamic mechanisms. For the identification of the best suited location in the country, the SWSR-project benefited from the profound knowledge on the local hydrogeology acquired by the Consortium through continuous work in the groundwater field of Abu Dhabi since 1995. Overall, some 36,000 existing third party wells had been visited in the field, some 4,000 vertical electrical soundings were conducted and more than 1,300 own water wells, drilled to maximum depth of 1,100 m, had been completed including some 50 core pre-drillings. For more than 300 drillings, geophysical borehole logging was carried out. All wells were state-of-the-art tested, comprising well performance test, long-term constant discharge test and dynamic fluid-logging. Moreover, at selected locations, infiltration and tracer tests were conducted. For all own water wells and a significant number of third party wells complete hydrochemical and environmental isotopes analyses were conducted.

A monitoring network for measuring groundwater level fluctuations and water quality parameters was established, comprising 325 continuously automatically recorded wells and more than 1,600 periodically measured wells. Furthermore, 20 rain gauge stations and five climate stations had been installed and operated.

Introducing all this information, sophisticated numerical groundwater models were developed for the entire Emirate's area as an indispensible tool for all later ASR-related simulations.

Brief description of the local aquifer systems

The available hard rock aquifers in the subsurface of the Abu Dhabi Emirate are unsuited for large-scale ASR-measures, because they are found mostly in very deep position and contain consistently highly saline groundwater. Hence, the investigations concentrated on the shallow aquifer system, which is developed over most of the Abu Dhabi Emirate's area.

The shallow aquifer consists of Holocene and Pleistocene sediments, predominantly of fine to medium grained, well sorted, loose to semi-consolidated Aeolian dune sands, while in the easternmost part of the country, west of the Hajar mountain ridge (Oman), semi-consolidated gravel and conglomerates prevail. Over large parts of the country, the shallow water table aquifer is confined at the bottom by low permeable sediments (mudstones,

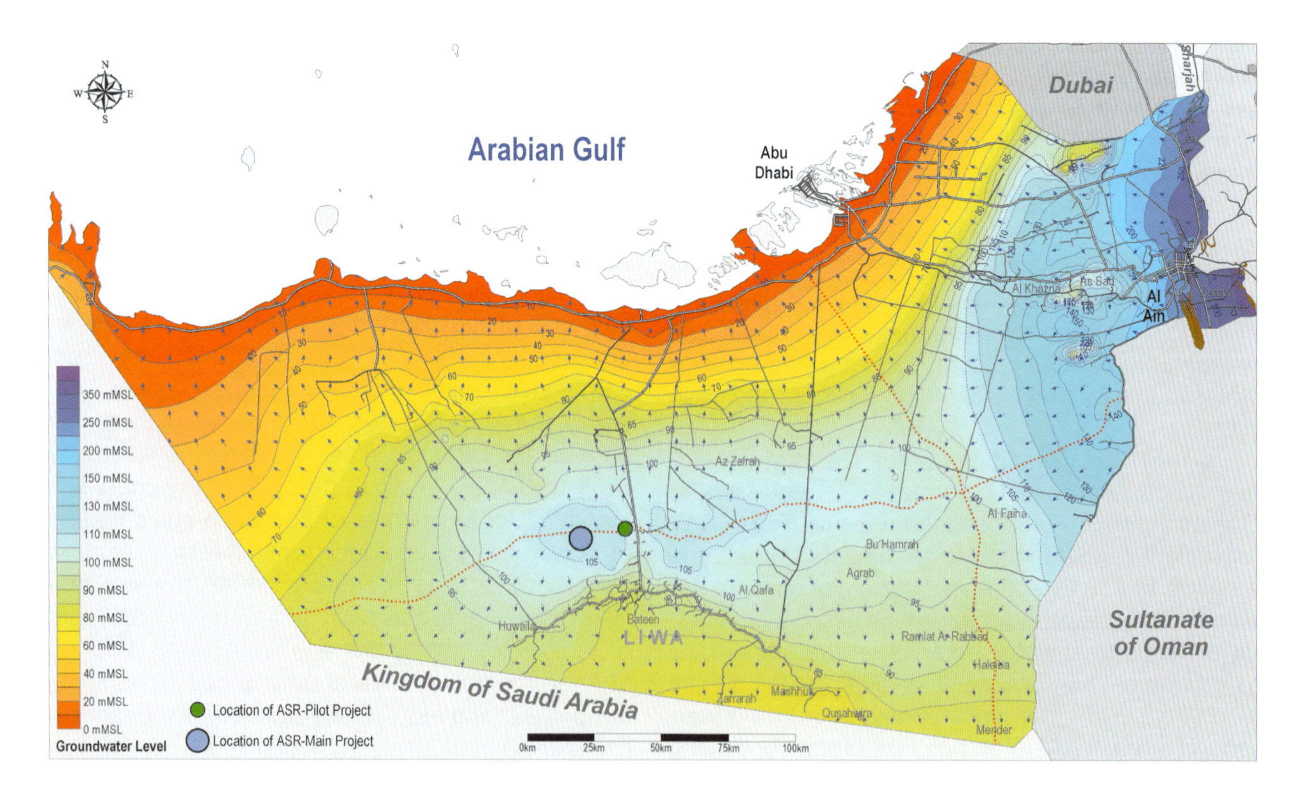

Figure 1: Groundwater level contours of the shallow aquifer in the Abu Dhabi Emirate.

evaporites and clastics) of Miocene age. The saturated aquifer thickness reaches about 150 m in the Western Region and more than 300 m in parts of the Eastern Region.

Figure 1 shows the groundwater flow pattern of the shallow aquifer system with highest level elevation along the eastern border to Oman, while lowest level elevation is found along the coastline. An approximately west-east stretching groundwater shed divides the Emirate's area into a northern and a southern part; all shallow to medium deep-seated groundwater north of it flows towards the Arabian Gulf as the receiving body, while all groundwater south of it flows towards an extended low elevated inland Sabkha structure in Saudi Arabia. Furthermore, an about north-south running groundwater shed separates the Eastern from the Western Region. A pronounced groundwater mound forms the central part of the Western Region, while large areas in particular in the Eastern Region exhibit significant groundwater depletion due to large-scale groundwater abstraction for irrigation purpose.

Wells completed in the dune sands can yield up to 250 m^3/h, while wells tapping the gravel/sand aquifer produce in exceptional cases more than 400 m^3/h. The hydraulic conductivity ranges from around 2 m/d to 60 m/d, generally decreasing with depth. Effective porosity and specific yield are mostly high, in places reaching 30%. The thickness of the vadose zone varies widely from almost zero in the coastal shallow plains to more than 250 m in parts of the Eastern Region, with an average of some 45 m in the central Western Region.

Although the total stored groundwater in the Emirate's shallow aquifer system is considerably high with an estimated 745 km^3, most of the groundwater is unsuited for domestic and irrigation uses. Figure 2 indicates that over half of the Emirate's area, the groundwater is saline showing maximum concentration of around 300 g/kg. Locations where the aquifer bears fresh native groundwater cover only some 4% (≤1,500 ppm) and less than 2% (≤1,000 ppm) of the total area, respectively. The remaining areas are taken by brackish groundwater (salinity range between 1,500 ppm and 10,000 ppm). Low mineralised groundwater in noticeable quantities occurs only along the eastern border to the Omani mountain ridge and in the central desert of the Western Region.

The fresh groundwater along the eastern border predominantly originates from young recharge from the Hadjar mountain ridge, while fresh groundwater in the Western Region is solely attributed to recent aquifer recharge due to infiltrating local precipitation. Significant tritium levels confirm that the local groundwater is not exclusively fossil but has

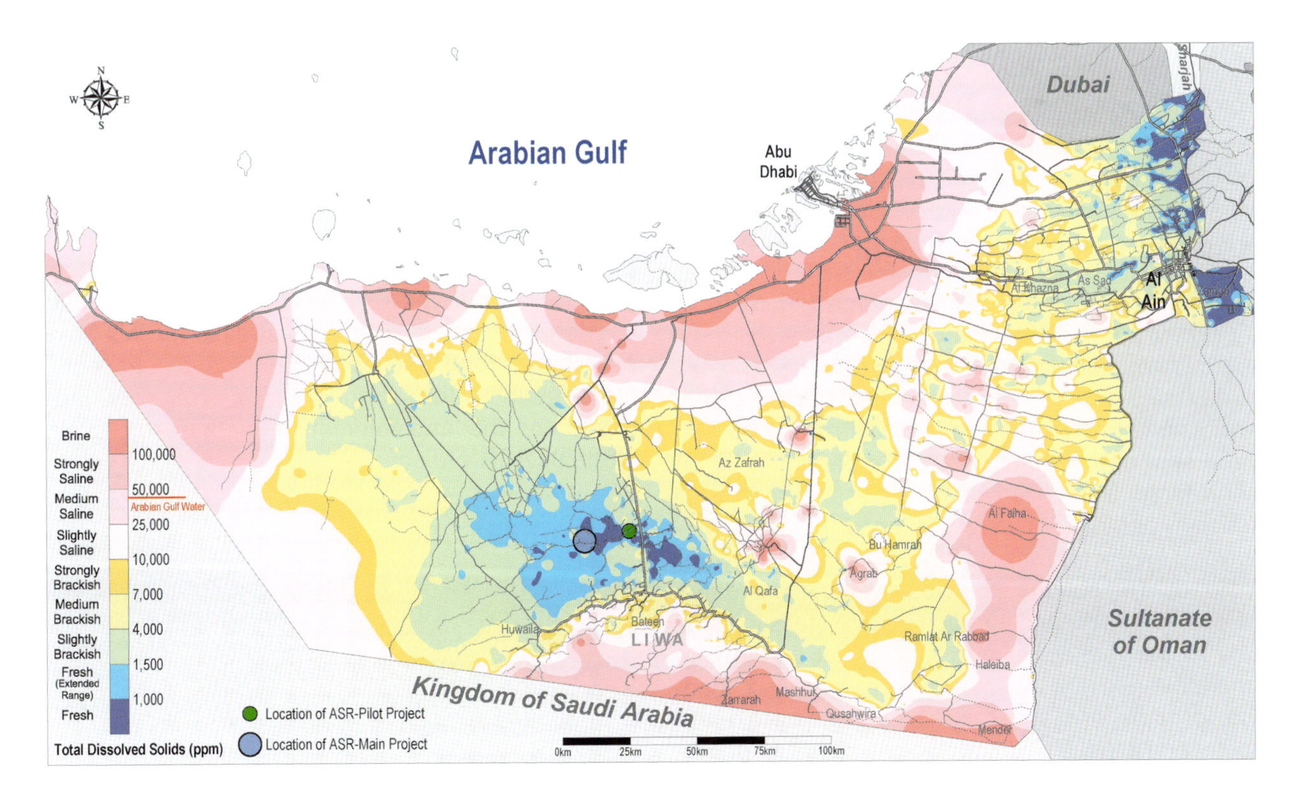

Figure 2: Salinity of the shallow groundwater in the Abu Dhabi Emirate.

Figures 3 and 4: The Liwa area in Abu Dhabi's Western Region after heavy rains on 12.05.2002 (45 mm within one hour).

a significant young component. This finding is substantiated by the hydrochemical character of such groundwater, identical to rain water, as well as the dynamic groundwater flow pattern of the shallow aquifer. Numerical simulations indicate that the areal groundwater recharge is about 10 mm/a on average. Although local rainfall is scarce (on average ranging from some 40 mm/a in the Western Region to 200 mm/a in the mountainous part of the Eastern Region), recharge is provided through occasional torrential storm floods (see Figs. 3 and 4). However, the natural recharge compensates only for a fraction of the groundwater volume, which is abstracted for irrigation purpose in the Emirate (about 3.4 million m^3/a).

ASR-suitability assessment

In order to assess the Emirate's area regarding its suitability for ASR-measures, first and foremost the hydrogeological settings govern the suitability but also a range of other aspects, like present and planned future land use and infrastructure, environment and general strategic aspects had to be considered. Thematic hydrogeological maps for all relevant parameters were produced, comprising lithology, geohydraulics, hydrochemistry and hydroisotopes. The country then was divided into squares of 100 km^2 each, and for each square statistical values of the respective hydrogeological parameters and other relevant information were assigned, leading to maps with suitability scoring for each parameter and aspect. Hydrogeological criteria for general exclusion of an individual square were insufficient thickness of aquifer or vadose zone, for instance.

Figure 5 shows the resulting map for hydrogeological criteria, clearly indicating that the centre of the Western Region, north of the Liwa Crescent is by far the best suited location for the implementation of large-scale ASR-schemes.

Here, along the west-east running groundwater shed, on top of the pronounced groundwater mound, the geological settings of the remote, virgin area offer a vast natural storage capacity and an excellent protection of the groundwater resources. The saturated aquifer thickness exceeds 100 m and the minimum thickness of the vadose zone is 35 m. On average, the upper 40 m of the saturated aquifer contain native fresh groundwater. Hence, quality losses due to mixing of the introduced DSW with local groundwater are considered minimal. The natural groundwater flow gradient is very low (less than 0.2‰); flow simulations indicated that the lateral migration of the DSW-plume would be less than 5 m/a.

Moreover, there are no constraints for this area as regards land use and other relevant aspects. Consequently, the northern Liwa area was proposed for the implementation of the large-scale strategic ASR-project.

References

Deutsche Gesellschaft für Technische Zusammenarbeit (GTZ) GmbH & Dornier System Consult – Daimler-Chrysler Services (DSC), 2002: Groundwater Assessment Project Abu Dhabi: Feasibility study on artificial recharge and utilisation of the groundwater resource in the Liwa area. – Internal report for the Office of Presidential Affairs Abu Dhabi, unpublished, 101 pp.

Deutsche Gesellschaft für Technische Zusammenarbeit (GTZ) GmbH & Dornier Consulting (DCo), 2005:

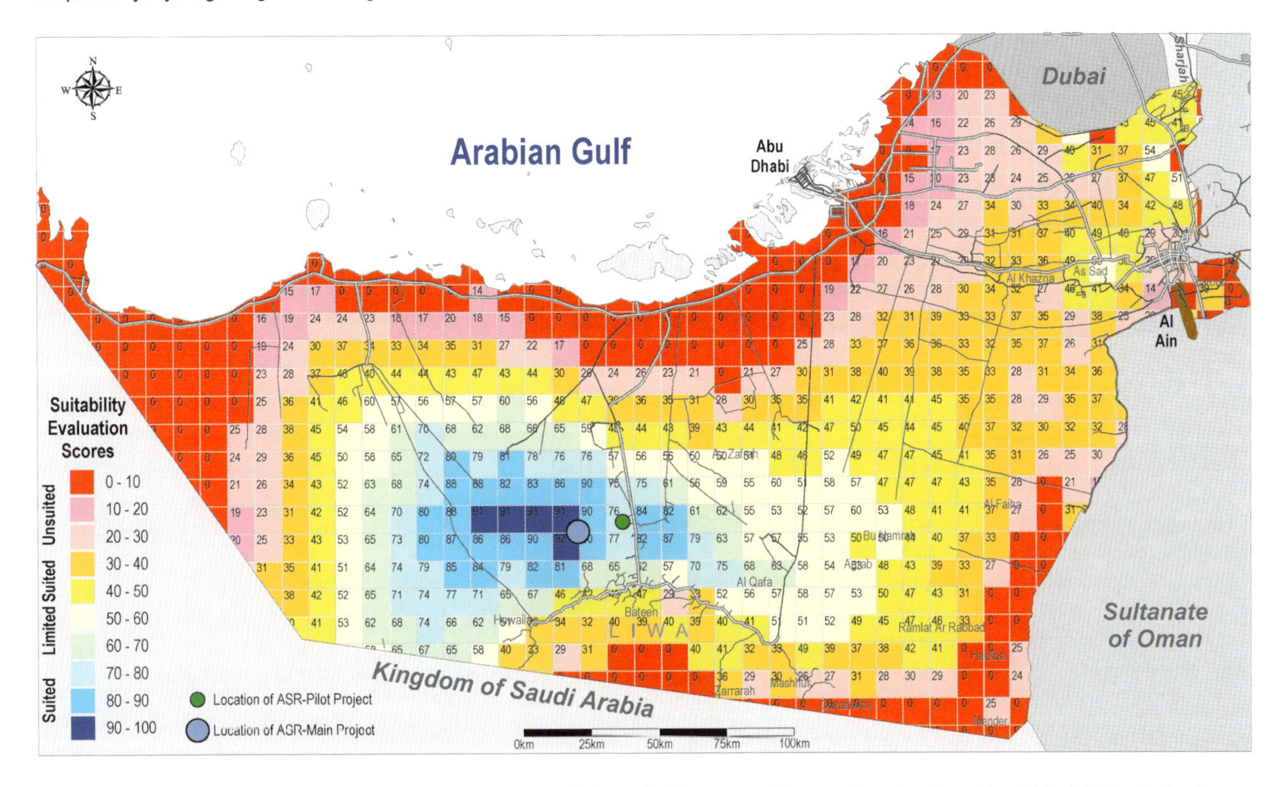

Figure 5: Spatial ASR-suitability assessment of the shallow aquifer system in the Abu Dhabi Emirate (synopsis of individual hydrogeological parameters).

Combined artificial recharge and utilisation of the groundwater resource in the Greater Liwa area (Western Region of the Emirate of Abu Dhabi), pilot project, final technical report. – Internal report for the Abu Dhabi National Oil Company (ADNOC), unpublished, 126 pp.

Deutsche Gesellschaft für Technische Zusammenarbeit (GTZ) GmbH & Dornier Consulting (DCo), 2005: Groundwater Assessment Project Abu Dhabi: Status Report – Phases IXa, IXb and IXc, Vol. 1 (1-2): Exploration. – Internal report for the Abu Dhabi National Oil Company (ADNOC), unpublished, 507 pp.

Deutsche Gesellschaft für Technische Zusammenarbeit (GTZ) GmbH & Dornier Consulting (DCo), 2005: Groundwater Assessment Project Abu Dhabi: Status Report – Phases IXa, IXb and IXc, Vol. 2: Geophysics. – Internal report for the Abu Dhabi National Oil Company (ADNOC), unpublished, 280 pp.

Deutsche Gesellschaft für Technische Zusammenarbeit (GTZ) GmbH & Dornier Consulting (DCo), 2005: Groundwater Assessment Project Abu Dhabi: Status Report – Phases IXa, IXb and IXc, Vol. 3 (1-2): Monitoring, Database and Modelling. – Internal report for the Abu Dhabi National Oil Company (ADNOC), unpublished, 629 pp.

Proceedings "Hydrogeology of Arid Environments", p. 76–80
Stuttgart, March 2012

Extended Abstract

The Khadin Water Harvesting System of Peru – An Ancient Example for Future Adaption to Climatic Change

Bertil Mächtle[1], Katharina Ross[2], Bernhard Eitel[1]

[1] Ruprecht-Karls-Universität Heidelberg, Institute of Geography, Im Neuenheimer Feld 348, D-69120 Heidelberg, email: bertil.maechtle@geog.uni-heidelberg.de
[2] Friedrich-Schiller-University Jena, Institute of Geosciences, Chair for Hydrogeology, Woellnitzer Str. 7, D-07749 Jena

Key words: water harvesting, southern Peruvian coastal desert, Khadin system, global warming, archaeology

In arid environments, the access to water resources is a crucial challenge of mankind since thousands of years. Singular runoff events of high magnitude and low frequency are characteristic of arid environments, and the loss of unexploited water by evaporation and seepage is overwhelming. Under these conditions, ancient cultures developed several strategies to fulfill their demands of water, which could be renewed in the context of adaption measures to the impacts of global warming.

One technique, which counts to the water harvesting strategies, is the concentration and storage of water using artificial sediment bodies. Building up an embankment to impound surface runoff, this system is supplied by feeders which collect runoff from a large catchment area and concentrate it on a small cultivated bed (Fig. 1). This system is called "Khadin" in India's Thar desert, where it was invented during the 15th century (the Khadin cultivation generally is assumed to be more than 5000 years old, Tiwari 1988). Recent studies show that the Khadin system works well under conditions, where bare rock in the catchment area provides a high proportion of surface runoff. Annual average rainfall (a.a.r.) ranges from 70 to 150 mm/yr and the catchment : concentration area ratio (CCAR) ranges from 25:1 to 15:1 (Kolarkar 1997). The water is good for cultivation of crops, for drinking or to water cattle due to a slow seepage of water, which prevents salinization and replenishes the aquifers. Furthermore, the increased soil moisture activates microbiological activity and increases the fertility of soils. Singular dry years are buffered by infiltration of water to the storage sediment, which supplies wells up to two years.

Elsewhere in the world, similar structures have been developed, namely in northern Africa and central Asia, and hereof independent in South America. In the coastal desert of southern Peru, near Ciudad Perdida de Huayuri (14°31'45"S, 75°16'15"W, 480 m a.s.l., close to the village of Palpa; see Fig. 1), present-day a.a.r. does not exceed 20 mm/yr due to currently less intense easterly air flow and a more southern position of the Intertropical Convergence Zone (ITCZ), which results in moisture transport across the Andes further to the south (Fig. 2b).

Therefore, the preconditions for rainwater harvesting are missing at present, and the existence of a Khadin system there gives evidence of more humid conditions during the past due to changes in atmospheric circulation (Mächtle & Eitel 2012, Fig. 2a). Reconstructed from paleoclimatic proxy data in the adjacent highlands , more humid conditions occurred in the study area during the period from 1150–1450 AD, interrupted by drier conditions around 1250 AD (Fig. 3). In archaeological terms, this was the "Late Intermediate Period" (LIP). Given that the settlement of Ciudad Peridida de Huayurí was founded some decades earlier and that fluvial sediment dynamics in this region showed that increased aridity took place during the time of Khadin construction, we infer that the innovation of the Khadin system was a human adaption forced by a climatic deterioration ("x" in Fig. 3).

www.borntraeger-cramer.de
2012/0076/$ 1.25

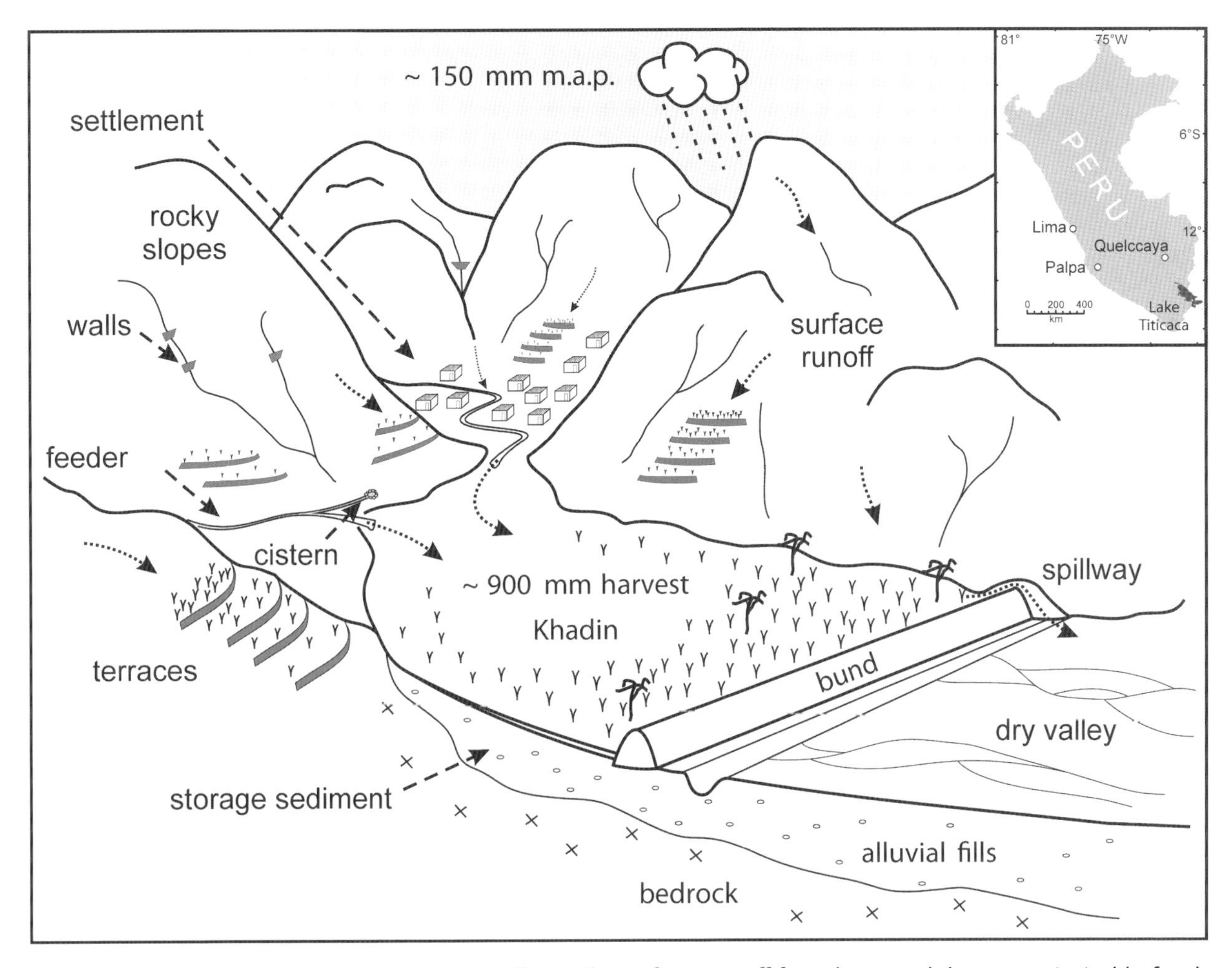

Figure 1: Khadin system in southern Peru: Episodic surface runoff from bare rock is concentrated in feeders, which end on an artificial sandy sediment body. Rapid infiltration minimizes evaporation loss, and a bund dams the runoff and fosters the growth of the sediment body and the storage capacity. From ~150 mm a.a.r., ~960 mm of water concentration results.

Usually, quantitative assessments of palaeorainfall are not possible from geomorphological archives, but the well known parameters of Khadins allow a precise reconstruction of its order. Considering a CCAR of 11:1, we assume an annual average rainfall of ~150 mm/yr for successful Khadin cultivation during the time of construction ~1260–1290 AD (Mächtle et al. 2009). Accounting for evaporation and seepage (for calculation see Prasad et al. 2004), the Khadin area received ~965 mm/m^2 water, which is sufficient to fulfill the water demands of maize growing. 10% loss due to evaporation loss on the Khadin should also be considered.

Unfortunately, crop yields for pre-Columbian maize cultivation are not available. Therefore, we calculated the yield in a first step based on modern data. Due to the sandy-silty soil structure, 62% of the harvested water is plant available (Ross 2007), which is good for a corn yield of 15 t/ha (15% water content, 344 l water demand for 1 kg of recent corn species, Benson 2011). Assuming a reduction of 66% in corn yield during pre-Columbian times, deduced from smaller maize cobs and taking into account less manuring, merely 5 t/ha seems to be more realistic for the past. The area of two Khadins in the vicinity of Ciudad Perdida de Huayurí was 2.5 ha, and the yield therefore was 12.5 t, good to feed nearly 180 dwellers with a diet of 33% maize (we assumed a calory requirement of 1800 kcal/day and a nutritional value of 3310 kcal/kg maize). This order agrees with the size of Ciudad Perdida de Huayurí, where the people also did gathering, pasture with camelids or used marine resources for dietary supplement. Albeit this is a rough estimate, it can be shown that the Peruvian Khadin enabled the people to adapt to climatic changes.

The Indian government realized the potential of the Khadin system, which uses local materials and

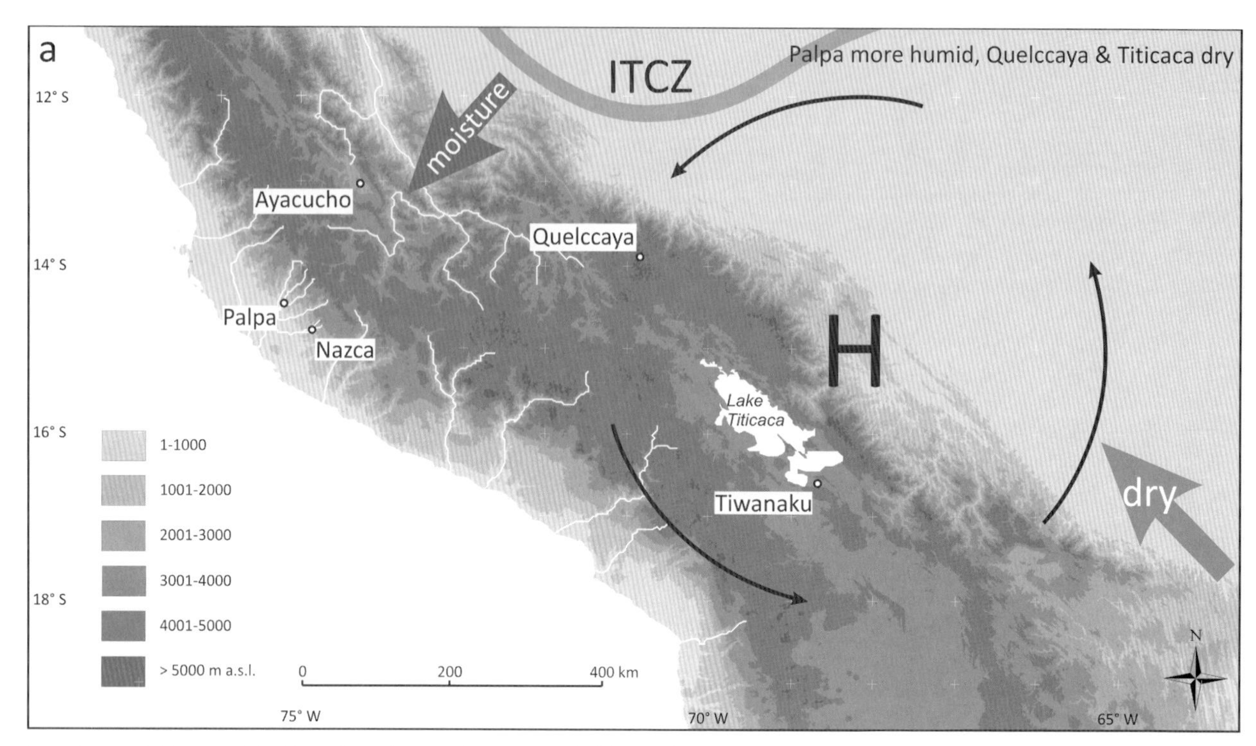

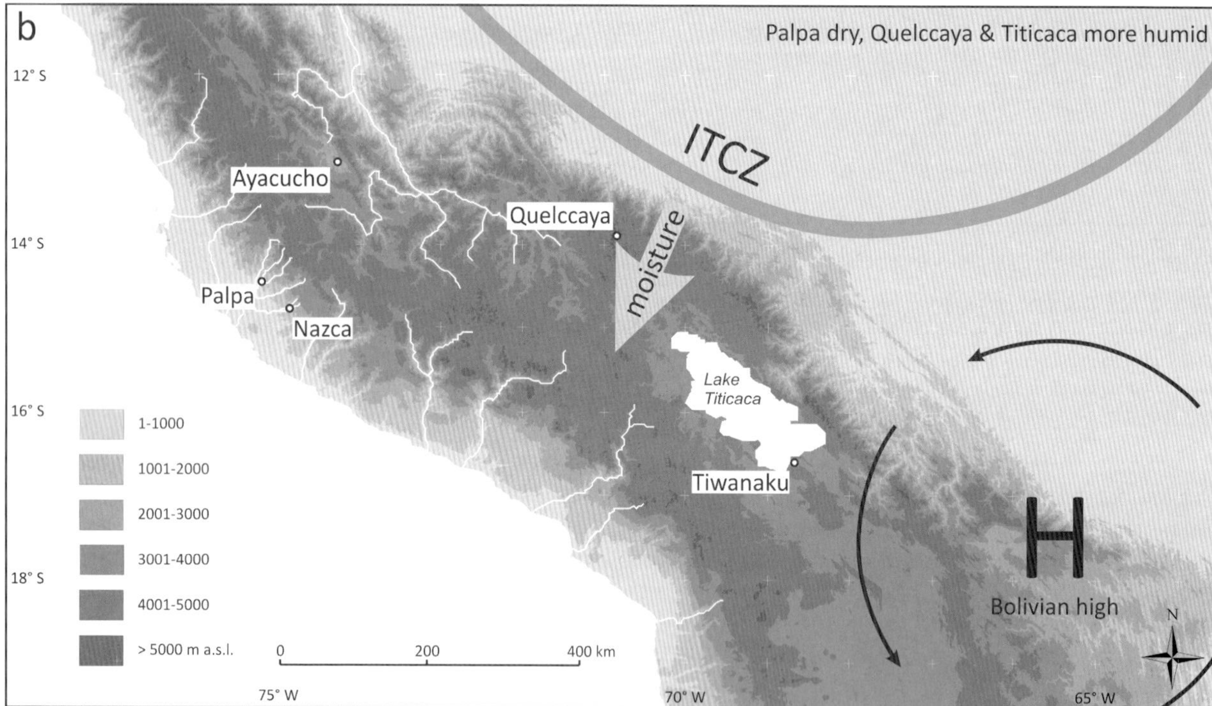

Figure 2: Moisture transport to the Palpa region due to a more northern position of the ITCZ (a) coincides with cultural boom times, the Titicaca region received less moisture and vice versa (b). The Khadin system was invented during short-term conditions of state "b" during a state "a" period of cultural boom (LIP, 1150–1450 AD).

is most cost-effective, starting various revival programs. In contrast to some modern techniques, traditional water management systems gain high acceptance by local people, which is a prerequisite for their sustainable maintenance. As long ago as 1980, more than 500 Khadins with an area of 12.000 ha have been rebuilt in the Jaisalmer district of northwestern India (100–200 mm annual average rainfall). The

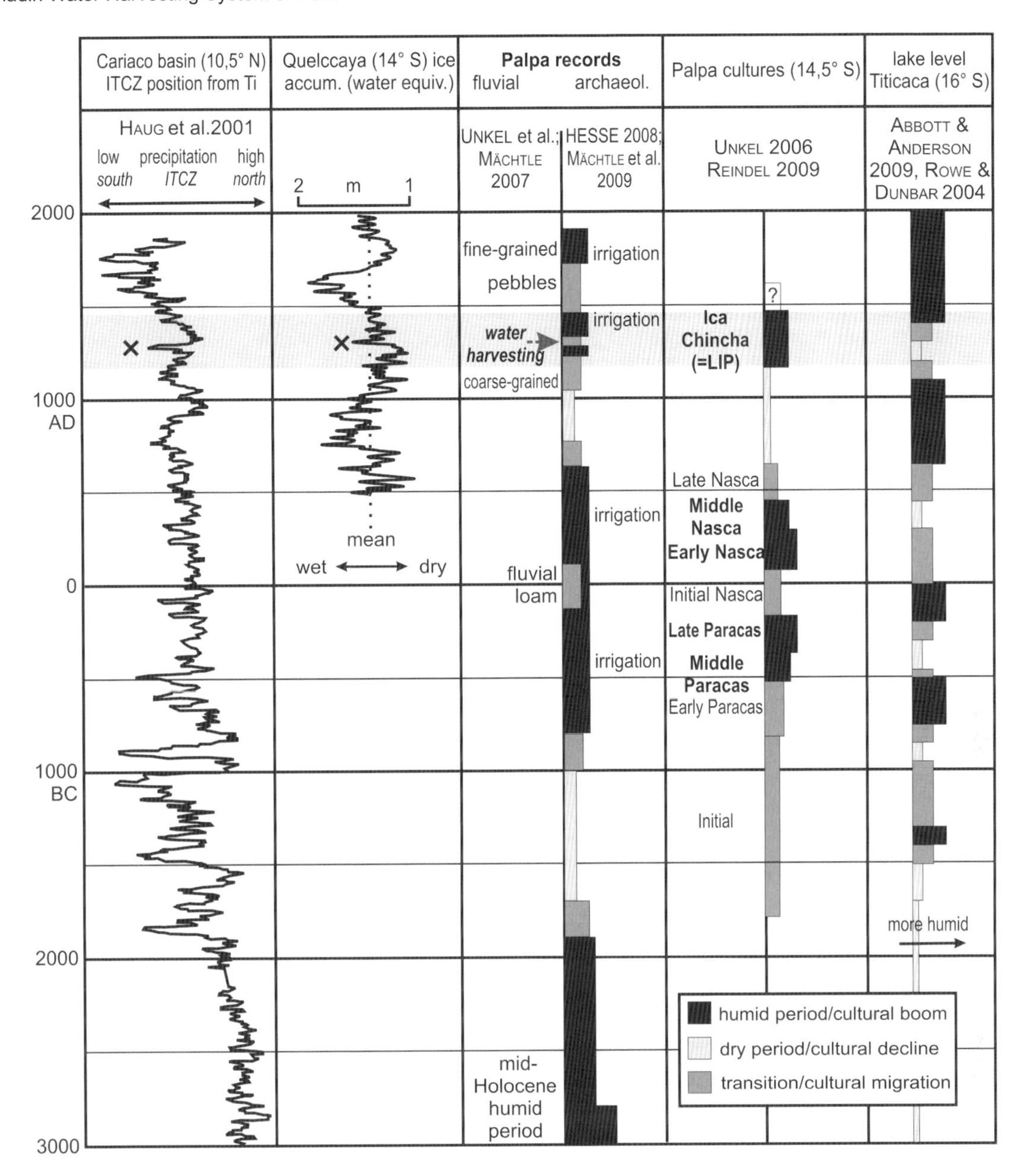

Figure 3: A southward shift of the Intertropical Convergence Zone (ITCZ) after 1250 AD may have triggered the innovation of water harvesting (x). The shift is confirmed by sediments from the Cariaco basin, an increase in ice accumulation at Quelccaya ice cap and coarse-grained fluvial sedimentation near Palpa, indicating drought and accentuated runoff events. For references, see Mächtle & Eitel 2012).

crop production from the Khadin beds is adequate to feed all people (Goyal et al. 2009). In dry years, the millet yield ranges from 300–500 kg/ha, in wet years wheat yield ranges from 1200–2000 kg/ha (Kolarkar & Singh 1984). In the context of global warming, the Khadin system will be a good chance to utilize water resources in regions which become more and more arid, or, on the other hand, to expand crop production in areas which are still too dry for cultivation, as it is expected for northwest India.

References

Benson, L. V., 2011: Factors controlling pre-Columbian and early historic maize productivity in the American southwest, part 1: The Southern Colorado plateau and Rio Grande regions. – J. Archaeol. Method. Theory **18**: 1–16.

Kolarkar, A. S. & Singh, N., 1984: An ancient system of water harvesting. – Ind. Farm. **34**: 21–22, 25.

Kolarkar, A., 1997: Traditional water harvesting systems – Thar desert – Khadins. – In: Agarwal, A. & Narain,

S. (eds.): Dying wisdom – Rise, fall and potential of India's traditional water harvesting systems. Centre of Science & Environment, New Dehli, = State of India's Environment Vol. 4, pp. 104–143.

Goyal, R. K., Angchok, D., Stobdan, T., Singh, S. B. & Kumar, H., 2009: Surface and Groundwater Resources of Arid Zone of India: Assessment and Management. – In: Kar, A., Garg, B. K., Singh, M. P. & Kathju, S. (eds.): Trends in Arid Zone Research in India. – Central Arid Zone Research Institute, Jodhpur, pp. 113–150.

Mächtle, B., Eitel, B., Schukraft, G. & Ross. K., 2009: Built on sand – climatic oscillation and water harvesting during the Late Intermediate Period. – In: Wagner, G. & Reindel, M. (eds.): New Technologies for Archaeology: Multidisciplinary Investigations in Palpa and Nasca, Peru. Springer, Heidelberg, pp. 39–46.

Mächtle, B. & Eitel, B., 2012: Fragile landscapes, fragile civilizations – how climate determined societies in the pre-Columbian south Peruvian Andes. – Catena, in revision.

Prasad, R., Mertia, R. S. & Narain, P., 2004: Khadin cultivation: a traditional runoff farming system in Indian Desert needs sustainable management. – J. Ar. Env. **58**: 87–96.

Ross, K., 2007: Geoarchäologisch-bodenkundliche Untersuchungen zu präkolumbischen Bewässerungstechniken im Raum Palpa (nördliche Atacama/Südperu). – Unpublished qualifying thesis, Universität Heidelberg, Institute of Geography, 130 p.

Tiwari, A. K., 1988: Revival of water harvesting methods in the Indian desert. – Ar. Lands Newsl. **26**: 3–8.

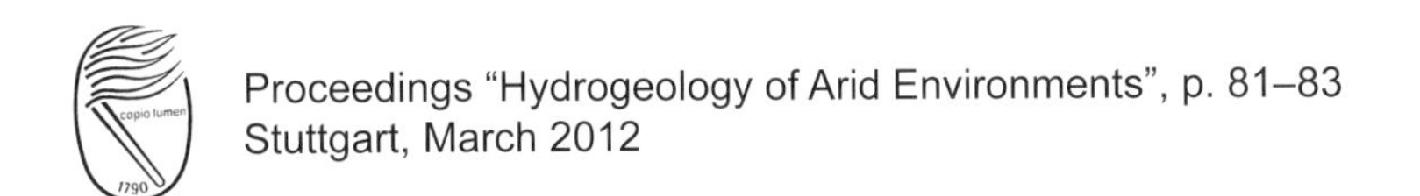

Proceedings "Hydrogeology of Arid Environments", p. 81–83
Stuttgart, March 2012

Extended Abstract

Managed Aquifer Recharge of Reclaimed Water: Storage and Treatment Opportunities in Arid Lands

Robert G. Maliva[1], Rolf Herrmann[2], and Frank P. Winslow[3]

[1] Schlumberger Water Services, 1567 Hayley Lane, Suite 202, Fort Myers, FL 33907, USA, email: rmaliva@slb.com
[2] Schlumberger Water Services, Abu-Dhabi, U.A.E., email: rherrmann@slb.com
[3] Schlumberger Water Services, 7634 Al Ma'ather-Al Sulaimania, Unit 1, Al Rashid Tower, Suite 201, Riyadh, 12621-2652, Kingdom of Saudi Arabia, email: fwinslow@slb.com

Key words: reclaimed water, managed aquifer recharge, aquifer heterogeneity, modeling

Introduction

Treated municipal wastewater and desalination are the only remaining large sources of additional water in most arid lands, as fresh groundwater and surface water resources are commonly already being utilized at quantities at or exceeding sustainable levels. Areas located near the coast can theoretically have unlimited supplies of water through seawater desalination. However, the high costs of desalination typically limit its use to potable and high-value industry water supply, and to nations that can afford this costly alternative. Treated wastewater will thus necessarily play an increasing role in the water supply of arid lands, because of the absence of economical alternative supplies for non-potable uses. In many parts of the world, using water once simply is no longer an option (Levine & Asano 2004).

Reuse of treated wastewater (reclaimed water) has several compelling benefits in that it avoids adverse environmental impacts associated with wastewater disposal, is the only source of additional water that increases as population grows, can provide needed plant nutrients and thus reduce fertilization costs, and is a reliable source of water with a relatively low inter-year and intra-year variation in availability. The use of reclaimed water for purposes that do not require high-quality water (e.g., landscape irrigation) may also allow for reservation of higher quality freshwater resources for higher-value uses, particularly potable supply. The key water management issue is, therefore, not whether or not to reuse treated water, but rather how to optimize its use. Optimization of the reuse of reclaimed water involves obtaining the maximum economic and societal value from the resource, while ensuring that public health and the environment are protected. It is important to also recognize that recharge of aquifers with wastewater occurs in many areas by unmanaged and unplanned means. For example, aquifer recharge may occur through irrigation return flows, sewage disposal systems, and wastewater pipe and canal leakage.

Dillon (2005) defined managed aquifer recharge (MAR) as the "intentional banking and treatment of waters in aquifers." The term MAR was introduced as an alternative to "artificial recharge", which has the connotation that the use of the water was in some way unnatural (Dillon 2005). MAR of reclaimed water can be an important tool in water management arid lands in three main manners, 1) by providing storage, 2) improving the quality of stored water through natural contaminant attenuation processes, and 3) protecting freshwater resources (e.g., salinity barrier systems). MAR systems can be used to store reclaimed water to balance supply and demand. Seasonally available excess reclaimed water can be stored underground using wells in an aquifer storage and recovery (ASR) system, for later use during peak demand (e.g., irrigation) periods. Although reclaimed water supply may generally be relatively constant, demand can be highly variable, and thus create a need for storage. Strategic recharge of reclaimed water by either injection or surface spreading can be used to create a salinity barrier to prevent, or even, reverse the landward migration of saline waters. ASR opportunities in the Middle East were reviewed by Maliva et al. (2010a) and Herrmann (2010).

www.borntraeger-cramer.de
2012/0081/$ 0.75

There is substantial documentation that reclaimed water can significantly improve in quality during aquifer recharge, transport, and storage, by a variety of physical, chemical and biological processes, which was reviewed by Maliva & Missimer (2010b). The placement of reclaimed or surface water into storage in natural groundwater environments increases the recycling time and thereby allows more time for the biodegradation of contaminants that degrade more slowly (Dillon et al. 2006a, b). The intentional use of natural attenuation processes to improve water quality is referred to as natural aquifer treatment (NAT). The greatest value of NAT is as a treatment element (polishing step) in a multiple barrier-approach to wastewater reuse. NAT can be particularly effective in the removal of pathogenic microorganisms and many chemical contaminants. NAT may also have great value in developing countries because it can be a more cost-effective wastewater treatment technology compared to advanced wastewater plants.

Performance of MAR Systems

MAR systems need to be evaluated on the basis of whether or not they meet both performance objectives and public heath, water resources, and environmental protection goals. Storage systems are typically evaluated in terms of recovery efficiency, which is quantified as the ratio of volumes of additional usable water recovered to recharged water. The important issue is that recharge should allow for the recovery of water of a quality suitable for its intended use at a quantity that would otherwise not be available. MAR systems with a treatment goal should be evaluated on the basis of the degree to which the recovered water meets target treatment or water quality goals. With respect to pathogens, the water quality targets could be either a specific numerical water quality standard (e.g., fecal coliform bacteria concentration) or a treatment level, such as specified number of $\log_{10}$ reductions in pathogen concentration. Salinity barrier systems can be evaluated in terms of their effectiveness in managing saline-water intrusion.

Protection of public health is a critical issue for wastewater MAR systems. Unplanned indirect potable reuse should be prevented. If indirect potable reuse is to occur, then it should be carefully planned and managed to ensure that there is no threat to public health. The primary health concern associated with reclaimed water is pathogens, because a one-time exposure to some water-borne pathogens may be sufficient to cause serious illness. Chemical contaminants are typically present at low concentrations in reclaimed water, which are insufficient to induce an acute response. The principal health concerns associated with chemical contaminants comes from long-term chronic exposure.

There are two main strategies for dealing with public health concerns associated with reclaimed water MAR. One option, referred to as "hyper-treatment," involves treating the wastewater to such a high degree that it poses essentially no significant health risks if consumed. Wastewater can be treated to any degree of purity desired. A well known example of the implementation of the hyper-treatment option is the Orange County Water District (California, USA) Groundwater Replenishment System (GRS), in which conventionally treated wastewater receives additional treatments including microfiltration (MF) followed by reverse osmosis (RO), and ultraviolet light and hydrogen peroxide (advanced oxidation) treatment. These treatment processes produce water of a quality far beyond that produced by many potable water systems (Markus 2009). GRS system water is used in the Talbert Gap Salinity Barrier. The hyper-treatment approach is employed because indirect potable reuse will eventually occur and it was reportedly necessary to obtain public support for the project.

An alternative option is to rely on physical and institutional controls to prevent recharged reclaimed water from unintentionally entering the potable water supply and to design and operate the system to take advantage of NAT processes. Physical controls involve the utilization of aquifers that are vertically and/or geographically separated from potable water wells to a sufficient degree so that there is essentially no possibility of the reclaimed water entering the potable water supply. Institutional controls include measures that either minimize the possibility of, or effectively prevent, consumption of reclaimed water in MAR systems. For example, the Destin Water Users Inc. (Florida, USA) reclaimed water ASR system takes advantage of a local ordinance that prohibits use of the storage zone (sand-and-gravel aquifer) for residential water supply. The storage zone is essentially reserved for residential landscape irrigation use, and public outreach further prevents incidental potable use (e.g. drinking from garden hoses).

Hydrogeology of Reclaimed Water MAR

A fundamental technical challenge for wastewater MAR systems is having an accurate understanding of the fate and transport of any contaminants of concern within the recharged water. Solute-transport modeling is thus a critical element of MAR system design in order to evaluate the extent and mixing of the recharged reclaimed water with native groundwater. With respect to pathogens, solute-transport-modeling can be used to estimate the geographic extent of recharged reclaimed water and the aquifer storage (retention) time before the recharged water reaches a sensitive receptor (e.g., a well). The later is important because pathogen inactivation (die off) is time dependent. Pathogen inactivation is usually expressed in terms of $\log_{10}$ inactivation per day or $\log_{10}$ reduction times (number of days for a 90% reduction in concentration). Using either site-specific data on pathogen inactivation rates or published inactivation rate data, and simulated retention times, the reduction in pathogen concentrations at sensitive receptors can be estimated.

Solute-transport modeling is much more complex than conventional groundwater flow modeling, which depends largely upon bulk aquifer hydraulic properties (e.g. transmissivity, storativity, and leakance). Aquifer heterogeneity (particularly with respect to hydraulic conductivity) strongly controls the rate, direction, and extent of solute transport. In cases where groundwater flow is dominated by one or more discrete flow zones or secondary porosity (e.g., fractures and karstic solution conduits), groundwater flow will be much more rapid and recharged water will have a greater areal extent and degree of mixing with native groundwater than would occur under homogenous (primary porosity-dominated) aquifer conditions. For example, ASR systems that store freshwater in brackish aquifers tend to have very poor recovery efficiencies where the storage zone has a high degree of aquifer heterogeneity (Maliva & Missimer 2010b). In MAR systems in which NAT is being relied upon a high-degree of aquifer heterogeneity may result in much shorter retention times and a lesser degree of contaminant attenuation than would occur in more homogenous aquifers.

Detailed aquifer characterization is thus a critical element of reclaimed water MAR projects. The aquifer testing program needs to characterize and quantify the type and degree of aquifer heterogeneity. The aquifer heterogeneity data, in turn, needs to be incorporated into detailed solute-transport models, to allow for accurate simulation of system performance. Standard aquifer pumping tests are needed to determine bulk aquifer properties. Aquifer heterogeneity in hydraulic conductivity can be evaluated through borehole logging (e.g. flow meters), coring and core analyses, packer (drill stem) testing, and tracer testing. Advanced borehole geophysical logs, such as nuclear magnetic resonance and micro-resistivity imaging can provide a continuous fine-scale record of total and effective porosity and pore size distribution, which can be processed to estimate hydraulic conductivity. Workflow software can then be used to process and analyze the data and then build and populate very detailed groundwater flow and solute-transport models.

Conclusions

MAR using reclaimed water has great potential towards optimizing the reuse of reclaimed water in arid lands, which will necessarily play an increasingly important role in overall water management. Reclaimed water MAR systems are inherently complex because of the need to predict and manage solute-transport, which is controlled by aquifer heterogeneity. The hydrogeological challenge is to quantitatively characterize aquifer heterogeneity and incorporate the heterogeneity into detailed solute-transport models. Solute transport modeling results are critical for predicting the movement of recharged reclaimed water and aquifer retention times, which can then be used to estimate pathogen attenuation.

References

Dillon, P., 2005: Future management of aquifer recharge. – Hydrogeol. J. **13**: 313–316.

Dillon, P., Pavelic, P., Toze, S., Rinck-Pfeiffer, S., Martin, R., Knapton, A. & Pidsley, D., 2006a: Role of aquifer storage in water reuse. – Desalination **188**: 123–134.

Dillon, R., Toze, S., Pavelic, P., Vanderzalm, J., Barry, K., Ying, G.-L., Kookana, R., Skjemstad, J., Nicholson, B., Miller, R., Correll, R., Prommer, H., Greskowiak, J. & Stuyfzand, P., 2006b: Water quality improvements during aquifer storage and recovery at ten sites. – In: Recharge systems for protecting and enhancing groundwater resources, Proceedings of the 5th International Symposium on Management of Aquifer Recharge, Berlin, Germany, 11–16 June 2005, Paris UNESCO, p. 85–94.

Herrmann, R., 2010: Aquifer Storage and Reuse in the Middle East. – GWI Global Water Intelligence, Global Water Summit 2010, Paris 26–27 April 2010.

Levine, A.D. & Asano, T., 2004: Recovering sustainable water from wastewater. – Env. Sci. & Tech. **38**: 201A–208A.

Maliva, R. G., Herrmann, R., Winslow, F. & Missimer, T. M., 2010a: Aquifer storage and recovery of treated sewage effluent in the Middle East – Arabian Journal for Science and Engineering **36**(1)1: 63–74.

Maliva, R. G. & Missimer, T. M., 2010b, Aquifer storage and recovery and managed aquifer recharge using wells: Planning, hydrogeology, design, and operation. – Schlumberger Water Services, Methods in Water Resources Evaluation Series No. 2, 578 pp.

Markus, M. R., 2009: Groundwater replenishment & water reuse. – The Water Report, n. **59** (January 15, 2009), p. 1–9.

Proceedings "Hydrogeology of Arid Environments", p. 84–88
Stuttgart, March 2012

Extended Abstract

Application of Thermal Data for Groundwater Studies in Arid Regions at the Example of the Dead Sea

U. Mallast[1,4], C. Siebert[1], F. Schwonke[2], B. Wagner[3], T. Rödiger[1], S. Geyer[1], R. Gloaguen[4], M. Sauter[3], F. Kühn[2], R. Merz[1]

[1] Helmholtz-Centre for Environmental Research (UFZ), Dept. Catchment Hydrology, 06120 Halle, Germany
[2] Federal Institute for Geosciences and Natural Resources (BGR), Sub-Department Geo-Hazard Assessment and Remote Sensing, 30655 Hannover, Germany
[3] Centre for Geosciences at the Georg-August-University Göttingen, Dept. Applied Geology, 37077 Göttingen, Germany
[4] Freiberg University of Mining and Technology, Institute of Geology, Remote Sensing Group, 09599 Freiberg, Germany

Key words: thermal, groundwater, Dead Sea, Landsat, satellite data, airborne data

1. Introduction

In arid regions like the Dead Sea (DS), water scarcity and groundwater salinisation is a well-known problem. There, water supply mostly relies on restricted groundwater resources. For a sustainable management it is essential to obtain detailed information concerning the genesis of this precious resource, one of the main aims within the BMBF-funded SUMAR (Sustainable Management of Arid and Semiarid Regions) project.

Along the DS, significant amounts of fresh groundwaters get lost by flushing through salty sediments and discharge via surface and submarine springs into the terminal lake. Hence, within this project, we study the applicability of satellite based and airborne thermal remote sensing data to provide information on groundwater in terms of discharge locations, temporal discharge variability and relative groundwater amounts.

Several studies have proven that thermal imagery offers a possibility to gain information on coastal aquifers and discharge-behaviour over a large scale. It generally assumes a temperature contrast between the fluids where groundwater causes thermal anomalies of the local sea surface temperature (SST) (Johnson et al. 2008). To detect and to map groundwater discharge airborne and spaceborne remote sensing platforms can provide useful information. Both contain individual advantages in either ground sampling distances (GSD) and hence the detection of small scale thermal anomalies or multiple overpasses resulting in a long-term analysis option of discharge behaviour.

In this study we analyse the advantages and disadvantages of both platforms shedding some light on general possibilities and limitations of groundwater detection. Within this context the focus lies on (i) the influence of surface water discharge mainly being unconsidered in other studies, (ii) the possibility to derive groundwater discharge and (iii) the discharge behaviour over time.

2. Study area

The DS is a terminal lake situated in the Jordan-Dead Sea Graben. Along its western shore, groundwaters originate mainly from Cretaceous limy Judea Group and from Quaternary alluvial coastal aquifers (Yechieli et al. 2010). It preferentially discharges in distinct locations of Ein Feshkha, Kane/Samar, Qedem and Ein Gedi with accumulated amounts of 80–150×10^6 m^{3}a^{-1}(Guttman 2000, Laronne Ben-Itzhak & Gvirtzman 2005). On its eastern flank, groundwaters emerge from Jurassic Zarqa limestone and Cretaceous Kurnub sandstone aquifers and the overlaying Upper Cretaceous Ajlun- and

www.borntraeger-cramer.de
2012/0084/$ 1.25

Belqa Group (Salameh & Bannayan 1994). Quantities for the entire eastern shore are estimated to approach 90×10^6 m^3a^{-1} (Salameh 1996), where it was already shown that the thermal localization is possible (Akawwi et al. 2008).

Springs can be generally classified in two types: i) terrestrial springs emerging along faults or sediment heterogeneities, which subsequently form erosion channels due to the lowering of the DS and ii) submarine springs that emerge on the lake's bottom establishing an upward (jet) flow due to density differences and which appears as a circular pattern on the DS surface (Munwes et al. 2010).

Surface runoff components that potentially cause similar thermal patterns as groundwater (Mallast et al. 2012a) are limited to the perennial Jordan River and ephemeral flash-floods generated after significant rainstorms in the rainy season (Oct-Apr) (Gertman & Hecht 2002). Flow durations of the latter range between 2 and 153 hours and exhibit temperatures of 10–15 °C (Ayalon et al. 1998, Greenbaum et al. 2006).

3. Satellite Data

Because of Becker (2006) who states that "the great difficulty in arid environments is that much of the groundwater flow is intermittent [...] why only multiple remote sensing images can provide secure information", we pursue our analysis on 19 Landsat ETM+ images (path 174 / row 38) using band 6.2. The images that are recorded at approximately 10 a.m. local time (GMT+2) cover the years 2000 to 2002 and are co-registered to UTM WGS 84 Zone 36N. Although the ground sampling distance (GSD) of band 6 is 60 m, all data delivered by the US Geological Survey are resampled to 30 m using cubic convolution (USGS, 2011). The image pre-processing is explained in Mallast et al. (2012a) and includes:

i) the exclusion of land pixels using a threshold of –0.1 of a normalized difference water index (NDWI) derived image of ETM+ band 4 and 2 from the earliest image of the series (15.02.2000) and
ii) the conversion from raw digital numbers (DN) of the ETM+ images to earth's surface temperatures using radiometric calibration coefficients according to Chander et al. (2009) and values on atmospheric transmissivity, upwelling and downwelling radiances from a web-based Atmospheric Correction Tool based on MODTRAN (Barsi et al. 2003).

The so gained sea surface temperature (SST) images of the DS have to be further processed as surface runoff also influences the SST distribution causing similar thermal anomalies. To identify surface runoff influence Mallast et al. (2012a) developed an influence factor (IF) where normalised temperatures (scale between 0 and 1) within a proximal area of a wadi outlets (SR) are subtracted from normalised temperatures of the central area (CA) of the DS).

As low tempered surface runoff occurs, SR is influenced which at the same time leads the normalised temperatures of SR (SRT) to approaches low values. Subtracting these low SRT values from continuously uninfluenced central area temperatures (CAT), results in strong negative IF values which directly indicates a surface runoff influence. With the help of the IF we found that influence time remains

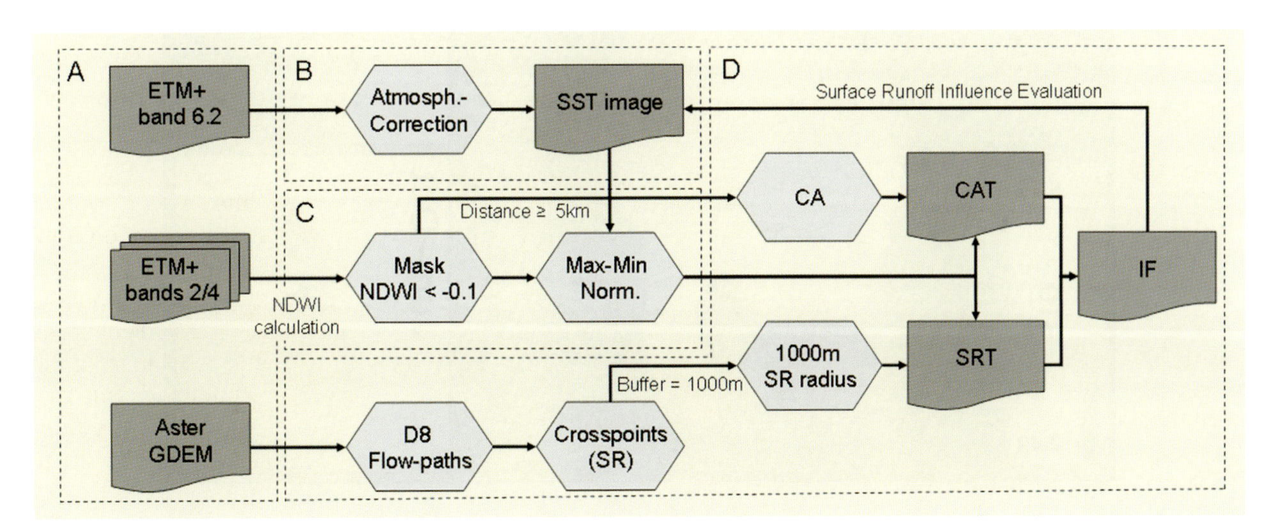

Figure 1: Influence factor (IF) flow chart for surface runoff identification.

for at least two days with an unknown maximum influence time. Excluding SST images with strong negative IF values leaves 12 images appropriate for groundwater related studies.

On the remaining images we calculate the standard deviation (STD) per pixel under the assumption that spatially and thermally persistent groundwater inflow stabilizes the water temperature at the inflow location resulting in low STD values. In contrast, areas with no groundwater inflow exhibits high STD values due to the daily and seasonally temperature variability's which is given as the covered time period of the remaining images spans from March until October. The result confirms the assumption. All spring areas on the western side can be distinguished as they exhibit small STD values (blue colours in Fig. 2). Same aspects account for the eastern side where three groundwater discharge locations can be observed that match those areas found by Akawwi et al. (2008). Moreover, the subsets on the right hand side of Fig. 2 illustrate that even if considering periods of a hydrological year a differentiation is achievable as the discharge amounts and likewise areas with low STD values decrease from the period of Mar-May (climax of the hydrological year with higher discharge) to the period of Jun-Oct (end of the hydrological year with lower discharge).

The STD values even represent a statistical significant clustering (verifiable groundwater occurrence) for 5 locations (Ein Feshka, Kane/Samar, Qedem (all in Israel), Zarka Ma'in, Zara (both in Jordan) with confidence values >90% if considering all images. Calculating the area [m^2] of these significant clusters along the shore results in areal distribution illustrated in Table 1. It is also possible to calculated relative and absolute groundwater discharge amounts taking Ein Feshkha as reference as

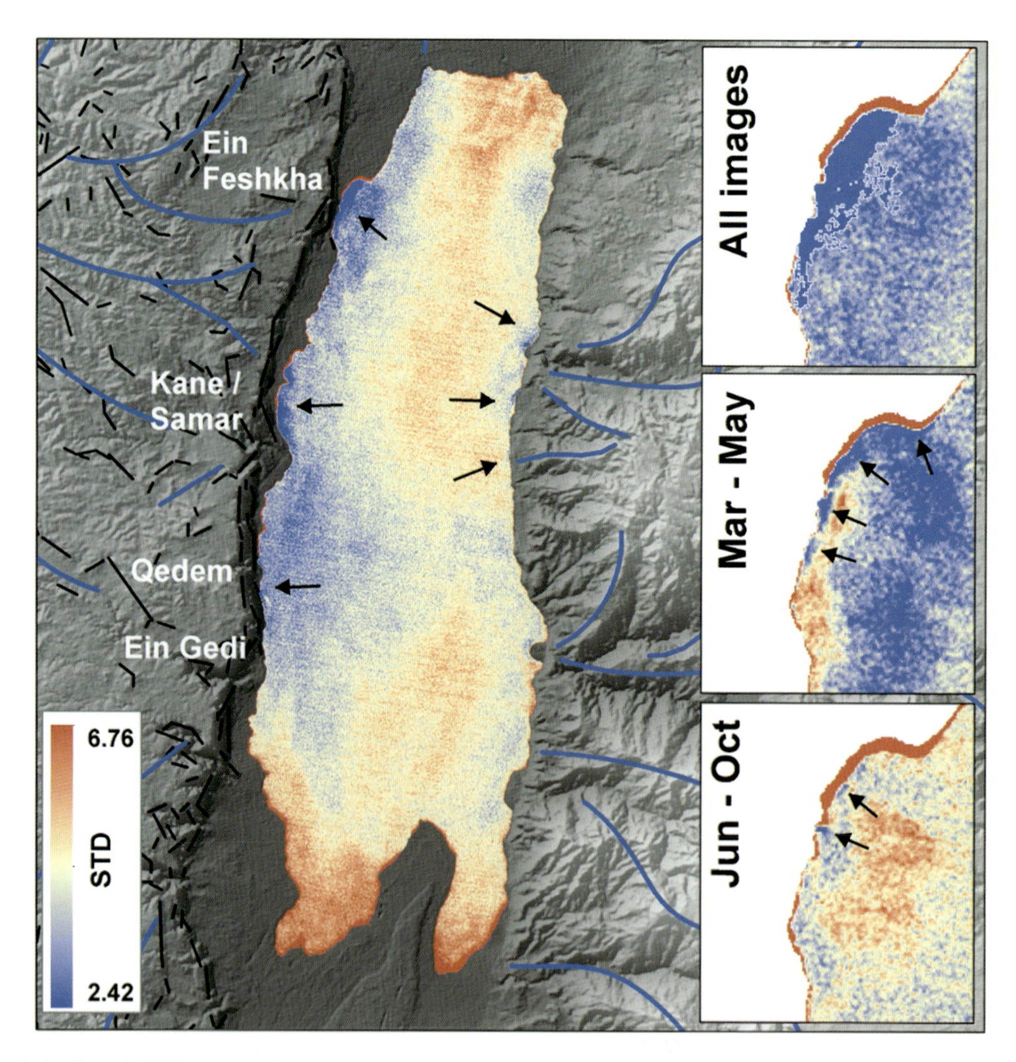

Figure 2: Calculated STD image based on only groundwater influenced images with blue solid lines indicating groundwater flow-paths – subsets are enlargements of Ein Feshkha spring area.

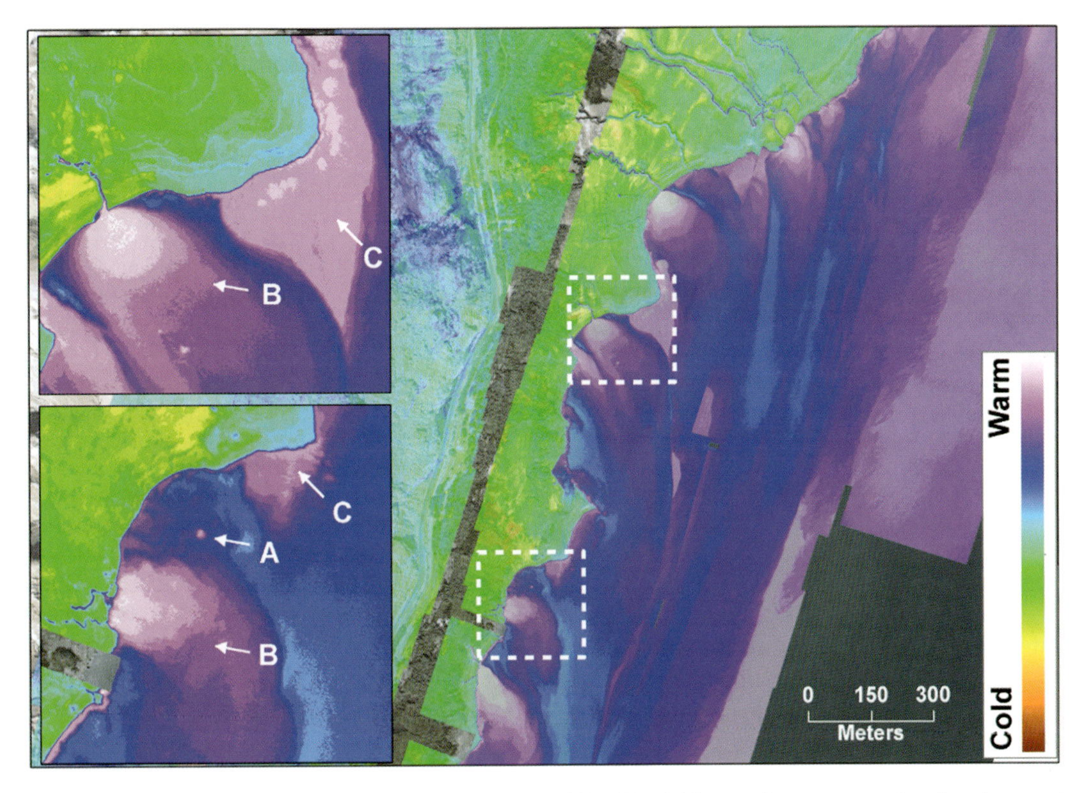

Figure 3: Airborne thermal image of the northern part of Ein Feshkha spring area and subsets emphasizing different spring types.

it represents the only spring location with measured amounts of $80*10^6m^3a^{-1}$ (Guttman 2000). Likewise, it appears that areas with lower discharge amounts (e.g. Ein Gedi) cannot be inferred with sufficient statistical significance (Mallast et al. 2012b).

Table 1. Areas, relative and absolute discharge amounts of significant clusters located at spring areas.

Location	Cluster Area [10^3m^2]	Relative Discharge	Absolute Discharge [$10^6m^3a^{-1}$]
Ein Feshkha	1976.4	1.00	80.00
Kane/Samar	1875.6	0.95	75.92
Qedem	175.5	0.09	7.10
Zarka	92.7	0.05	3.75
Zara	53.1	0.03	2.15

4. Airborne Data

Thermal images from airborne platforms contain two major advantages compared to satellite platforms: the possibility to define acquisition time by hour / by day to account for e.g. seasonally dependencies and the typically much better GSD. Fig. 3 shows a subset from the northern part of the Ein Feshkha area that was recorded during a campaign in January 2011 with a GSD of 0.5 m (Mallast et al. 2011). Groundwater discharge in this region was also detected using satellite images; however it becomes clear that while on the satellite images only general spring areas can be observed, airborne thermal data refine these areas showing numerous and differentiable discharge locations. These locations vary in size of their thermal plumes and hence in discharge amounts. Also visible are different spring types ranging from submarine springs that appear as circular pattern on the water surface (A) to terrestrial springs flowing via an erosion channel into the DS (B). Whereas both types are known the number was unknown. Also unknown was a third type of spring indicated by thermal anomalies along the shore without the existence of erosion channels or circular patterns (C). It appears that groundwater at these locations seeps through the sediment and discharges rather diffuse or emerges submarine and is forced to flow upwards. B and C represent the spring types with the highest discharge amounts where thermal plumes can be observed from a few to several hundred decimetres. The relationship between the 2D-area of the thermal plumes and actual discharge that was measured at 36 locations will be explored with a regression analysis. The result can then be extrapolated on all remaining discharge locations allowing to determine the total groundwater discharge.

5. Conclusion

Both platforms show significant potential for groundwater studies but also individual limitations. Whereas satellite images provide valuable information on general discharge locations, groundwater behaviour over time and conditioned information on groundwater amounts, they require several preprocessing steps and have to be evaluated against surface-runoff influence. The same aspect needs to be considered if analysing thermal images obtained from airborne platforms. Because of the fine GSD, it offers a great possibility to localize and to quantify groundwater discharge and even to differentiate between spring types. The major drawback is that due to monetary reasons thermal images from airborne platforms represent a one-time recorded temporal snap-shot that hinders to draw conclusions on temporal behaviour. On contrary, the recording of satellite images is only limited to cloud conditions, which results in a data basis of several images per year, which, in case of Landsat, are free of charge.

Being aware of these pros and cons, platforms can be chosen according to the scope of the respective study, to optimize the result or to shed light from different perspectives on groundwater issues.

References

Akawwi, E., Al-Zouabi, A., Kakish, M., Koehn, F. & Sauter, M., 2008: Using Thermal Infrared Imagery (TIR) for Illustrating the Submarine Groundwater Discharge into the Eastern Shoreline of the Dead Sea-Jordan. – American Journal of Environmental Sciences **4**(6): 693–700.

Ayalon, A., Bar-Matthews, M. & Sass, E., 1998: Rainfall-recharge relationships within a karstic terrain in the Eastern Mediterranean semi-arid region, Israel: δ18O and δD characteristics. – Journal of Hydrology **207**(1–2): 18–31.

Barsi, J., Barker, J. L. & Schott, J. R., 2003: An Atmospheric Correction Parameter Calculator for a single thermal band earth-sensing instrument. – Proceedings of IGARSS 2003, Centre de Congress Pierre Bandis, Toulouse, France.

Becker, M. W., 2006: Potential for Satellite Remote Sensing of Ground Water. – Ground Water **44**(2): 306–318.

Chander, G., Markham, B. L. & Helder, D. L., 2009: Summary of current radiometric calibration coefficients for Landsat MSS, TM, ETM+, and EO-1 ALI sensors. – Remote Sensing of Environment **113**(5): 893–903.

Gertman, I. & Hecht, A., 2002: The Dead Sea hydrography from 1992 to 2000. – Journal of Marine Systems **35**(3–4): 169–181.

Greenbaum, N., Ben-Zvi, A., Haviv, I. & Enzel, Y., 2006: The hydrology and paleohydrology of the Dead Sea tributaries. – In: Enzel, Y., Agnon, A. & Stein, M. (eds.): New Frontiers in the Dead Sea Paleoenvironmental Research. – The Geological Society of America, Boulder, CO, USA, pp. 254.

Guttman, Y., 2000: Hydrogeology of the Eastern Aquifer in the Judea Hills and Jordan Valley. – Mekorot.

Johnson, A. G., Glenn, C. R., Burnett, W. C., Peterson, R. N. & Lucey, P.G., 2008: Aerial infrared imaging reveals large nutrient-rich groundwater inputs to the ocean. – Geophys. Res. Lett. **35**(15): L15606.

Laronne Ben-Itzhak, L. & Gvirtzman, H., 2005: Groundwater flow along and across structural folding: an example from the Judean Desert, Israel. – Journal of Hydrology **312**(1–4): 51–69.

Mallast, U., Schwonke, F., Siebert, C., Maraschek, U., Kemper, G., Geyer, S., Kühn, F., 2011: Hochauflösende flugzeuggestützte Thermalfernerkundung zur Kartierung submariner und terrestrischer Grundwasserquellen in der hochsalinaren Uferzone des Toten Meeres. – DGPF-Jahrestagung, Mainz.

Mallast, U., Siebert, C., Gloaguen, R., Friesen, J., Rödiger, T., Geyer, S., Merz, R., 2012a: How to derive groundwater caused thermal anomalies in lakes based on long-term medium resolution images in semi-arid regions. – Journal of Arid Environments (submitted).

Mallast, U., Siebert, C., Wagner, B., Gloaguen, R., Sauter, M., Merz, R., 2012b: On groundwater detection and characterization using thermal long-term data. – Ground Water, (submitted).

Munwes, Y. et al., 2010: Direct measurement of submarine groundwater spring discharge upwelling into the Dead Sea. – IWRM, Karlsruhe.

Salameh, E., 1996: Water Quality Degradation in Jordan (Impacts on Environment, Economy and Future Generations Resources Base). – Friedrich Ebert Stiftung, Royal Society for the Conservation of Nature, Amman.

Salameh, E. & Bannayan, H., 1994: Water Resources of Jordan – Present Status and Future Potentials. – Friedrich Ebert Stiftung Amman.

USGS, 2011: Landsat ETM+ product description from the Earth Resources Observation and Science (EROS) Center. http://eros.usgs.gov/#/Find_Data/Products_and_Data_Available/ETM (17.11.2011).

Yechieli, Y., Shalev, E., Wollman, S., Kiro, Y. & Kafri, U., 2010: Response of the Mediterranean and Dead Sea coastal aquifers to sea level variations. – Water Resour. Res. **46**(12): W12550.

Proceedings "Hydrogeology of Arid Environments", p. 89–91
Stuttgart, March 2012

Extended Abstract

Water Resources Protection for the Water Supply of Beirut

Armin Margane[1], Ismail Makki [2]

[1] Federal Institute for Geosciences and Natural Resources, BGR, email: armin.margane@bgr.de
[2] Council for Development and Reconstruction, CDR, email: ismailm@cdr.gov.lb

Key words: karst, groundwater protection, wastewater, Lebanon

Abstract

Within the framework of technical cooperation, the German Federal Institute for Geosciences and Natural Resources (BGR) currently supports the Lebanese Council for Development and Reconstruction (CDR) in protecting the water resources of Jeita spring. A better protection of this important spring is aimed to be achieved by a) assistance in the planning of wastewater schemes in the groundwater contribution zone and b) delineation and implementation of groundwater protection zones. Tracer tests were used to delineate the extent of the groundwater contribution zone, previously unknown and to characterize the groundwater flow mechanism in a highly karstified limestone aquifer system.

Introduction

After the end of the civil war in Lebanon, a rapid and uncontrolled development has taken place. The Government has started only recently to establish collection and treatment systems for wastewater. The water supply of the capital of Lebanon, Beirut, depends to a large degree on springs that emerge from a nearby limestone aquifer, which is highly karstified. The Jeita and Kashkoush springs, located some 15 km northeast of Beirut, discharge on average around 200 Mm^3/a and provide around 75% of the water supply for the Greater Beirut area. Both springs are polluted since many years, mainly by wastewater.

Activities to Reduce Pollution Risks

Two projects, funded by German development aid (Ministry of Economic Cooperation and Development), aim to reduce the pollution risk for Jeita spring. The technical cooperation project (TC), implemented by the Federal Institute for Geosciences and Natural Resources (BGR) works hand in hand with a financial cooperation project (FC), implemented by KfW Development Bank, to achieve this objective. The TC project provides advice to the FC project, especially concerning geoscientific aspects related to the site selection for wastewater treatment plants (WWTPs), collector lines, effluent discharge locations, wastewater reuse, and concerning environmental impact assessments (EIAs) for the wastewater facilities to be established by the FC projects.

Based on groundwater investigations, carried out by the TC project, proposals have been made how to collect, where to treat and where to potentially reuse wastewater in the groundwater contribution zone of Jeita spring (Margane 2011). A criteria catalogue was proposed to improve planning in the wastewater sector. High-risk zones were delineated, indicating where investments in the wastewater sector should have priority. Groundwater protection zones, being the first in Lebanon, will soon be established in the Jeita catchment for all major springs and wells used for drinking water supply. The related land use restrictions in the most sensitive area, protection zone 2, will include special regulations for the construction of wastewater facilities in declared protection zones, among others with the obligation to connect households to existing sewer lines or to build closed septic tanks and regularly empty them.

www.borntraeger-cramer.de
2012/0089/$ 0.75

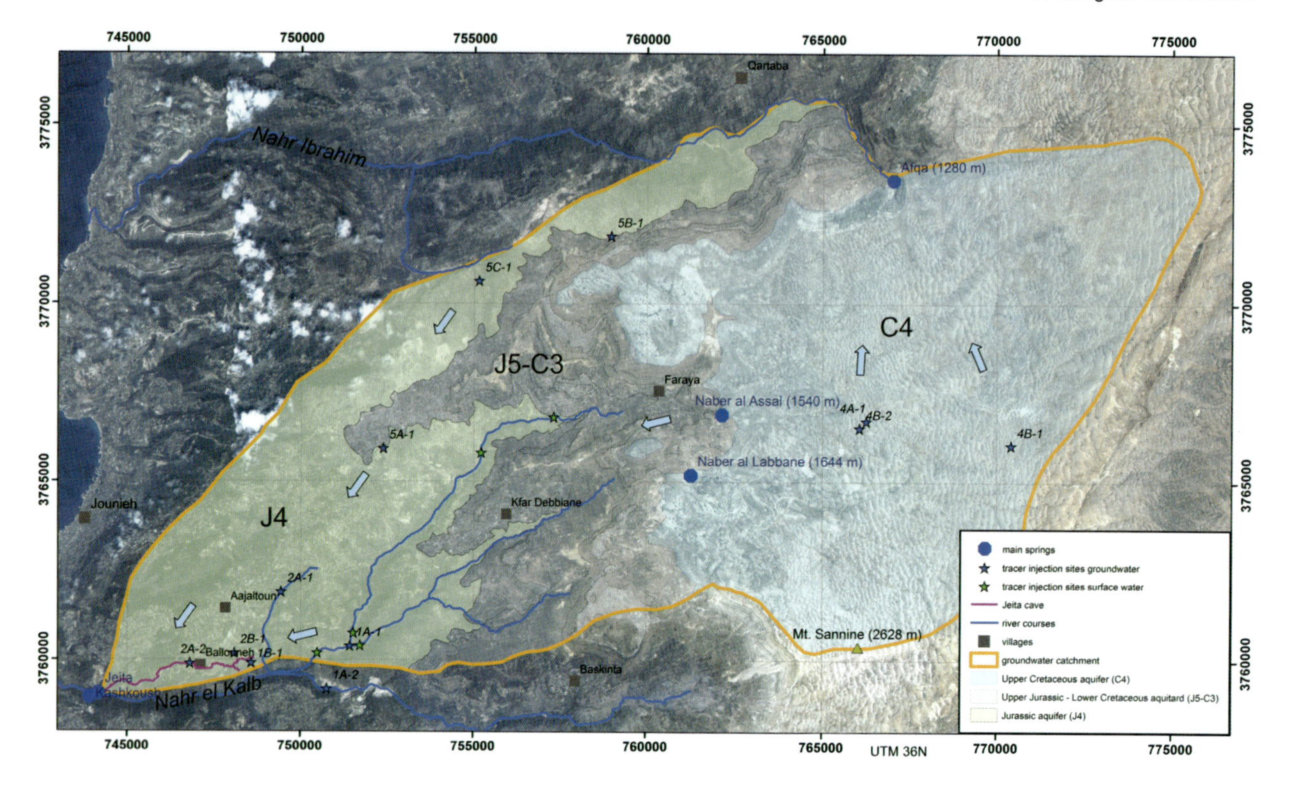

Figure 1: Groundwater contribution zone of Jeita spring and location of tracer injection sites.

The TC project assists municipalities in the catchment area in integrating water resources protection aspects into their local land use plans, which is currently not the case.

Tracer tests proved to be highly valuable for the delineation of the groundwater contribution zone (Figure 1), for the delineation of groundwater protection zones (using groundwater vulnerability maps) and for site selection of wastewater facilities (Figure 2). They showed where direct and fast infiltration pathways exist and indicated that in certain areas flow velocities in the saturated part of the groundwater system are extremely high (up to 2,000 m/h). Based on the results of tracer tests the initial planning of wastewater facilities had to be changed (Margane 2011). According to current planning most collected wastewater would be conveyed towards the coastal zone and treated there.

In order to improve water resources protection and environmental practices, a best management practice guideline for wastewater facilities in karstic areas (Steinel & Margane 2011), a standard for treated wastewater reuse (Margane & Steinel 2011) and a guidance document for environmental impact assessments (EIA) in the wastewater sector (Margane & Abi Rizk 2011) have been prepared by the TC project. The EIA guideline lists which investigations need to be done in order to properly address all issues of the EIA dealing with the impact on water resources and georisks (tectonic movements, earthquakes, landslides, rock falls, flooding, soil stability).

Recommendation

In many countries water resources protection is often neglected in the planning and design process for wastewater projects, and water supply and wastewater issues are not dealt with in an integrated approach. Therefore the investment often fails because the objective of water resources protection is not sufficiently met. The combination of financial cooperation projects, which establish wastewater and geotechnical facilities, and technical cooperation projects, which provide advice to the former in all geoscientific aspects, is a new approach of German development aid that aims to reach a better protection of water resources.

EIAs for wastewater schemes in less developed countries do often not sufficiently address all geoscientific aspects. It is therefore recommended using a standard EIA guideline, such a proposed by the project, which includes all potential impacts on water resources and impacts from georisks.

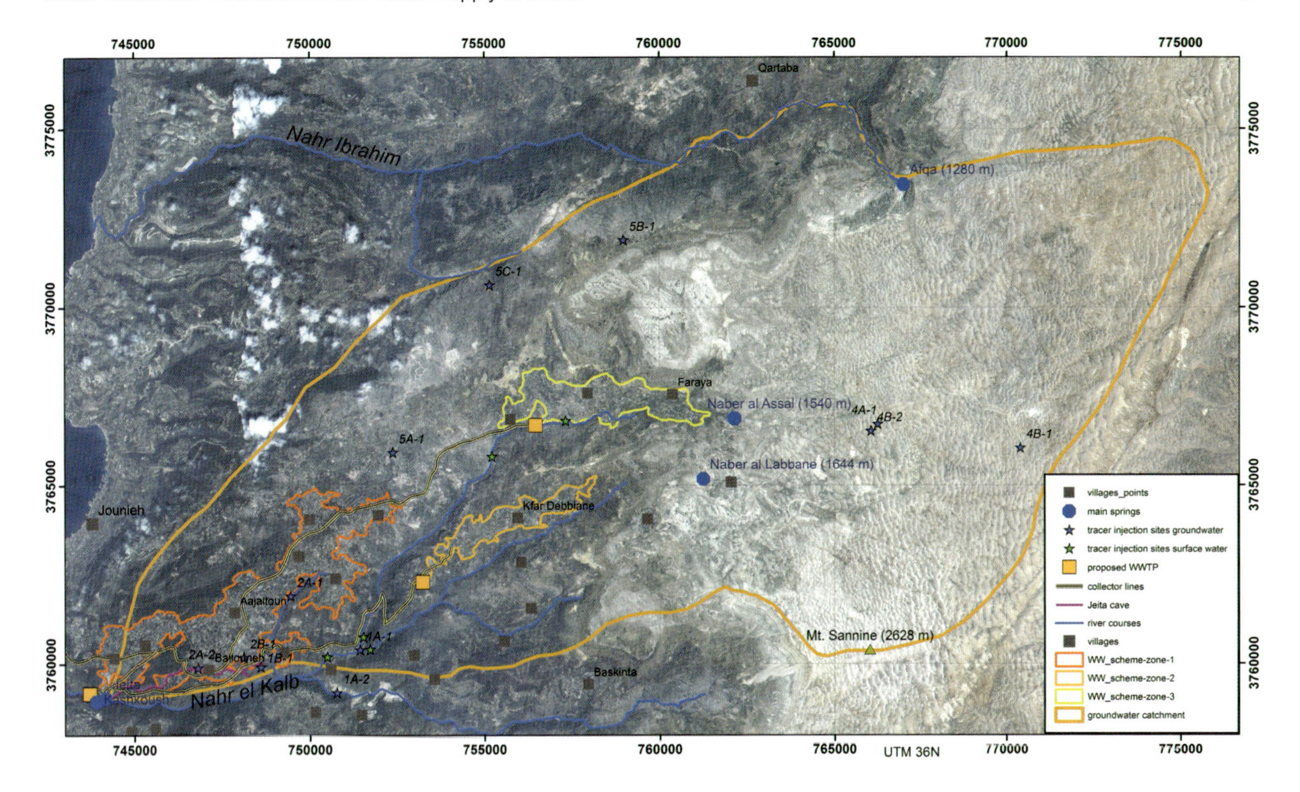

Figure 2: Proposed wastewater schemes for the Jeita catchment.

References

Margane, A., 2011: Site Selection for Wastewater Facilities in the Nahr el Kalb Catchment – General Recommendations from the Perspective of Groundwater Resources Protection. – German-Lebanese Technical Cooperation Project 'Protection of Jeita Spring', Technical Report No. 1, 155 p., Ballouneh.

Margane, A. & Abi Rizk, J., 2011: Guideline for Environmental Impact Assessments for Wastewater Facilities in Lebanon. – German-Lebanese Technical Cooperation Project 'Protection of Jeita Spring', Technical Report No. 3, 27 p., Ballouneh.

Margane, A. & Steinel, A., 2011: Proposed National Standard for Treated Domestic Wastewater Reuse for Irrigation. – German-Lebanese Technical Cooperation Project 'Protection of Jeita Spring', Special Report No. 4, 42 p., Ballouneh.

Steinel, A. & Margane, A., 2011: Best Management Practice Guideline for Wastewater Facilities in Karstic Areas of Lebanon – with special respect to the protection of ground- and surface waters. – German-Lebanese Technical Cooperation Project 'Protection of Jeita Spring', Technical Report No. 2, 147 p., Ballouneh.

Proceedings “Hydrogeology of Arid Environments”, p. 92–96
Stuttgart, March 2012

Extended Abstract

Integrated Remote and in situ Assessment of a Playa Lake Groundwater System in Northern Chile

K. H. Markovich[1], S. A. Pierce[2]

[1] Dept. of Geological Sciences, email: khmarkovich@gmail.com
[2] Center for International Energy and Environmental Policy, The University of Texas at Austin, 1 University Station, C9000, Austin, TX 78712-0254

Key words: Remote sensing, playa lake, water resources, Chile

Introduction

Playa lakes or internally drained, evaporative basins in the Atacama Desert of Chile, such as Salar de Ascotán and Salar de Carcote, are important water resource features for both human and environmental needs. Spring-fed perennial surface water supports diverse flora and fauna in the salar basins, including flamingo, vicuña, and the endemic fish species *Orestias ascotanensis*. Mining projects in the region also depend on the playa lakes as the most economically viable source of fresh water, and global demand for Chile's mining exports such as copper will inevitably increase with concomitant stresses on water resources. This comparative study uses satellite imagery to detect changes in surface water extent in the two salars and evaluate the results for possible correlation with regional recharge and/or anthropogenic factors.

Study Site

Salar de Ascotán and Salar de Carcote are located 200 km northeast of Antofogasta in Region II, Chile, at about 3700 m.a.s.l. (Fig. 1). Regional geology is controlled by active volcanism and N-S trending compressional faults from the oceanic-continental subduction margin (Hartley et al. 2000). The regional faulting plays a major role in governing the size and morphology of the intravolcanic salar basins, and in particular the springs at Ascotán are proposed to occur along a fault plane. Local stratigraphy of the salars includes Ca-SO_4 salt crusts which overlie interbedded Cenozoic tuffs and sedimentary strata, with a deep carbonate sequence that could play host to the regional groundwater system (Stoertz & Ericksen 1962).

Regional average annual precipitation ranges from 50 to 300 millimeters, with potential evapotranspiration greatly exceeding that (Risacher et al. 2003). The majority of precipitation occurs as snowfall on the volcanic peaks during the summer months of November to February, commonly referred to as the “Altiplanic Winter.” Salar de Ascotán and Salar de Carcote are part of a regional groundwater system that recharges in the adjacent uplands to the east and terminates in the regional topographic low at Salar de Uyuni, Bolivia. This regional groundwater system is discharged locally as spring-fed perennial surface water that flows across

Table 1. Selected features of study salars (modified from Risacher et al. 2003).

	Latitude (South)	Longitude (West)	Altitude Salar (m)	Temperature mean (°C)	Precipitation (mm/year)	Evaporation (mm/year)	Area Basin (km^2)	Area Salar (km^2)
Salar de Ascotán	21°33'01	68°16'58	3716	5.8	125	1630	1757	243
Salar de Carcote	21°22'36	68°20'55	3690	5.8	125	1630	561	108

www.borntraeger-cramer.de
2012/0092/$ 1.25

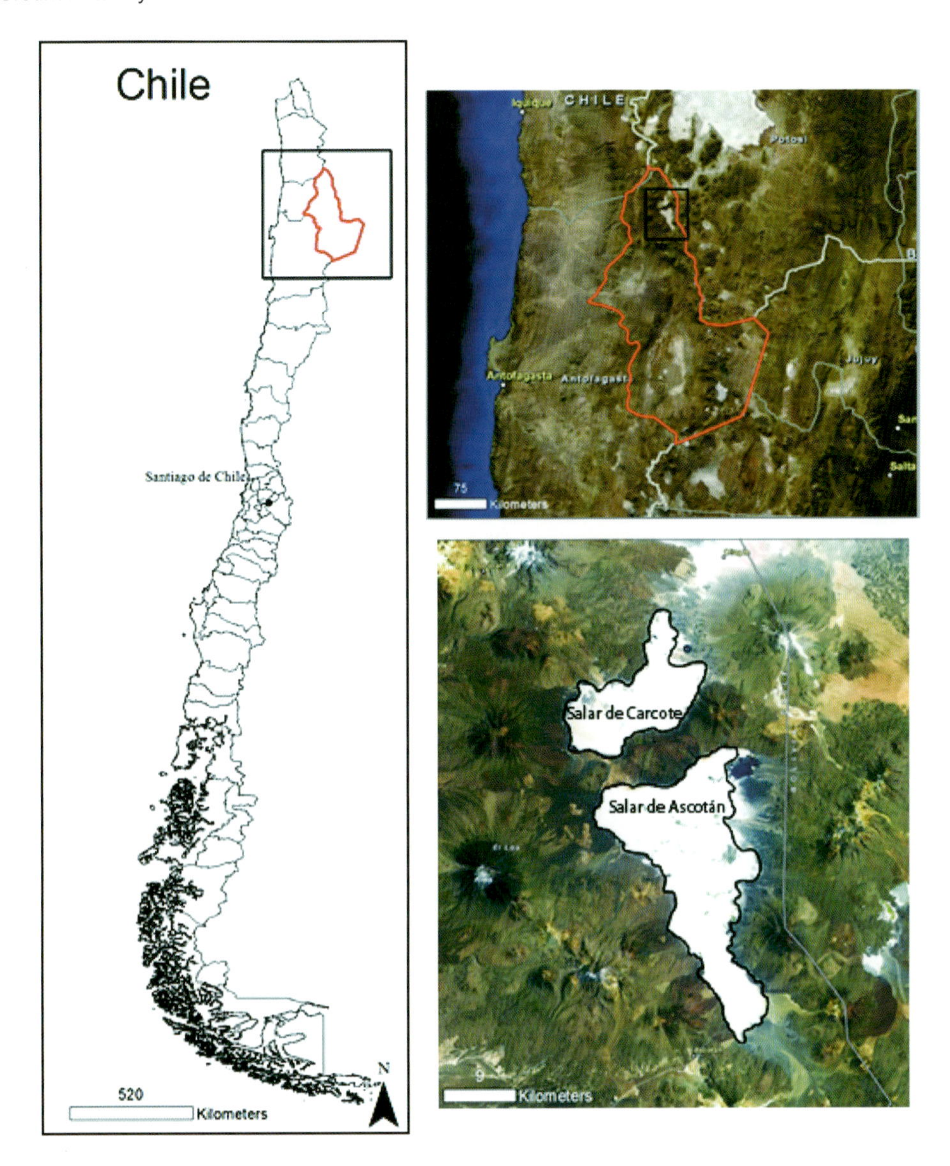

Figure 1: Map of El Loa Province in Chile, with regional and local inset satellite imagery of Salar de Ascotán and Salar de Carcote.

the salar surface and either evaporates, or reinfiltrates, in lagoon-like environments.

Production wells were established in the Ascotán basin in the mid-1990's, leading to concern about the preservation of spring-fed surface flows as a function of drawdown. While hydrologic and ecologic monitoring efforts have been coordinated since mining activity began, antecedent data collection precede extraction by approximately six months. Remote sensing can provide a means for large scale monitoring of the salars, as well as providing additional historical data for understanding antecedent conditions and support environmental management of the systems.

Water Budget

Groundwater abstraction in Ascotán currently supports operations for one of the largest copper mines in the world. The balance between regional recharge and drawdown of springflows due to pumping is critical to understanding the potential for ecological impacts to groundwater dependent systems within the basin. Rodriguez-Rodriguez et al. (2006) describes the water budget for a playa lake as,

$$\Delta V = (P + I_{GW} + I_{SW}) - (ET + O_{GW} + O_{SW}) \quad (1)$$

where ΔV is change in storage, P is precipitation, ET is evapotranspiration, I_{GW} is the groundwater input, O_{GW} is the groundwater output, I_{SW} is the surface water input, and O_{SW} is the surface water output.

The hydrologic and climate regimes affecting these salars allows for several assumptions to simplify the budget. First, the basins are internally drained and the only source of water is from the springs, thus the surface water input and output term can be eliminated. Further, direct precipitation to the surface is expected to be negligible and snow accumulation on the surrounding peaks sublimates before reaching the surface, and so the precipitation term can be eliminated to result in Equation 2. Finally, if evaporation and infiltration rates are not changing significantly over time, a change in volume of water is indicative of changes to the groundwater storage over time.

$$\Delta V = (I_{GW}) - (E + O_{GW}) \qquad (2)$$

Remote sensing of water extent is useful then because it can be used to quantify a change in area, and since the salars possess relatively low topographic relief, this can be used as a rough analog to the groundwater system.

Methods and Data

A multi-temporal remote sensing analysis was performed using Landsat 4-5 Thematic Mapper and 7 Enhanced Thematic Mapper Plus orthoimagery. Thirteen scenes (Row: 75/Path: 233) over the time period of 1985-2011 were downloaded from the USGS Landsat Archive, which provides cloud-free, georeferenced, and orthorectified images in a GeoTIFF format. 5 of the scenes were chosen over the course of 2009 to establish seasonal response to wet/dry periods, as well as to determine the best window of which to assess long-term changes. The scenes were pre-processed and corrected using ERDAS Imagine 2011, and then multiple classification methods were applied to quantify surface water extent. Of these include optical analysis of single band and "false" composite images, Normalized Difference Water Index (Xu 2006), unsupervised classification (Casteneda et al. 2005), and supervised classification (Lillesand & Kieffer 1987). Resultant classified images were imported into a geographic information system (ESRI ArcGIS 10) to quantify the total water extent, as well as zonal estimates. Precipitation data was downloaded from the NASA Tropical Rainfall Monitoring Mission (TRMM) Global Precipitation dataset (GPCP), which provides monthly average rain rates for the region from 1979-present. The horizontal resolution for this data is 2.5° and the region selected is 20-22S and 60-68W.

While remote sensing is an economically viable tool for monitoring surface water response to climate and groundwater abstraction over time, in situ data can be integrated to establish hydrologic connections within and between the salar basins. This has implications for determining if groundwater abstraction in the southern portion of Ascotán captures spring-fed discharge in the rest of the system. In situ data such as surface water depth (m), water temperature (°C), conductivity (μS/cm), and pH were taken during a sampling trip in late-September of 2011.

Results

Early remote sensing results indicate that surface water extent on the Salar de Ascotán surface diminished following the onset of pumping, but the rate decreases until reversing slightly in the mid-2000's as a result of artificial recharge and environmental management efforts (Fig. 2). This suggests a pre-pumping, non-optimal pumping, and mitigation pumping stage, where the mitigation pumping stage

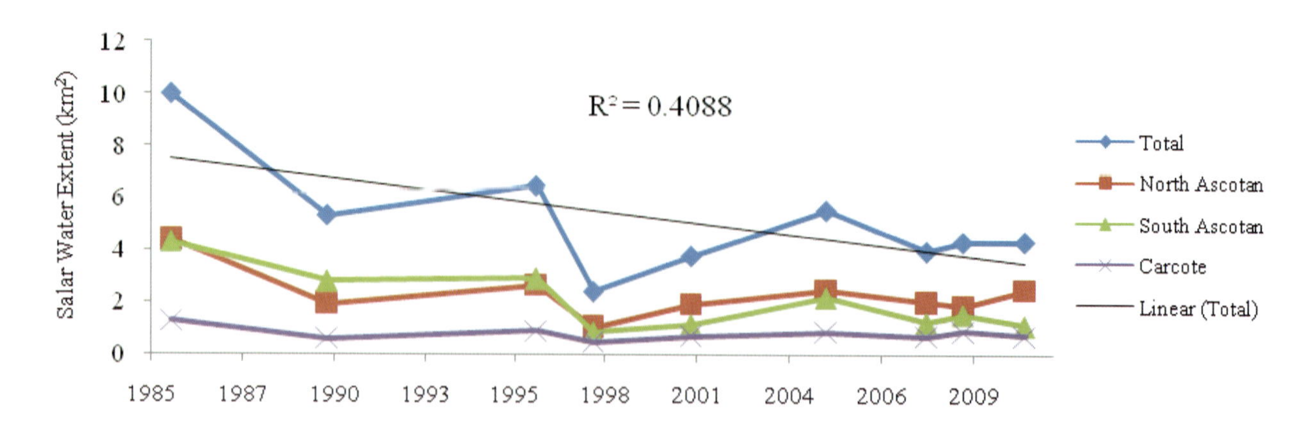

Figure 2: Multi-temporal remote sensing results for the dynamic playa lake groundwater system.

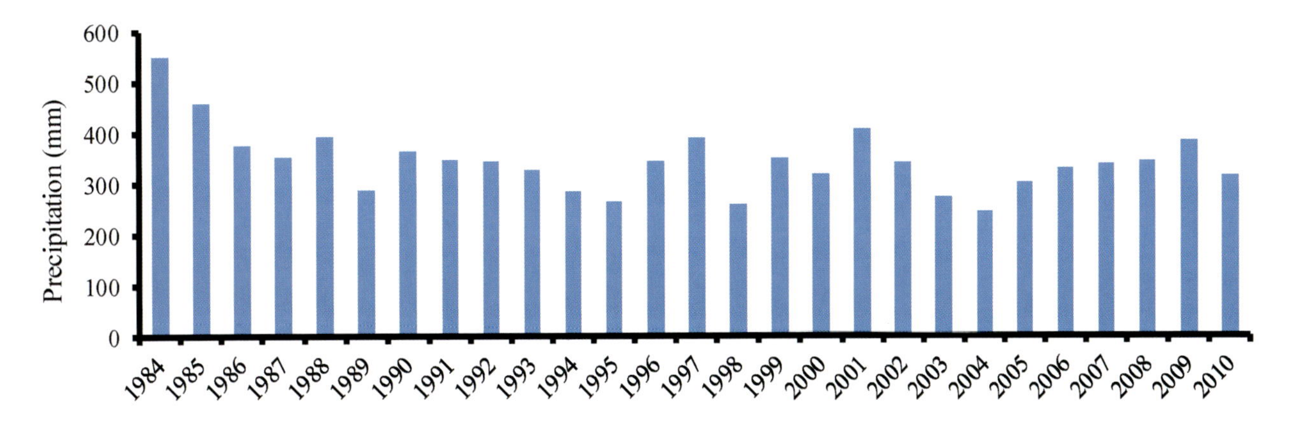

Figure 3: Cumulative annual precipitation derived from NASA TRMM monthly values.

can potentially reach equilibrium between economic and environmental dependency on the groundwater system in the playa lake basins. Early results also show that the water extent on the Salar de Carcote surface remains relatively unchanged over the same time period, suggesting that pumping in the Ascotán basin has not significantly affected the regional flowpaths. A comparison with precipitation and in situ data over the same time period suggests that the playa lake groundwater system is sensitive to climatic shifts, which is shown by the correlation between decreasing cumulative annual precipitation and decreasing surface water extent on the salars during the years preceding abstraction (Figs. 2; 3).

Discussion

In a natural state, this playa lake groundwater system is in dynamic equilibrium, where the long-term average discharge equals the long-term average recharge to the system. This is confirmed by the presence of the endemic fish species *Orestias ascotanensis*, which evolved in the perennial supply of fresh (<10,000 µS/cm) surface water since the last glacial maximum (Keller & Soto 1998). Zhou (2009) suggests that groundwater abstraction can be sustained over time, and may even result in the system reaching a new equilibrium state if pumping does not exceed the natural and induced recharge. However, this induced recharge comes at the cost of captured natural discharge, such as reduced amounts of spring-fed surface water, which threatens the delicate ecological system of this hyper-arid playa lake environment. Integrated remote and in situ results show the initial capture from groundwater abstraction, however precipitation records suggest that this system is largely responsive to climate, and in situ data suggests that springs in the southern part of Ascotán are hydrologically isolated from of the northern portion. Systems dynamics modelling can be utilized with future abstraction and climate scenarios to provide an optimized pumping regime. Future research is expected to focus on understanding the regional recharge component, establish a systems model, and evaluate scientific assumptions using scenario analysis.

References

Briere, P., 2000: Playa, playa lake, sabkha: Proposed definitions for old terms. – Journal of Arid Environments **45**: 1–7.

Casteñeda, C., Herrero, J. & Casterad, M. A., 2005: Landsat monitoring of playa-lakes in the Spanish Monegros desert. – Journal of Arid Environments **63**: 497–516.

Hartley, A., May, G., Chong, G., Turner, P., Kape, S. J. & Jolley, E. J., 2000: Development of a continental forearc: A Cenozoic example from the Central Andes, northern Chile. – Geology **28**: 331–334.

Keller, B. & Soto, D., 1998: Hydrogeologic influences on the preservation of *Orestias ascotanensis* (Teleostei: Cyprinodontidae) in Salar de Ascotán, northern Chile. – Revista Chilena de Historia Natural. **71**: 147–156.

Lillesand, T. M. & Kieffer, R., 1987: Remote Sensing and Image Interpretation. – 2nd ed. (New York: John Wiley & Sons).

Risacher, F., Alonso, H. & Salazar, C., 2003: The origin of brines and salts in Chilean salars: a hydrochemical review. – Earth-Science Reviews **63**: 249–293.

Rodriguez-Rodriguez, M., Benavente, J., Cruz-San Julian, J. J. & Moral Martos, F., 2006: Estimation of ground-water exchange with semi-arid playa lakes (Antequera region, southern Spain. – Journal of Arid Environments **66**: 272–289.

Stoertz, G. E. & Ericksen, G. E., 1974: Geology of salars in northern Chile. – U.S. Geological Survey Professional Paper 811, 65 pp.

Xu, H., 2006: Modification of normalised difference water index (NDWI) to enhance open water features in remotely sensed imagery. – International Journal of Remote Sensing **27**(14): 3025–3033.

Zhou, Y., 2009: A critical review of the groundwater budget myth, safe yield, and sustainability. – Journal of Hydrology **370**: 207–213.

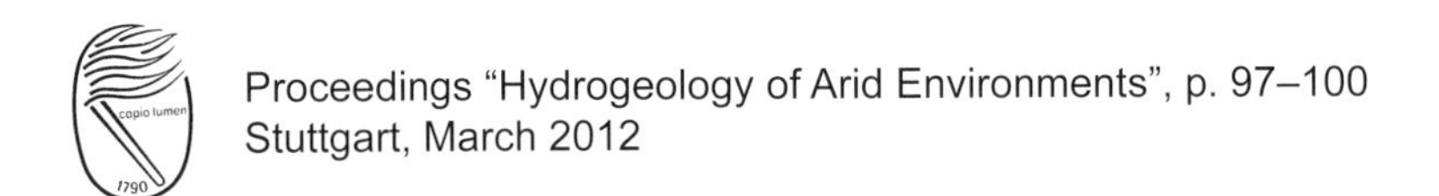
Proceedings “Hydrogeology of Arid Environments”, p. 97–100
Stuttgart, March 2012

Extended Abstract

The Water Resources of the Eastern Mediterranean: Present and Future Conditions

Lucas Menzel[1], Tobias Törnros[1]

[1] Department of Geography, Heidelberg University, Im Neuenheimer Feld 348, 69120 Heidelberg, Germany, email: lucas.menzel@geog.uni-heidelberg.de, tobias.toernros@geog.uni-heidelberg.de

Key words: Eastern Mediterranean, water resources, TRAIN model, scenarios

Introduction

The Jordan River Region is ranking among the most water poor regions of the world. Over a distance of only 300–400 km, the climatic conditions range from sub-humid in the upper north of the basin (mean annual precipitation above 800 mm) to hyper-arid at the Gulf of Aqaba in the south (mean annual precipitation below 50 mm, potential evaporation well above 2000 mm). The scarce water resources are competitively shared among several nations and different water use sectors, with irrigation agriculture as one of the major water users. An increasing population, rising water demands, reduced precipitation totals with increasing precipitation intensities make the region susceptible to frequent droughts (Törnros 2010). In the framework of the GLOWA-Jordan project (www.glowa-jordan-river.de), an interdisciplinary and multinational group of researchers investigate the impact of climate and land-use change on the water resources of the Jordan Region. Within the sub-project “Climate and Hydrology” simulation tools have been developed and applied in order to convert observations and information from the scenarios into hydrological data which serves as input for possible water management options, regional development and co-operation.

Material and methods

A principal tool applied in the project is the hydrological model TRAIN (Menzel 1997, Menzel et al. 2009). TRAIN is a physically-based, spatially distributed model which includes information from comprehensive field studies of the water and energy balance of different types of land-cover and land-use. It is designed to simulate the spatial pattern of the individual water budget components at different spatial and temporal scales. Typical applications are at the point and the regional scales, with a temporal resolution of one hour or one day. Special emphasis is on the processes at the soil-vegetation-atmosphere interface. TRAIN can be applied in two modes, the “bottom up” and the “top down” approaches. In the “bottom up” mode, processes at the point scale, such as evapotranspiration or soil moisture dynamics are simulated, driven by data measured at experimental sites. This aims at further developing and validating the model. In the GLOWA-Jordan project, TRAIN has been applied with data from the Yatir Forest, located at the northern edge of the Negev in the semi-arid part of Israel (mean annual precipitation 285 mm). There, the Weizmann Institute of Science (www.weizmann.ac.il/ESER/People/Yakir/YATIR/) operates a flux tower with a detailed determination of the radiation balance components and of evapotranspiration based on the Eddy Correlation method. Further, soil moisture data are regularly measured at different soil depths and estimations of rainfall interception are available. Hausinger (2009) found that TRAIN was able to generally predict the overall dynamics of measured evapotranspiration at Yatir. For the validation period a Nash-Sutcliffe Index of 0.6 and a relative bias of 11.1% were obtained. The simulated canopy interception of approx 6% of total precipitation is consistent with findings from experimental studies at Yatir (Shachnovich et al. 2008).

In the “top down” approach, TRAIN simulates the water balance of a large spatial domain. Occasional model improvements are based on findings from the “bottom up” applications at experimental sites.

www.borntraeger-cramer.de
2012/0097/$ 1.00

This approach delivers policy-relevant information for whole catchments or administrative units, including the impact of environmental change scenarios on the water conditions as a base for planning and mitigation purposes. In the GLOWA-Jordan project, a square has been selected which covers an area of about 90,000 km^2, ranging in north-south direction from the Mount Hermon to the Gulf of Aqaba, and from the Mediterranean in the western part to the Jordanian Highlands in the East. This area is subdivided into 1x1 km grid cells for which TRAIN has been applied in daily resolution over the period 1961–1990. Model input is a Digital Elevation Model, raster maps of land-use/land-cover and soils as well as daily grids of interpolated meteorological data (for details see Menzel et al. 2009). For each grid cell the model simulates daily evapotranspiration and its individual components, possible irrigation water requirements (in case of irrigated agriculture) as well as water availability, which is an aggregated measure of deep percolation and surface runoff. The output maps can be aggregated to any period.

Results

The model based investigations give for the first time a coherent overview of the water conditions of the Eastern Mediterranean region, with the highest possible spatial detail of 1x1 km. The simulation of water availability for the reference period 1961–1990 demonstrates the scarcity of water resources in major parts of the region, with extremely low values of water availability in the semi-arid and arid parts (Figure 1).

The spatial course of water availability does not clearly follow the precipitation distribution but also reflects impacts e.g. from soils with different water

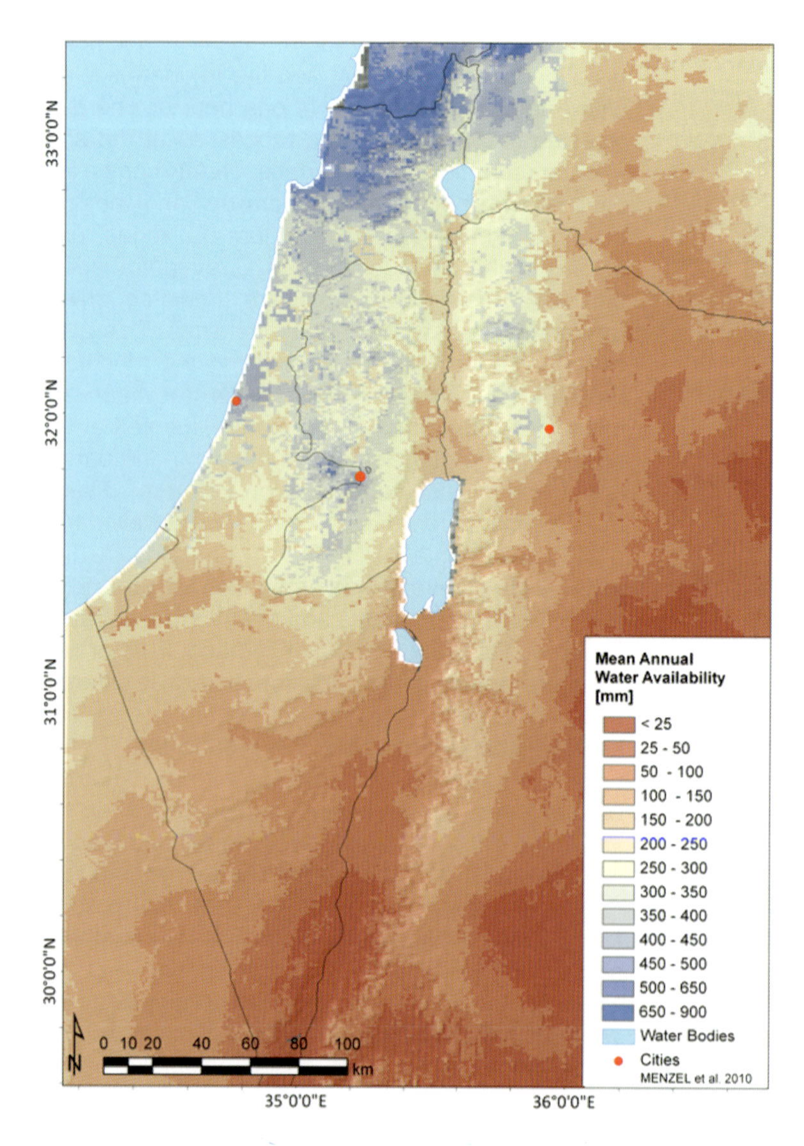

Figure 1: Mean annual water availability (percolation and surface runoff) for the reference period 1961–1990. The map is based on daily simulation steps over the individual 1x1 km grid cells.

holding capacities and the spatial variation of evapotranspiration which is both driven by meteorological and physiographic conditions.

In a next step, several combinations of Regional Climate Models (RCM) and Global Climate Models (GCM) have been derived based on the IPCC emission scenario A1B (Smiatek et al. 2011). The climate data from the scenarios ECHAM5-RegCM3, ECHAM5-MM5 and HadCM3-MM5 were delivered to the hydrological model TRAIN. Then, the future (2031–2060) relative changes [%] (with regard to the reference period) in evapotranspiration, water availability and irrigation water demand were determined. For the comparison, the project region was subdivided into three precipitation zones, representing sub-humid (>450 mm), semi-arid (250–450 mm) and arid (<250 mm) conditions. Figure 2 shows that precipitation is projected to drop considerably over the three sub-regions. As a consequence, water availability shows an over-proportional decrease, with percentage changes in a range between ca. –10% and –30%. Hence, in order to sustain agriculture at its current extent and intensity, an additional water amount of 15–50% would be needed.

Finally, TRAIN has been applied with land-use / land-cover scenarios (Menzel et al. 2009) in order to investigate possible, regional developments and land management options for a mitigation of the expected adverse impact of climate change on water availability. For our study, two intermediate scenarios were selected, with projected land-use developments ranging between optimistic and pessimistic futures (with regard to social and economic conditions in the region). Relevant drivers for land-use / land-cover change were fed into the land-use model LandSHIFT (Schaldach et al. 2011) which produced land-use / land-cover maps for the different scenarios. These maps were then used as input to TRAIN in order to generate respective hydrological scenarios for the region. Given that climate condi-

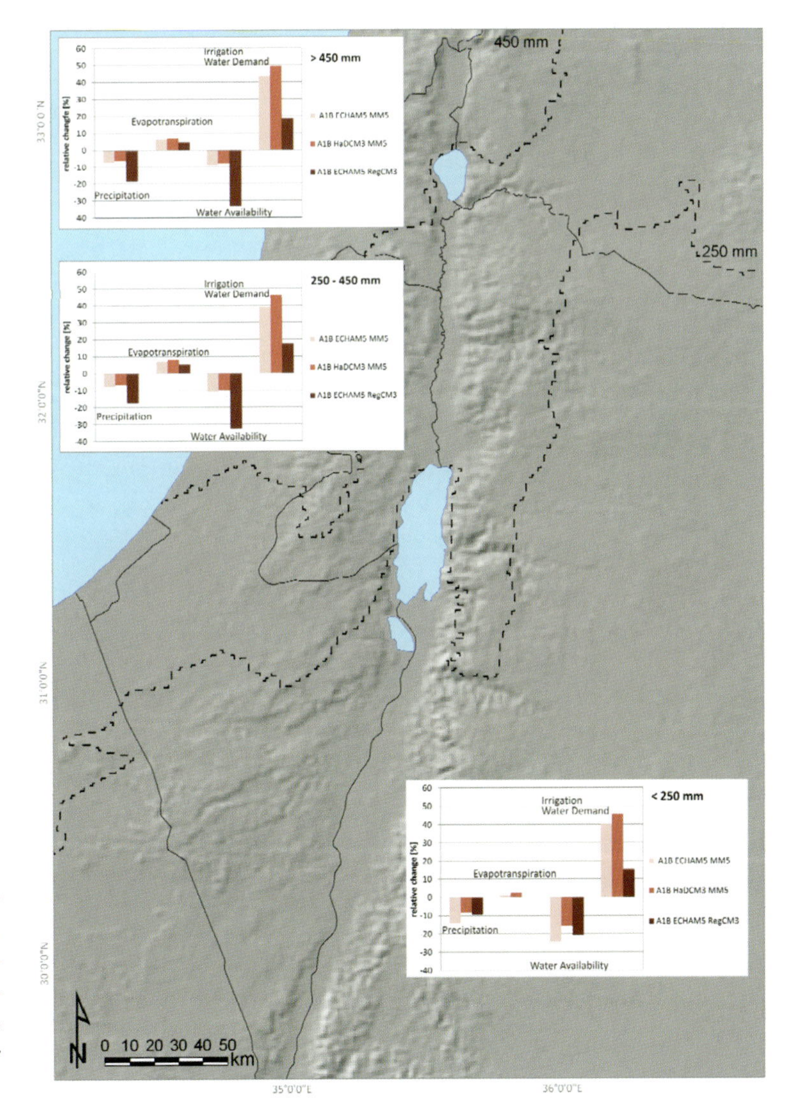

Figure 2: Relative changes [%] in precipitation, irrigation water demand, evapotranspiration and water availability between current and future conditions, given for three GCM-RCM combinations. Results are presented for three different precipitation zones.

tions remain unchanged in our model experiment, the simulations show both increases and decreases in water availability, depending on the future pattern of natural and agricultural vegetation and the related dominance of hydrological processes. The simulation study shows that land-use and land-cover changes have an obvious impact on the amount and distribution of water. However, the predicted climate change and its control over water resources is projected to clearly overbalance any efforts to conserve water through land management measures.

References

Hausinger, I., 2009: Application and validation of TRAIN at the Yatir forest. – Master thesis at the Ecological Modelling Department, University of Bayreuth (unpublished), 93 p.

Menzel, L., 1997: Modellierung der Evapotranspiration im System Boden-Pflanze-Atmosphäre. – Zürcher Geographische Schriften **67**, ETH Zurich, 128 p.

Menzel, L., Koch, J., Onigkeit, J. & Schaldach, R., 2009: Modelling the effects of land-use and land-cover change on water availability in the Jordan River region. Adv. Geosci. **21**: 73–80, doi:10.5194/adgeo-21-73-2009.

Schaldach, R., Alcamo, J., Koch, J., Kölking, C., Lapola, D., Schüngel, J. & Priess, J., 2011: An integrated approach to modelling land-use change on continental and global scales. – Environmental Modelling and Software **26**: 1041–1051.

Shachnovich Y., Berliner, P. & Bar, P., 2008: Rainfall interception and spatial distribution of throughfall in a pine forest planted in an arid zone. – J. Hydrol. **349**: 168–177, doi:10.1016/j.jhydrol.2007.10.051.

Smiatek, G., Kunstmann, H. & Heckl, A., 2011: High-resolution climate change simulations fort he Jordan River area. – J. Geophys. Res. **116**, D16111, doi:10.1029/2010JD015313.

Törnros, T., 2010: Precipitation trends and suitable drought index in the arid/semi-arid southeastern Mediterranean region. – In: Servat, E., Demuth, S., Dezetter, A. & Daniell, T. (eds.): Global Change: Facing Risks and Threats to Water Resources (Proc. of the Sixth World FRIEND Conference, Fez, Morocco, 25–29 October 2010). IAHS Publ. **340**: 157–163.

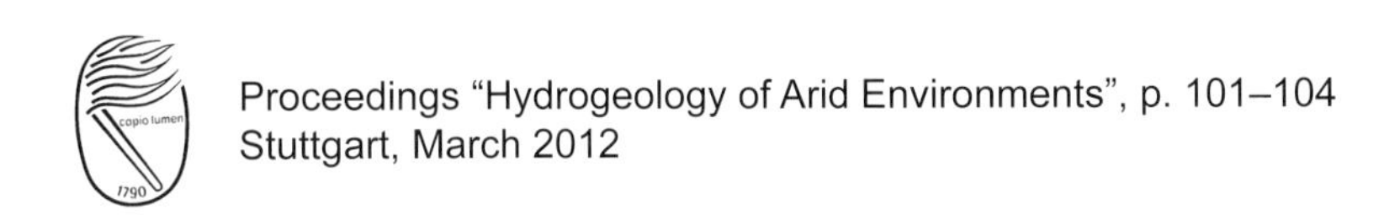
Proceedings "Hydrogeology of Arid Environments", p. 101–104
Stuttgart, March 2012

Extended Abstract

Assessment of Natural Radioactivity Occurring in Saudi Arabian Sandstone Aquifers

Nils Michelsen[1], Michael Schubert[2], Christoph Schüth[1], Randolf Rausch[3], Mohammed Al Saud[4]

[1] TU Darmstadt, Institute for Applied Geosciences, Schnittspahnstraße 9, 64287 Darmstadt, Germany, email: michelsen@geo.tu-darmstadt.de, schueth@geo.tu-darmstadt.de
[2] Helmholtz-Centre for Environmental Research – UFZ, Permoserstraße 15, 04318 Leipzig, Germany, email: michael.schubert@ufz.de
[3] GIZ IS, P.O. Box 2730, Riyadh 11461, Kingdom of Saudi Arabia, email: randolf.rausch@gizdco.com
[4] Ministry of Water & Electricity – MoWE, Saud Mall Center, Riyadh 11233, Kingdom of Saudi Arabia, email: malsaud@mowe.gov.sa

Key words: groundwater radioactivity, sandstone aquifers, radium, Saudi Arabia

Introduction

In many semi-arid and arid areas, such as the Kingdom of Saudi Arabia, renewable water resources are scarce or non-existent. As seawater desalination and the subsequent water transportation via pipelines are energy-intensive and thus costly, areas that can be supplied by this source are limited and mainly restricted to coastal regions. Hence, water supply in semi-arid and arid regions mostly depends on non-renewable groundwater resources that have been recharged during pluvial times thousands or even ten thousands of years ago. As sustainable management of such non-renewable resources is not possible, the fossil groundwater should at least be "mined" in a responsible way. An essential prerequisite for this is the sound understanding of the concerned aquifer systems – in terms of water quantities and qualities.

Motivation

In Saudi Arabia, only the so-called Arabian Shield in the west of the country receives enough precipitation to allow a water supply that is at least partly based on renewable water resources. The remaining part of the Kingdom relies on fossil groundwater originating from several large-scale aquifers located on the Arabian Platform east of the Shield (see Al Saud & Rausch, this publication). Isotope investigations have demonstrated that most of the waters in these deep aquifers are fossil. Due to the currently low recharge rates and the thick unsaturated zone, these waters are usually not prone to anthropogenic impact. Yet, groundwater quality can be deteriorated by natural contaminants. A good example is the elevated groundwater salinity frequently encountered in Saudi Arabia. However, saline waters can be easily identified and constituents of concern such as nitrate or fluoride are usually part of standard water analyses. A more critical issue is natural groundwater radioactivity, which is usually not addressed in routine water analyses. The problem is particularly relevant to sandstone aquifers derived from granites, the latter being known to contain substantial amounts of incompatible elements such as uranium and thorium.

For the principal sandstone aquifers on the Arabian Platform, two main provenances can be assumed. They either represent the direct weathering products of the granite-bearing Arabian-Nubian Shield or they developed through erosion of such sandstones and re-deposition of the weathering products, i.e., they are indirectly derived from the Shield. The presence of radiometric anomalies has previously been demonstrated in airborne gamma-ray surveys and associated follow-up ground studies (e.g. Pitkin & Huffman 1989) and geophysical borehole logging campaigns (e.g. GTZ/DCo 2010a).

www.borntraeger-cramer.de
2012/0101/$ 1.00

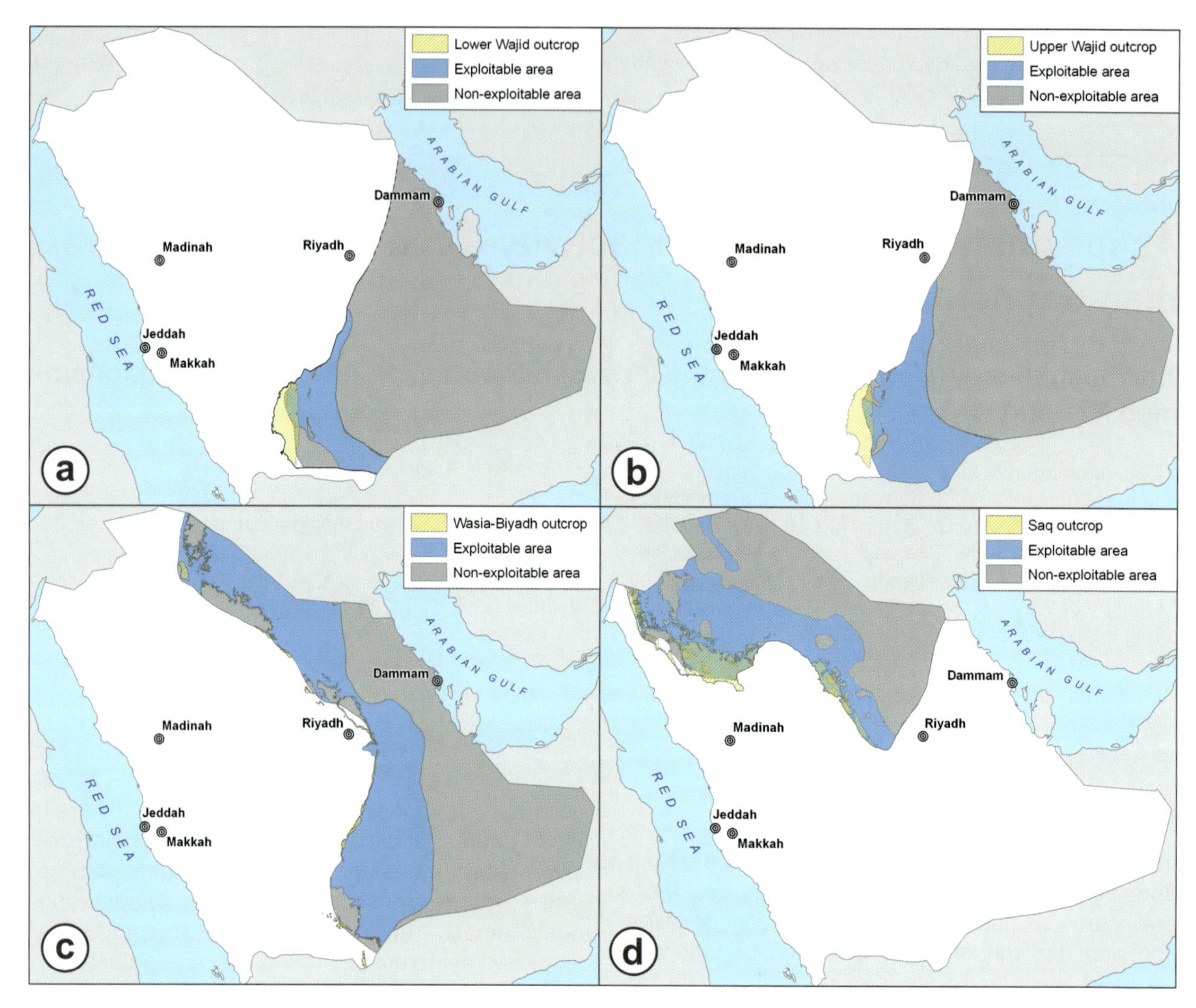

Figure 1: Maps showing the outcrop and exploitable as well as non-exploitable areas of major Saudi Arabian sandstone aquifers (GTZ/DCo 2010b): a) Lower Wajid, b) Upper Wajid, c) Wasia-Biyadh, and d) Saq; assessment of exploitability based on water salinity (<2,000 mg/l), drilling depth (<2,000 m), and pumping height (<300 m).

In view of that fact and considering the long groundwater residence times possibly favouring the mobilisation of nuclides from the aquifer matrix, it seems plausible that these sandstones contain groundwaters with elevated radionuclide concentrations. In Jordan, such waters have recently been described by Vengosh et al. (2009).

Methods

In order to find out whether Saudi Arabian aquifers are subject to the mobilisation of natural radionuclides, groundwater samples were obtained from selected sandstone aquifers, with emphasis being placed on the aquifers Lower Wajid (Cambrian-Ordovician), Upper Wajid (Silurian-Permian), Biyadh (Early Cretaceous), and Wasia (Middle Cretaceous). Moreover, existing data on the Saq aquifer (Cambrian-Ordovician) by BRGM/ATC (2008) was considered. The outcrops of the mentioned aquifers and their approximate extents on Saudi Arabian territory are shown in Figure 1.

If the quality of a groundwater is to be studied with respect to its radiochemistry, an undifferentiated analysis for all radionuclides potentially present in the aquifer and/or groundwater is not appropriate. Instead, various nuclides can be neglected, either due to their poor solubility in water or because of their scarce natural occurrence. In addition, isotopes with very short half-lives (<5 days) are only of minor importance because they are represented by their parent nuclides (Fig. 2). Hence, the collected

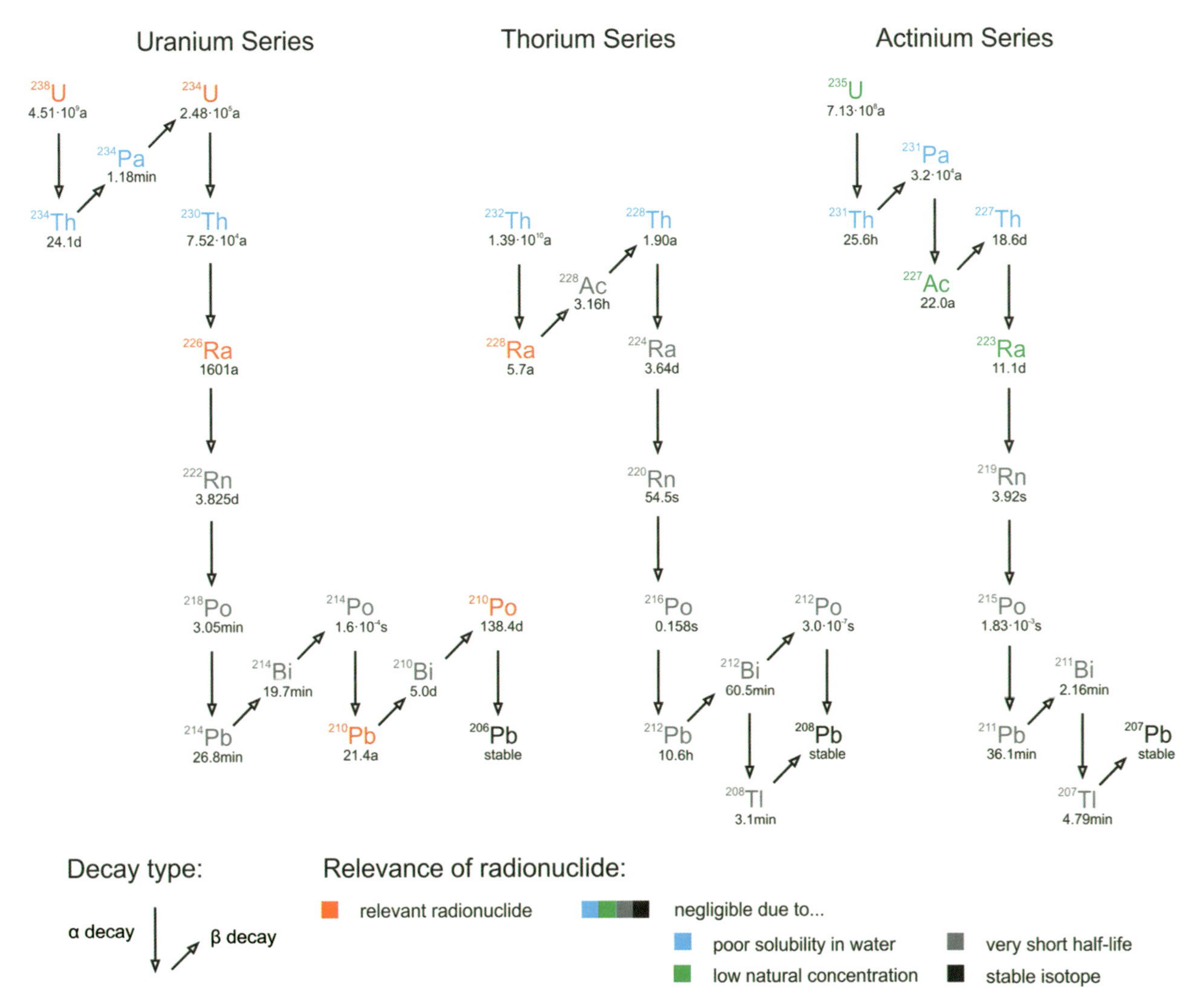

Figure 2: Natural decay chains (incl. half-lives, Kraemer & Genereux 1998) with evaluation of nuclides with respect to their radio-ecological relevance.

water samples were analysed for uranium-234, uranium-238, radium-226, and radium-228. Partly, also polonium-210, lead-210, and/or radon-222 activities were determined. Moreover, the major ion chemistry of the waters was investigated.

Results and Conclusions

Although uranium is present in the aquifer matrix, the obtained uranium nuclide activities are mostly low to moderate and range well below their corresponding WHO (2011) guideline values accounting for 1,000 mBq/l (uranium-234) and 10,000 mBq/l (uranium-238). With respect to the studied aquifers, the uranium isotopes are thus only of minor importance from a radio-ecological perspective.

With a few exceptions, also radium-226 shows activities that do not violate their nuclide-specific guideline level of 1,000 mBq/l. Radium-228, by contrast, frequently occurs in significant concentrations, partly exceeding the corresponding threshold value of 100 mBq/l. Distribution maps created for the studied aquifers do mostly not reveal clear spatial trends, i.e., radium-228 appears erratically. An inter-aquifer comparison however shows that elevated radioactivities are more common in the Saq, Lower Wajid, and Upper Wajid aquifers, in other words the older clastic strata cropping out in the eastern part of the country. This indicates that the depositional setting of the aquifer, particularly its distance to the Arabian Shield, plays a significant role. Considering these findings – erratic occurrence of elevated radium-228 values within a specific aquifer but overall clustering in the aquifers adjacent to the Shield – thorium-bearing heavy minerals in the sandstone aquifers, originating from the crystalline basement complex, seem to be a plausible source for the encountered radium-228. In this context, it is noteworthy that preliminary datasets on limestone aquifers (e.g. Late Cretaceous Aruma aquifer) do not exhibit elevated radium-228 activity concentrations, thus corroborating the heavy mineral hypothesis. This, in turn, might provide an additional tool to identify characteristic fingerprints for aquifers on the Arabian Peninsula.

References

BRGM/ATC – Bureau de Recherches Géologiques et Minières/Abunayyan Trading Corporation, 2008: Investigations for Updating the Groundwater Mathematical Model(s) for the Saq and Overlying Aquifers – Groundwater Quality, Volume 8. – Ministry of Water and Electricity, Riyadh, unpublished.

GTZ/DCo – Gesellschaft für Technische Zusammenarbeit/Dornier Consulting, 2010a: Detailed Water Resources Studies of Wajid and Overlying Aquifers. – Volume 19 – Geophysical Borehole Logging of Existing Wells; Ministry of Water and Electricity, Riyadh, unpublished.

GTZ/DCo – Gesellschaft für Technische Zusammenarbeit/Dornier Consulting, 2010b: Kingdom of Saudi Arabia – Assessment and Strategic Plan of the Water Sector. – Final Report; Ministry of Economy and Planning, Riyadh, unpublished.

Kraemer, T. F. & Genereux, D. P., 1998: Applications of Uranium- and Thorium-Series Radionuclides in Catchment Hydrology Studies. – In: Kendall, C. & McDonnell, J. J. (eds.): Isotope Tracers in Catchment Hydrology. Elsevier, Amsterdam, pp. 679–722.

Pitkin, J. A. & Huffman, A. C., 1989: Geophysical and Geological Investigations of Aerial Radiometric Anomalies in the Phanerozoic Cover Rocks of Southern Saudi Arabia. – Open-File Report USGS-OF-07-7; Ministry of Petroleum and Mineral Resources, Jiddah.

Vengosh, A., Hirschfeld, D., Vinson, D., Dwyer, G., Raanan, H., Rimawi, O., Al-Zoubi, A., Akkawi, E., Marie, A., Haquin, G., Zaarur, S. & Ganor, J., 2009: High Naturally Occurring Radioactivity in Fossil Groundwater from the Middle East. – Environ. Sci. Technol. **43**: 1769-1775.

WHO – World Health Organization (2011): Guidelines for Drinking-water Quality. – 4th Edition. WHO, Geneva, 541 pp.

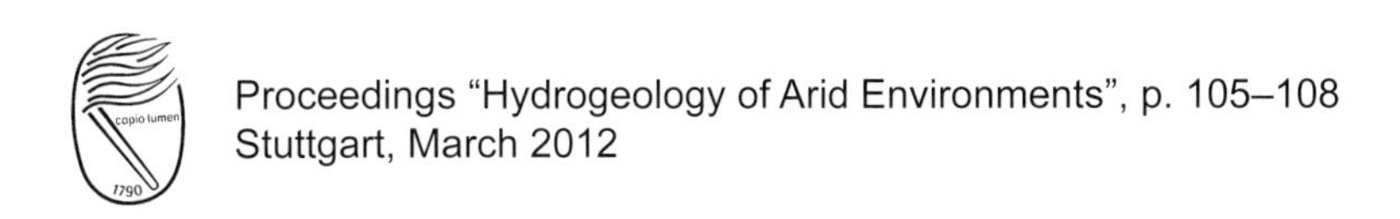

Proceedings “Hydrogeology of Arid Environments”, p. 105–108
Stuttgart, March 2012

Extended Abstract

Hydrological and Hydrochemical Process of the Sebkha Oum El Khialette, South East of Tunisia

N. Nesrine[1], B. Rachida

[1] Ecole Nationale d’Ingénieur de Tunis. BP 37, Le Belvedere.1002 Tunis. Tunisie, email: nasri.hydro@gmail.com

Key words: sabkha; brine; hydrological processes; geochemistry; modelling

Introduction

Saline systems have a significant importance hydrological, economic importance. Brines in sedimentary basins are related to evaporites (dissolution of evaporite minerals, incongruent alteration of hydrous evaporites minerals and interstitial fluids) (Carpenter 1978). Evaporites are sediment deposited from natural waters that have been concentrated as a result of evaporation (Ingebritsen et al. 1998) in semi arid to arid area .Their deposit is mined for salt which constitutes economically important minerals like gypsum, halite, natron, mirabilite and other.

The genesis of these deposits is controlled by many aspects such us hydrological condition, the sedimentological framework and chemistry and mineralogy of brine (Eugster 1980). For continental brine, inflow compositions are control by solute acquisition which includes rain water and weathering reaction. Studying the accumulation of brines in closed system will permit to understand their origin. The solutes in the dilute inflow will explain the final composition of the brine and the mineralogy of the sediment.

Description of the Study Area

Sebkha Oum El Khialate is closed basin situated in South East of Tunisia, 55 km from Tataouine and 65 km South Ouest from Ben Guerdene. (Fig. 1). It comprises an area of about 75 km^2 and an altitude between 32°45’00’’N and 10°53’00’. It is characterised by an arid type climate with an annual average precipitation 104 mm during 1956–2006.

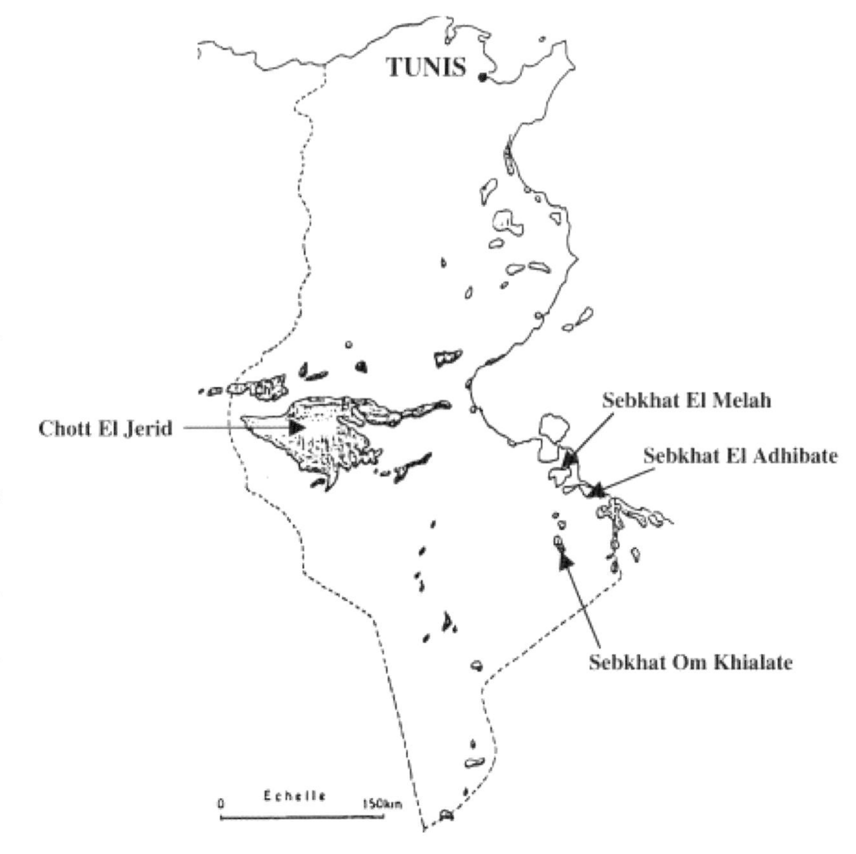

Figure 1: Map of the study zone.

The mean annual temperature is 20.7 °C while potential evaporation exceeds 1106 mm/year and rainfall is about 100 mm/year (Baccar et al. 2007). The catchment system of sebkha Om El Khialate receives 2 Mm^3 coming from the river and 3.5 Mm^3 from direct rain fall contribution.

The catchment area occupies the central place of the province of Jbel Rehach Sidi Toui and the substratum is characterized by clay of Mhira Ouled Oun. The mineralogical association of the different deposit in the study area according to the X-ray analyses is made up mainly by Gypsum, Halite, Kaolinite, Quartz and Dolomite (Table 1).

Table 1: Mineralogy of the evaporate outcrop in the catchment area of the sebkha (Baccar & Louhaichi 2007)

Formations	Mineralogy composition
Dune deposits	quartz, gypsum, dolomite, illite and kaolinite
Quaternary encrusting	gypsum, quartz, calcite
Clay of Mhira	illite and kaolinite, quartz, gypsum, halite, feldspars
Mestaoua evaporitic	gypsum, dolomite, quartz

Hydrochemistry

The results show the relative concentration of the different ions of analysed water samples. Solutions present high concentrations in sodium, sulphate and chloride while it is poor in carbonate, magnesium and calcium. pH ranges from 7.05 to 8.1 and it has salinities between 9500 and 178000 mg/l samples in central playa but at the margin the dry residue is between 2480 and 5580 mg/l.

The hydrochemical properties show that is Na-SO_4-Cl and Na-Mg-Cl-SO_4 rich water for samples situated in central playa but for the other in the borders Na-Mg-Ca-Cl-SO_4 and Ca-Na-Mg-SO_4-Cl water type. The chemical analysis results from samples are given in Fig. 2.

The samples show almost similar hydrochemical characteristics (Fig. 2) for basin center but the inflow groundwater show elevated proportion of Na, sulphate and chloride. The sample 7751/5 show only high proportion of Ca and SO_4^{2-}, which is probably derived from Triassic evaporites.

Modelling

Modelling of saturation and speciation of solutions was completed by the PHREEQC geochemical modelling codes using data base PITZER (Parkhurst & Appelo 1999).

During the evaporation process of water in order we have followed the mineral precipitation sequence in sabkha of Oum El Khialate. Saturation index indicated elevation in minerals: Calcite and Aragonite. Solutions are rich principally in Dolomite, Magnesite, Mirabilite and sometimes in Halite.

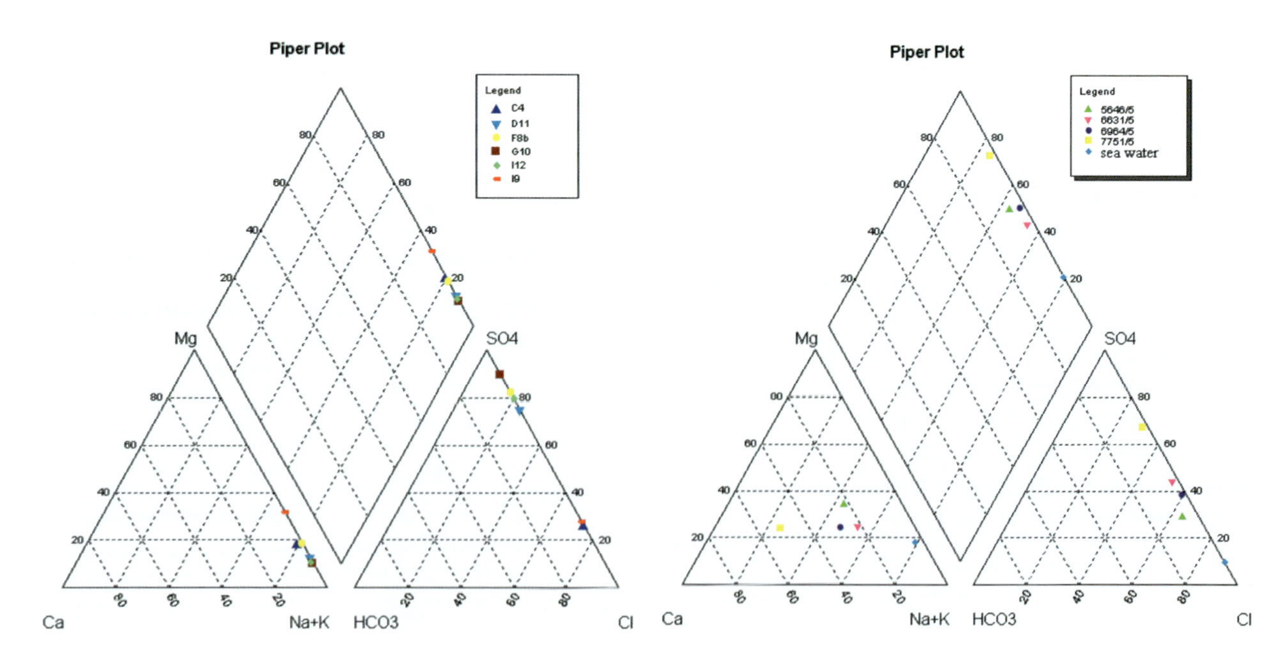

Figure 2: Piper plot of groundwater samples in central and at the margin of sebkha.

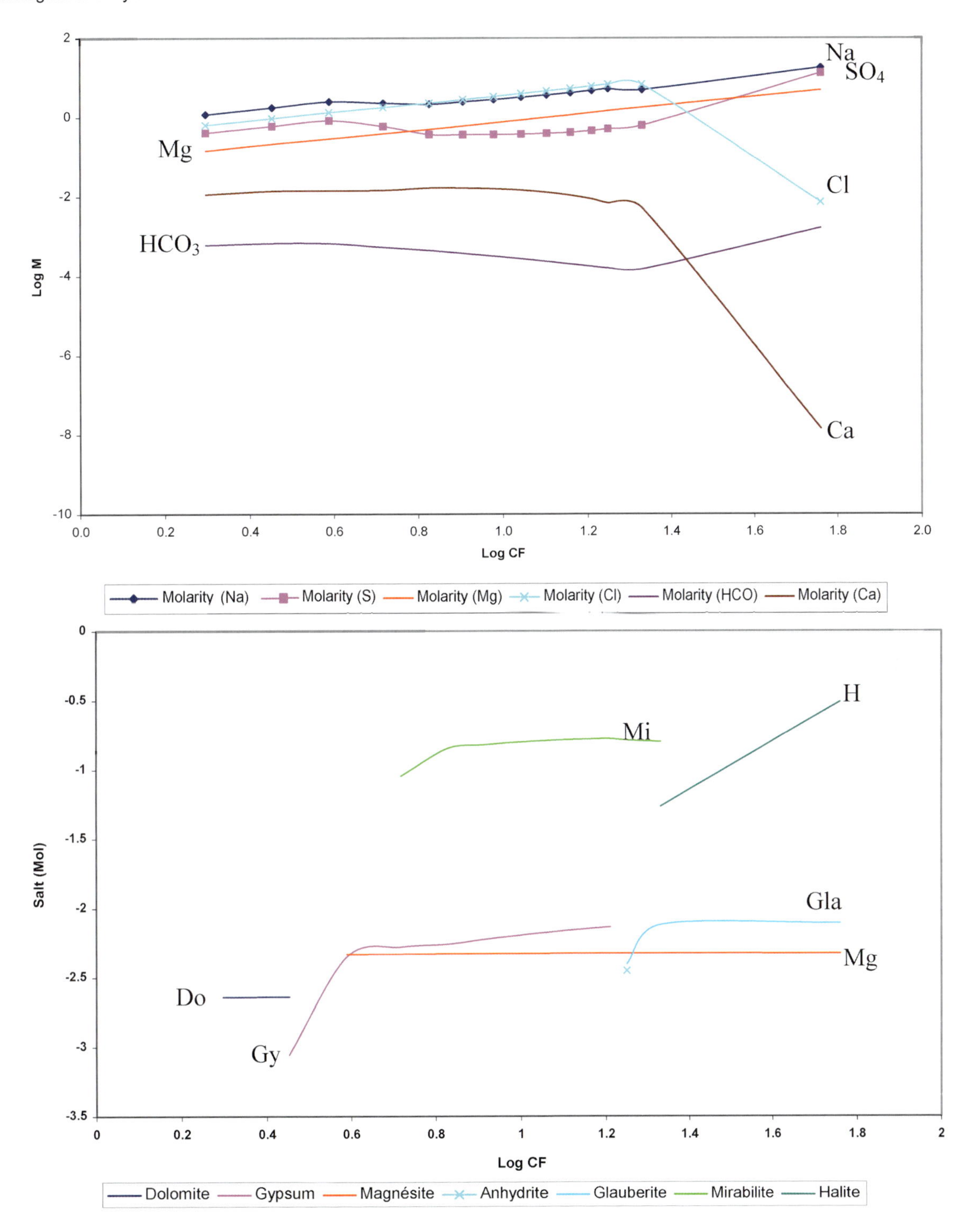

Figure 3: Example of Simulated evaporation of brine.

Conclusion

The Hydrogeochemical composition of different samples in sebkha of Oum EL Khielate show a major water type Na-SO_4-Cl in central of sebkha and Na-Mg-Cl-SO_4 or Na-Mg-Ca for samples situated in the margin in the playa.

The simulation of the brine evaporation process was carried out using PHEEQC included data base Pitzer geochemical equilibrium-reaction model. The sequence of precipitating minerals is controlled mainly by evaporation. Also mineral-brine reactions are important during the process of evaporation.

References

Baccar, L. & Louhaichi, M. A., 2007: Hydrogeological survey of Sebkha Oum El Khialate for the evaluation of the mining stok of Na_2SO_4. – Final Report. 90p.

Carpenter, A. B., 1978: Origin and chemical evolution of brines in sedimentary basins. – Proceedings of the 13th annual meeting Forum on the geology of industrial minerals. 60–77.

Eugster, H. P., 1980: Geochemistry of evaporative lacustrine deposits. – Annual Review Earth and Planetary Science **8**: 35–63.

Ingebritsen, S. E., Sanford, W. E. & Neuzil, C., 1998: Evaporites. Groundwater in Geologic Processes. – Cambridge Univ. Press.

Parkhurst, D. L. & Appelo, C. A. J., 1999: User's guide to PHREEQC (version 2) – A computer program for speciation, batch-reaction, one-dimensional transport, and inverse geochemical calculations: U.S. Geological Survey Water-Resources Investigations Report 99-4259, 312 p.

Proceedings "Hydrogeology of Arid Environments", p. 109–111
Stuttgart, March 2012

Extended Abstract

Constraints for Managed Aquifer Recharge in Arid and Semi-Arid Regions

S. A. Prathapar

International Water Management Institute, NASC Complex, DPS Marg, Pusa, New Delhi, India.
email: S.prathapar@cgiar.org.

Key words: aquifer, recharge, arid, semi-arid, constraints, groundwater

Introduction

In many arid and semi-arid regions in South Asia and the Middle East, large amounts of water annually flood out to the sea or to deserts during extreme rainfall events. Frequently, runoff from rainfall accumulates in low lying areas, and evaporates. Storing these waters during extreme events will increase the supply of water when required. In principle, groundwater storage, commonly referred in literature as Managed Aquifer Recharge (MAR), is far superior to conventional dams because 1) it only requires limited land surface, 2) it does not cause environmental damage, 3) evaporation losses are minimal and 4) it is much less expensive.

Groundwater is a common resource, it is available to all as long it is managed (recharged and extracted) in a sustainable manner. Where access is uncontrolled, there is no incentive for individual users to conserve the resource. In fact there is incentive to use as much of the resource as they are able as someone else pays the degradation cost. This is often rational behavior when viewed in isolation. The approach to managing long term degradation is to impose some form of limit on the resource to ensure sustainable use, which will be the responsibility of the government and the local community. The main criticism by users of this approach is that their property right to the resource is being curtailed. This is a false notion in that a property right in an unmanaged common resource is not defined and has no certainty. In other words, there is no guarantee in perpetuity that the resource will exist in the future. In a fully managed system where a limit to extraction can be imposed to allow sustainable development, each user has a right in perpetuity to extract. The tragedy is, in most cases it is not managed sustainably, because the irrigators need for profitability often supersede the need to extract groundwater sustainably.

For example, India, where the western part is arid or semi-arid, uses around 230 km^3 of groundwater per year and is the largest groundwater user in the world, accounting for more than a quarter of the global total. Over 60% of water needs for agriculture and 85% of drinking water are met from groundwater sources. Groundwater use in many instances has reached its limit and aquifers are at unsustainable levels of exploitation. Some estimates indicate that one-third of groundwater blocks in India are in the semi-critical, critical or overexploited categories with the situation deteriorating rapidly. The master plan for artificial recharge of groundwater in India indicated that about 0.5 million km^2 (15% of the total land area of India) is suitable for artificial recharge and that about 36,000 million m^3 of water can be recharged through MAR – nearly 20% of the current groundwater use. As a consequence there has been significant interest and investments in watershed development programs, particularly through in-situ moisture conservation, rainwater harvesting and storage and groundwater recharge.

In summary, considering the sporadic availability of excess water, advantages of aquifers as a storage media, and self-regarding attitude of groundwater irrigators, the challenge for water resources managers is to store excess water in aquifers when possible through Managed Aquifer Recharge (MAR), and then promote sustainable and equitable levels of withdrawals by the irrigators.

www.borntraeger-cramer.de
2012/0109/$ 0.75

The types and scale of MAR structures vary widely across water scarce regions. Gulf countries and Balochistan province of Pakistan for example, where flood waters are sporadic and in large volumes, have constructed recharge dams (also known as delay action dams) across wadis. In contrast, in western India, excess water for recharge is available during specified monsoon months, and therefore there is limited temporal variability in its availability. Furthermore, topography is often mildly undulating, if not flat, and therefore run-off water tend to accumulate in low lying areas, and often lost due to evaporation. Many of these ponds had been modified into recharge structures in Gujarat and Punjab, states where water scarcity is acute.

The impacts of these structures are varied. In Salalah, Oman, MAR has largely contained decline in groundwater levels, and mitigated sea water intrusion. But in India and Pakistan, their impact is limited either because of poor management of these structures and/or continuous withdrawal of large volumes of water for irrigation. There are several constraints to MAR in arid and semi-arid regions, which are discussed below.

Water

The prime reason for water scarcity in arid and semi-arid regions is their land use is often incompatible with rainfall. These regions receive annual average rainfall ranging from 100–600 mm, but grow crops throughout the year requiring evapotranspiration need between 1000–1200 mm. In addition, surface irrigation being the common method of irrigation, water is lost through non-beneficial evaporation. Therefore, even in an average year, water, either from groundwater reserves or from rivers will be required to meet irrigation demand. Sustainable solution to these regions, is to change land use either by growing crops during winter months and/or and reduce irrigated areas to be in balance with rainfall of the region. Since such changes are not foreseeable, the interim solutions will be to maximize recharge to groundwater during extreme events, and to minimize groundwater pumping through improvement in irrigation efficiencies. However, there are aspects related to rainfall and land use, which warrant adoption of MAR in arid and semi-arid regions.

Often the area under irrigation is less than the area which receives rainfall. Surface condition unirrigated areas include partially covered ground catchments, sealed roads and roof tops. Their run-off coefficients range from 0.3 for soil surface to 0.9 for roof surfaces. Hence rainfall runoff from non-agricultural land may be channeled into MAR structures.

Although the average annual rainfall in arid and semi-arid regions ranges from 100–600 mm, the annual variation had been wide. For example, the Sangrur District in Indian Punjab receives 590 mm of average annual rainfall. But records show that, the annual rainfall has ranged between 121 mm in 2002 and 1012 mm in 1983 in thirty years prior to 2007. During the 17 year period from 1954 to 1970, the annual rainfall was less than 80 per cent of the average in five years, and was between 450 and 750 mm in 10 years (CGWB 2007). This implies that there had been excessively wet years occasionally, resulting in runoff suitable for recharge.

Occasionally, the arid and semi-arid regions have experienced rainfall events of very high intensities, capable of resulting in large volumes of run-off, even from irrigated fields. For example, Batinah region of Oman received and average of 300 mm rainfall, which included 800 mm rainfall in certain parts, during the tropical cyclone, Gonu in 2007.

In summary, it may be concluded that although water is a constraint in arid and semi-arid regions, there is still scope to increase recharge to groundwater.

Land

Ideal sites to locate MAR structures have been difficult to find. For example, the Sangrur District in Punjab is 5024 km^2 in extent, but 4110 km^2, is under cultivation. Only 624 km^2 within the district is not under agriculture. The lands not under agriculture are not contiguous, not necessarily in low lying areas to construct MAR structures, and often privately owned. Owners of the uncultivated lands are unwilling to install a MAR structure, which would benefit the community as a whole. Often urban development and constructions are made across wadis and drain, obstructing the flow of runoff water to low lying areas suitable for recharge.

Capital

Construction of Recharge dams has been very expensive. In Oman, the government has built 15 recharge dams in the coastal Batinah region since 1985. These dams detain runoff water from being discharged to the sea, de-silt flood waters, and increase opportunity time for infiltration upstream of the dam. Furthermore, when released in a controlled manner when they help in recharging aquifers. The total cost of these dams is estimated to $188 Million at constant price of 2008 with a total annual capacity of 52 M m^3. Between 1990 and 2005 more than 600 M m^3 of water were collected in these dams.

Assuming that 75% of the water detained infiltrated the soil surface either upstream or downstream (Al-Ajmi & Abdel Rahman 2001), total recharge is thus estimated to be 450 M m^3. The estimated cost of groundwater recharge through dams varies from $ 0.21 to $ 0.78 per m^3 with an average of $ 0.49/m^3 (personnel communication Zaher Al-Sulaimani). The lowest cost of recharge is $ 0.21 per m^3 far below the highest marginal benefit from water $ 0.143 per m^3. This indicates that any use of recharged groundwater for agricultural purposes results in a net economic loss of at least $ 0.067 per m^3 (Zekri & Abdalla 2008). This example illustrate the cost of recharging is often prohibitive for developing countries to invest is large recharge dams.

Labor

In capital-scarce labor-rich countries like India, MAR activities remain labor intensive. In order to provide livelihood security to rural unemployed, Government of India has enacted the Mahatma Gandhi National Rural Employment Guarantee Act (MNREGA) in 2005. This Act provides for 100 days of guaranteed employment to every rural house hold in a financial year for unskilled manual labor. It covers 99% of districts in India and benefit 55 million households in 2009 (Sharma 2009). Works related to water & soil conservation, afforestation, and land development were given top priority under MNREGAs. Their success so far is limited because of a lack of guidelines to the labor available at village level on how best to integrate hydrology with civil engineering aspects of MAR structures. A new research supported by the Australian Council for International Agricultural Research is aiming to develop required guidelines to fill this void. In certain villages, differences among castes in villages have impeded effectiveness of MAR.

Maintenance

Almost exclusively, all wadis carry considerable amounts of silt and clay during runoff events and hence sedimentation is a common issue in all dams in Oman. Prathapar & Bawain (2011) reported that approximately 5–6% of runoff volume consists of sediments. Soil and rock transported from hillslopes and other land surfaces into watercourses are the major sources of sediment. These sediments settle upstream of the dam during detention and significantly reduce the rate of infiltration and recharge to aquifers. Increase in the thickness of sediments delays commencement of recharge and recharge volumes. Therefore, periodic removal of sediments not only restores storage capacity, but also accelerates recharge.

At Sahalnowt dam in Salalah, Oman, Infiltration capacity prior to the construction of the dam has decreased by at least two orders of magnitude (16 cm/h to 0.18 cm/h) in 17 years (Prathapar & Bawain, 2001). It appears that macro pores are being clogged due to movement of fine sediments with percolating water. Therefore, the maintenance activities should include removal of sediments and breaking up of soil surface to improve hydraulic conductivity of clogged profiles upstream of the dam after removal of sediments.

Conclusions

In arid and semi-arid regions Governments are investing large amounts directly or indirectly to increase recharge to aquifers. However, their success is limited because of constraints on water, land, labor and maintenance. Often, socio-political reasons supersede genuine hydrological and/or financial reasons when decisions to invest in MAR are made. The primary reason for water scarcity in arid and semi-arid regions is that the land use is often incompatible with renewable water resources of the region. The secondary reason is poor care taken by irrigators in using this valuable resource. Despite these major shortcomings, there is no option but to pursue MAR, because water is a vital resource, and aquifer storage of water is superior to surface storage.

References

Al-Ajmi, H., Abdel Rahman, H., 2001: Water Management Intricacies in the Sultanate of Oman: the Augmenting Conservation Conundrum. – Water Int. **26**: 68–79.

CGWB, 2007: Groundwater Information Booklet. Sangrur District, Punjab. – CGWB, MWR, GoI, North West Region, Chandigarh, 17 pp.

Prathapar, S. A. & Bawain, A., 2011: Impact of sedimentation to groundwater recharge at Sahalnoot dam, Salalah. – International Conference on Drought Management in Arid and Semi Arid Regions, Muscat, Oman.

Sharma, R., 2009: National Rural Employment Guarantee Act. – National Workshop on NREGA, Delhi, India.

Zekri, S. & Abdalla, O., 2008: Groundwater management in the Sultanate of Oman. – Oman Country Report to FAO. Rome, Italy.

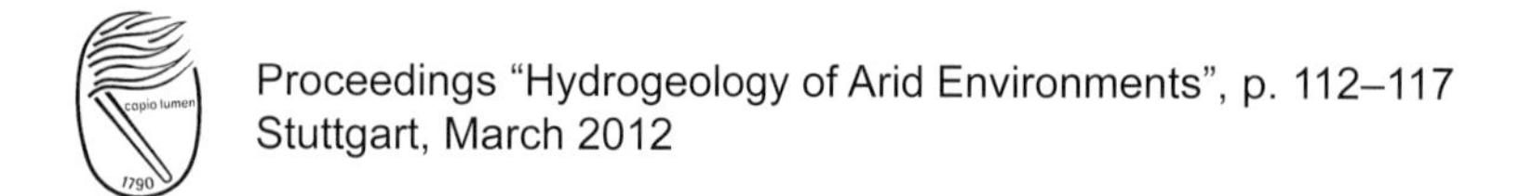

Proceedings "Hydrogeology of Arid Environments", p. 112–117
Stuttgart, March 2012

Extended Abstract

Groundwater Modeling Considering Climate Change and Significant Change in Recharge by Urban Development Along the Western Coast of Saudi Arabia

Angela Prein[1], Johannes Weiß[1], Rabih Samir Makhoul[2]

[1] CDM Consult GmbH, Alsbach, Neue Bergstraße 9-13, email: preina@cdm.com, weissj@cdm.com
[2] HUTA Group Ltd., email: rabih@huta.com.sa)

Key words: hydrology of arid environments, groundwater recharge, land-use development, water quality aspects, 3D groundwater modeling

The urbanization in former low or not developed arid areas results in a rising groundwater level. The rapid development of the City Jeddah in the Kingdom of Saudi Arabia brings the necessity to protect the city not only from flash flooding but also from rising groundwater.

The geohydrology of the area, the distribution of the ground-water recharge in time and area and also the aspects of the groundwater quality have been examined in order to plan preventive measures against rising groundwater and to control groundwater rise.

The primary objective of the study is to develop methods to keep groundwater elevations below an acceptable level over a 100-year time-horizon. Significant concerns triggered the need for the study. These were the measured groundwater table rise, existing basements and utilities that are below the groundwater table, recent intense rainfall events, severe flash flooding and climate change, the associated sea level rise, and its impact on groundwater levels.

To support this effort an analysis of available data was carried out regarding the measured groundwater elevations, measured rainfall data, the measured chemical groundwater data and the development of land-use.

The area of investigation is the City Jeddah located on the West coast of the Kingdom of Saudi Arabia. The figures 1.1 to 1.4 show the development of land-use in this area between the years of 1975 (Ortho-photomap 1975) and 2029 (Jeddah Municipality 2009) based on the evaluation of satellite imageries exemplary (Google Earth 2011). It shows the rapid development of the city area caused by increasing of the total population from approx. 0.5 million in the year 1975 to 3.2 million inhabitants today.

Figure 2 shows the groundwater elevations at three selected locations in the coastal plain area. Because data are not continuously available at nearby measuring points were summarized to get data over a time period of ten years. It shows a trend to rise up.

Figure 3 shows the annual total rainfall between years 1961 to 2009 (Saudi Oger et al.) with an average of 70 mm/year and with no trend but much more extreme values in the last years. The last rainfall events in November 2009 and January 2011 led to flooding of large city areas. The connection with the urban development is obvious.

Figure 4 show the recent distribution of chlorides in the investigation area measured in the year 2011. The analysis of the measured chemical groundwater data results in the conclusion that the influence by the seawater intrusion from the Red Sea was less than the anthropogenic influence and cannot be differentiated clearly. The mineralizing of the groundwater is a consequence of evaporation and precipitation. The dissolution of carbonate minerals is the most important process. The groundwater pollution from landfills is also of great importance for the area.

For the development of a numerical groundwater model the following processes were considered: recharge from bare land, recharge from irrigation, leakage from ponds and lakes, lekage from septic

www.borntraeger-cramer.de
2012/0112/$ 1.50

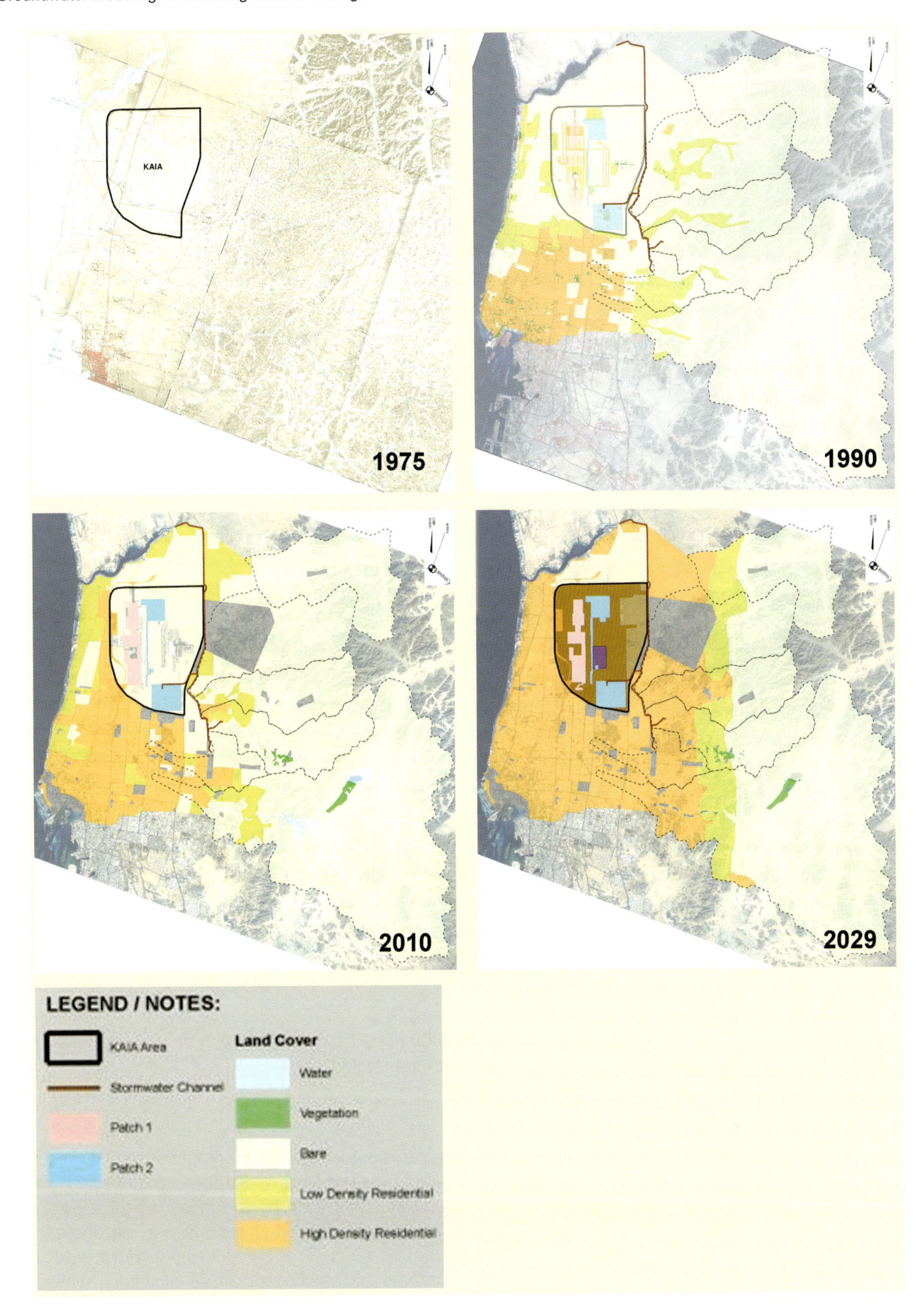

Figure 1: The development of land-use in this area for the years between 1975 (Ortho-photomap 1975) and 2029 (Jeddah Municipality 2009) based on the evaluation of satellite imageries exemplary (Google Earth 2011).

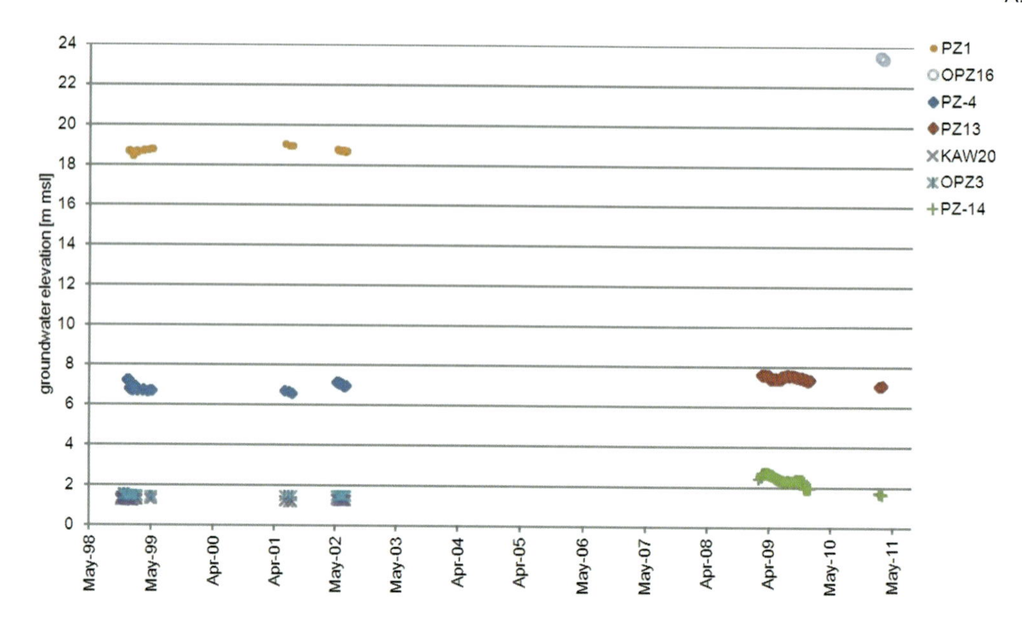

Figure 2: The groundwater elevations at three selected locations in the coastal plain area.

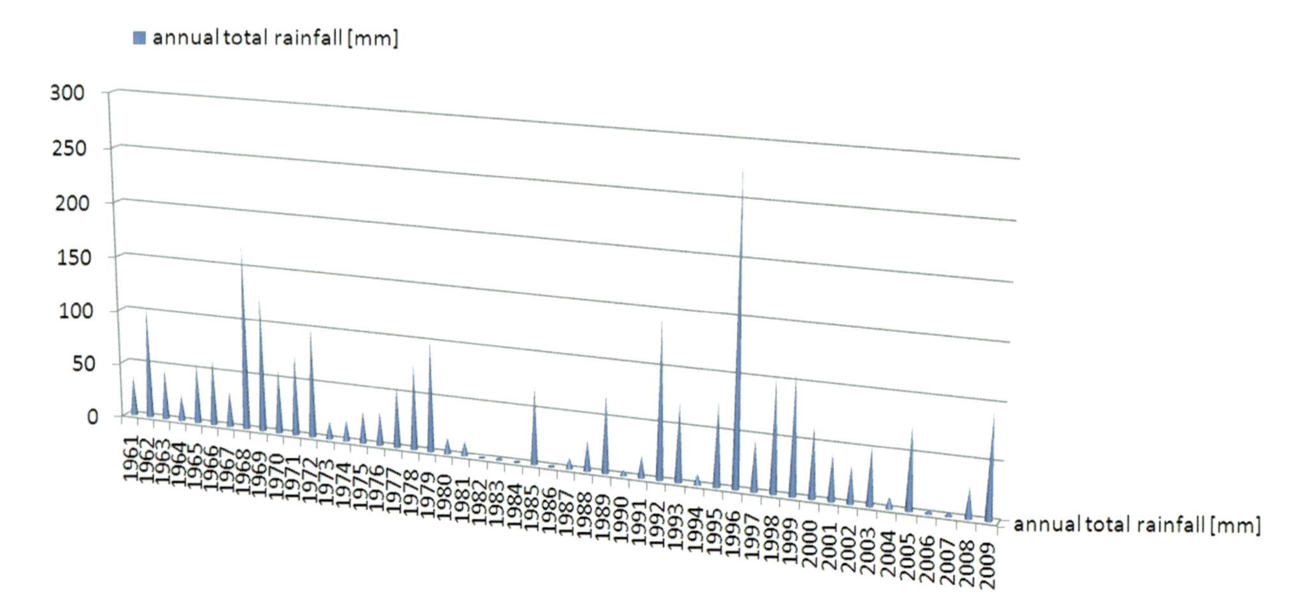

Figure 3: The annual total rainfall for the years 1961 to 2009 (Saudi Oger 2010) with an average of 70 mm/year and with no trend but much more extreme values in the last years.

tanks, leakage from residential and commercial areas based on land-use, evaporation depending on depth to groundwater elevation, temporary discharges.

The model area contains two areas, the coastal plain aquifer of approx. 340 km² and the eastern wadi area of approx. 600 km².

The western boundary of the coastal plain aquifer is located at the Red Sea and the eastern boundary is located at the morphological boundary to the mountains. The geological map (Ministry of Petroleum and Mineral Resources 1989) shows the geology completed by a number of geological and hydrogeological investigations like boreholes and pumping tests. The considered eastern wadi area contains six catchment areas of wadis, which are tribute the coastal plain aquifer. The thickness of the alluvium in the wadis was estimate by the assumption that the slopes of the hills continue below the alluvium. The maximum thickness of the alluvium

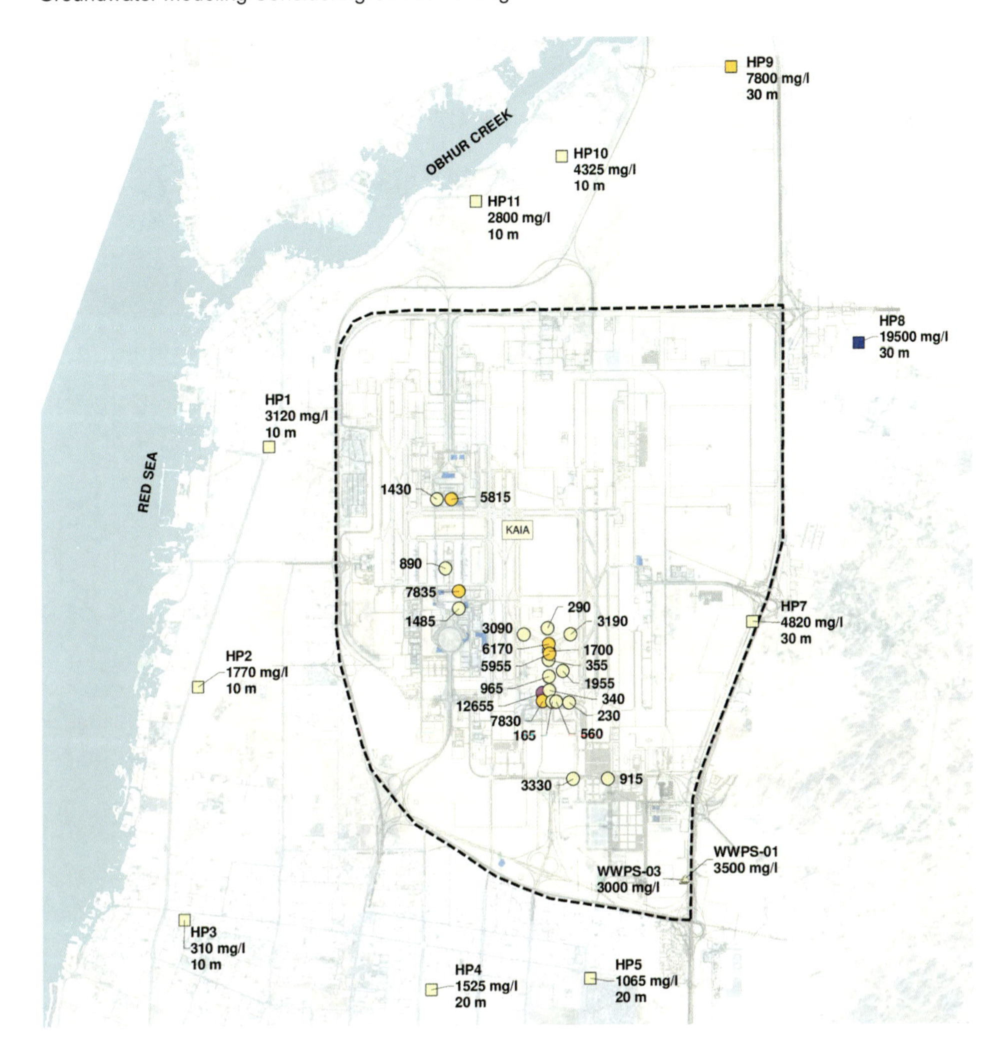

Figure 4: The recent distribution of chlorides in the investigation area measured in the year 2011.

was determined for typical cross sections of the wadis channels.
The numerical groundwater model differentiates between three main layers: alluvium, weathered rock and fractured rock. In hydrogeological investigations the bedrock was found deeper than 170 under ground level. For modeling of groundwater flow the thickness of the whole aquifer of 90 m was sufficient following the assumption that the consideration of a larger depth has no impact on the groundwater flow.

The recharge from bare land and the recharge from irrigation were estimated by using TALSIM-procedures and was found to be zero caused by available storage capacity of the soil of 100 mm/m. Also the leakage from ponds and lakes is lower than the available storage capacity.

The leakage from septic tanks assumed as included in the leakage from residential and commercial areas based on land-use. The development of land-use is characterized by increasing the total population from zero in 1975 to 15,000 inhabitants per km² today. So the resulting leakage based on the water consumption grew from zero to 102 mm/year today. For areas with commercial land-use the leakage of the network-systems was estimated by using the water consumption as well as the length and location of the pipeline. The leakage from networks estimated locally between 240 and 380 mm/year. The dumping of sewage water and retention of rainwater were considered for the time-period 1990 to 2010 in the south of the eastern wadi area.

The leakage caused by land-use reduced by evaporation. The evaporation is a function of the depth to groundwater and assumed at 2190 mm/year in average. The rate of evaporation is also a function of the soil type. The maximum extinction depth in clay is approx. 6.2 m, for sandy clay 2.1 m and for sand/gravel 0.5 m (Shah et al. 2007). More than 50 % of the water evaporates from the upper 20 percent of total influenced depth. For the calculation of the recharge caused by the network the location of pipes were assumed at a depth of 2 m below the ground level.

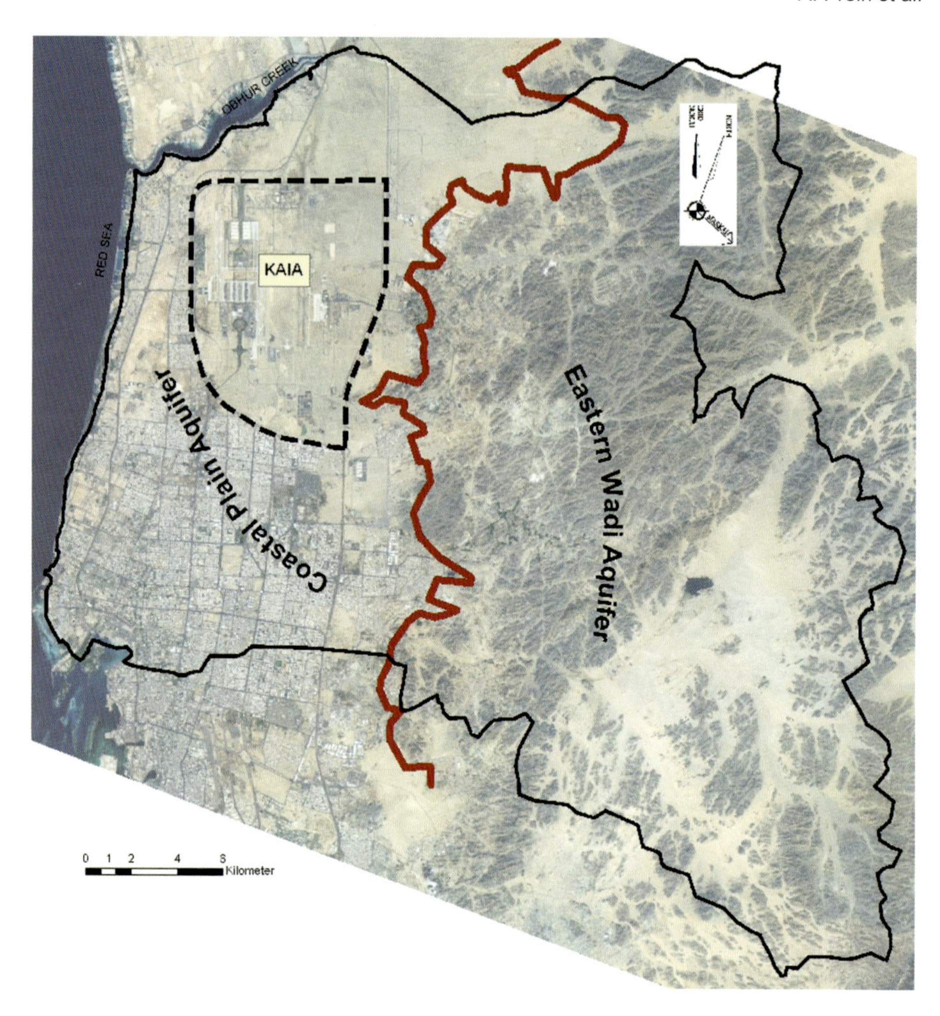

Figure 5: The model extent with coastal plain aquifer and eastern wadi area. Legend: black line: model boundary, red line: morphological boundary, dotted line: investigation area.

Temporary groundwater discharges were neglected if the duration was smaller than half a year. The model is built in GMS7.1/Modflow and it contains 6 layers, 529 rows and 456 columns. The model calibration was conducted almost exclusively through steady-state simulations. The overall model development process also included incorporation of information and data from hydraulic tests and from hydrographs of groundwater levels showing responses to natural climatic and hydrologic changes. The model-testing and sensitivity analysis identified the factor land-use as well as the extinction depth for key model parameters. The uncertainty associated with the available data characterized by non-continuous data for groundwater elevations and discharges as well as not available groundwater data in the eastern wadi area. This affects prediction of future water table elevation changes but was balanced by conservative assumptions.

Several parameters and factors were incorporated in the groundwater balancing like recharge caused by rainfall, but especially recharge caused by leakage of pipelines. The last is an important factor for the rising groundwater table due to the rapid development of the residential and industrial areas despite the high evaporation rate.

The water balance reflects the understanding of current conditions, as facilitated by the conceptual model development and subsequent implementation of a numerical model for each historic condition simulated for calibration purposes (1975–80, 1990 and 2002–2011), and for each future-condition scenario simulated (2029 and 2111). The future-condition modeling incorporated several factors likely to influence groundwater elevations, including Red Sea water level rise due to climate change, and potentially significant changes in land-use, water distribution systems, wastewater management, stormwater and flood control, irrigation and agricultural practices, and water conservation and reuse (Jeddah Municipality 2009).

The rise of the seawater level has been estimated for the region of Jeddah over the next 100 years at 1.6 m (Report: Climate Change 2007). Assuming a linear seawater rise that results in a rise of 0.29 m in the year 2029 when the study area will be fully developed.

The simulation results provided the basis for assisting the formulation of field-based groundwater and surface water monitoring, and for the decision making for groundwater control at facilities, which are to be built.

References

Saudi Oger / Murray & Roberts, J. V., 2010: S23-REO-04. KAIA (GSE) Tunnel: 100 Year Rainfall Intensity Curves.

Ministry of Petroleum and Mineral Resources, 1989: Geologic Map of the Makkah Quadrangle, Sheet 21D, Kingdom of Saudi Arabia (1986). Compiled by Thomas A. Moore and Mohammed H. Al-Rehaili, A.H. 1410, A.D.

IPCC Fourth Assessment Report: Climate Change, 2007: www.ipcc.ch/publications_and_data/yrr4/wg1/en/ch6s6-4-3-2.html.

Ortho-photomap 3921-43 North Jiddah. 1975.

Jeddah Municipality, April 2009: Jeddah Strategic Plan (Draft) – Section 1-8. Urban Territory and Land-use patterns. www.jeddah.gov.sa/strategy/English/JSP/index.php

Shah, N., Nachabe, M. & Ross, M., 2007: Extinction Depth and Evapotranspiration from Ground Water under Selected Land Covers. – Ground Water Journal **45** (3): 329–338.

Google Earth, 2011: US Dept. of State Geographer.

Proceedings "Hydrogeology of Arid Environments", p. 118–123
Stuttgart, March 2012

Extended Abstract

Linkage of WEAP and MODFLOW Models for the Azraq Basin

Raphael Reuss[1], Kai Zoßeder[1], Jobst Maßmann[2], Markus Huber[3], Klaus Schelkes[2], Johannes Stork[4], Oliver Priestly-Leach[5], Ali Subah[6]

[1] Technische Universität München, Germany, email: raphael.reuss@mytum.de
[2] Federal Institute for Geosciences and Natural Resources (BGR), Hannover, Germany
[3] geo:tools, Munich, Germany
[4] Centre for international Migration (CIM) / Ministry of Water and Irrigation, Amman, Jordan
[5] Deutscher Entwicklungsdienst (DED) / Ministry of Water and Irrigation, Amman, Jordan
[6] Ministry of Water and Irrigation, Amman, Jordan

Key words: MODFLOW, WEAP, Azraq Basin

1 Introduction

Water planning systems and hence groundwater flow models are gaining importance for a sustainable water management in arid regions as these face new challenges induced by growing domestic and irrigation demands, shrivelling water resources and the effects of climate change. This case study linkage between the water resources planning software WEAP and the finite differences groundwater flow model MODFLOW is implemented for a semiarid region in Jordan.

In the course of the recently revised National Water Master Plan of the Hashemite Kingdom of Jordan, WEAP models were developed and applied for each surface water basins in Jordan. A linked WEAP-MODFLOW approach is chosen, where the surface water–groundwater interaction is the crucial point in the basin. This is true for the Azraq Basin, which is one of the major ground and surface water basins of Jordan.

2 WEAP-MODFLOW Decision Support System

A WEAP-MODFLOW Decision Support System (DSS) will be implemented for a number of Jordan's surface water basins to ensure a sustainable water management. The DSS itself is a software product that enables the user to calculate and visualize the time-dependent behavior of a hydraulic system, if one or many of the system's parameters change. The WEAP-MODFLOW DSS is the combination of two pre-existing software products: the finite differences groundwater flow model MODFLOW developed by the U.S. Geological Survey (Harbaugh et al. 2001), and the water planning software WEAP, developed by the Stockholm Environment Institute (SEI) (Yates et al. 2005). MODFLOW is based on Darcy's law for laminar flow and the conservation of the water volume. WEAP calculates both current and future demands and the development of the resources, based on groundwater and surface water balances at catchment, landuse class level.

Since MODFLOW takes spatial relationships into account and WEAP does not, a dynamic link has been developed by BGR, SEI and ACSAD, which provides some sort of a dictionary enabling the translation of results between the two models. This ensures that WEAP results address to MODFLOW cells correctly and vice versa.

A dynamic link between the two models enables the transfer of results at each time-step (Fig. 1). WEAP calculates boundary conditions (i.e. groundwater recharge, abstraction rates, and river stages) for MODFLOW, which calculates hydraulic heads, storage volumes and in- and outflows in the subsurface. These values are used in turn by WEAP.

Thus, river-groundwater interaction, spring discharge or recharge as well as management constraints regarding the groundwater head or discharge can be considered (Al-Sibai et al. 2009).

www.borntraeger-cramer.de
2012/0118/$ 1.50

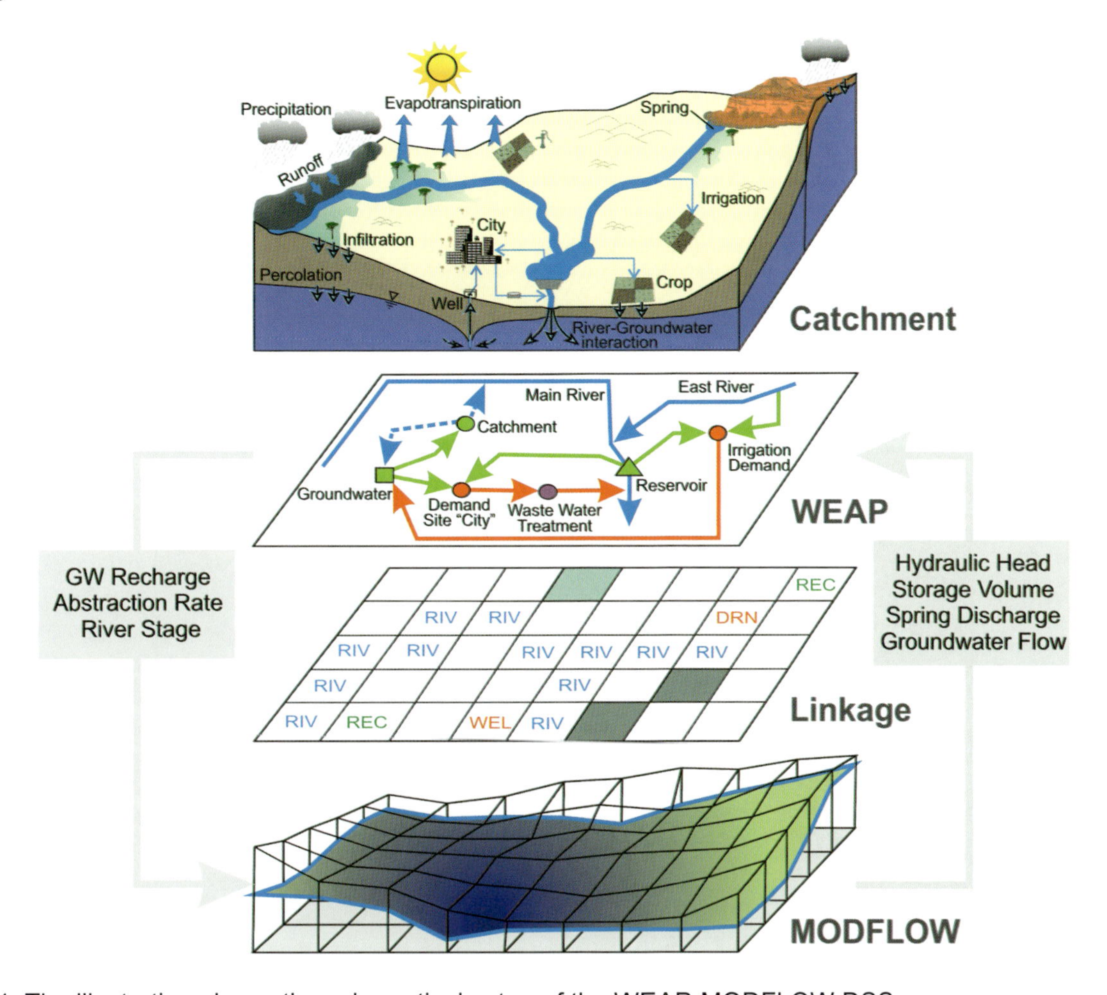

Fig. 1: The illustration shows the schematical setup of the WEAP-MODFLOW DSS.

A calibrated DSS allows the user to investigate, compare and evaluate various water management scenarios. Future constraints, such as changes in demography, economy, climate, land use, irrigation efficiency, or return flow, may easily be integrated into this model. The advantage of the linkage is that not only the impact of different scenarios in terms of hydraulic heads, flow rates or irrigation amounts can be shown, but also the reactions and dynamics of the groundwater system, discretized in time and space, can be predicted and evaluated.

3 Case study Azraq Basin

The Azraq Basin is located in the north central part of Jordan; an area of about 12,700 km², partly extending to Syria and Saudi Arabia. Hot and dry summers and fairly wet and mostly cold winters characterize the semiarid to arid climate in the Azraq Basin. Precipitation, at a weighted annual average of 88 mm, mainly occurs in the form of thunderstorm rainfalls caused by cold air masses coming from the Mediterranean Sea. Cyclonic rainfall can however reach the area from December to October. In general, the occurrence of precipitation is limited to the period from October until May (University of Jordan 1996).

The two dominant topographical features of central north Jordan are the gently undulating relief of the southward sloping Harrat-Ash Shaam basaltic province and the elongated low of the Al´Azraq depression. Six major wadis drain into the Azraq depression. None of the streams are perennial (NWMP 2004).

The Azraq oasis, which is centrally located in the basin, forms a unique habitat with a great biodiversity and numerous endemic species. However, as this region not only plays an important role for local agriculture but also supplies the capital Amman with water, the Azraq Basin has shown a negative water budget for decades.

Beside the Tertiary to Quaternary basalts (BA) in the north, Upper Cretaceous formations (B3 and B4/B5) are exposed within the Azraq Basin (Fig. 2). The marly B3 formation separates the overlying

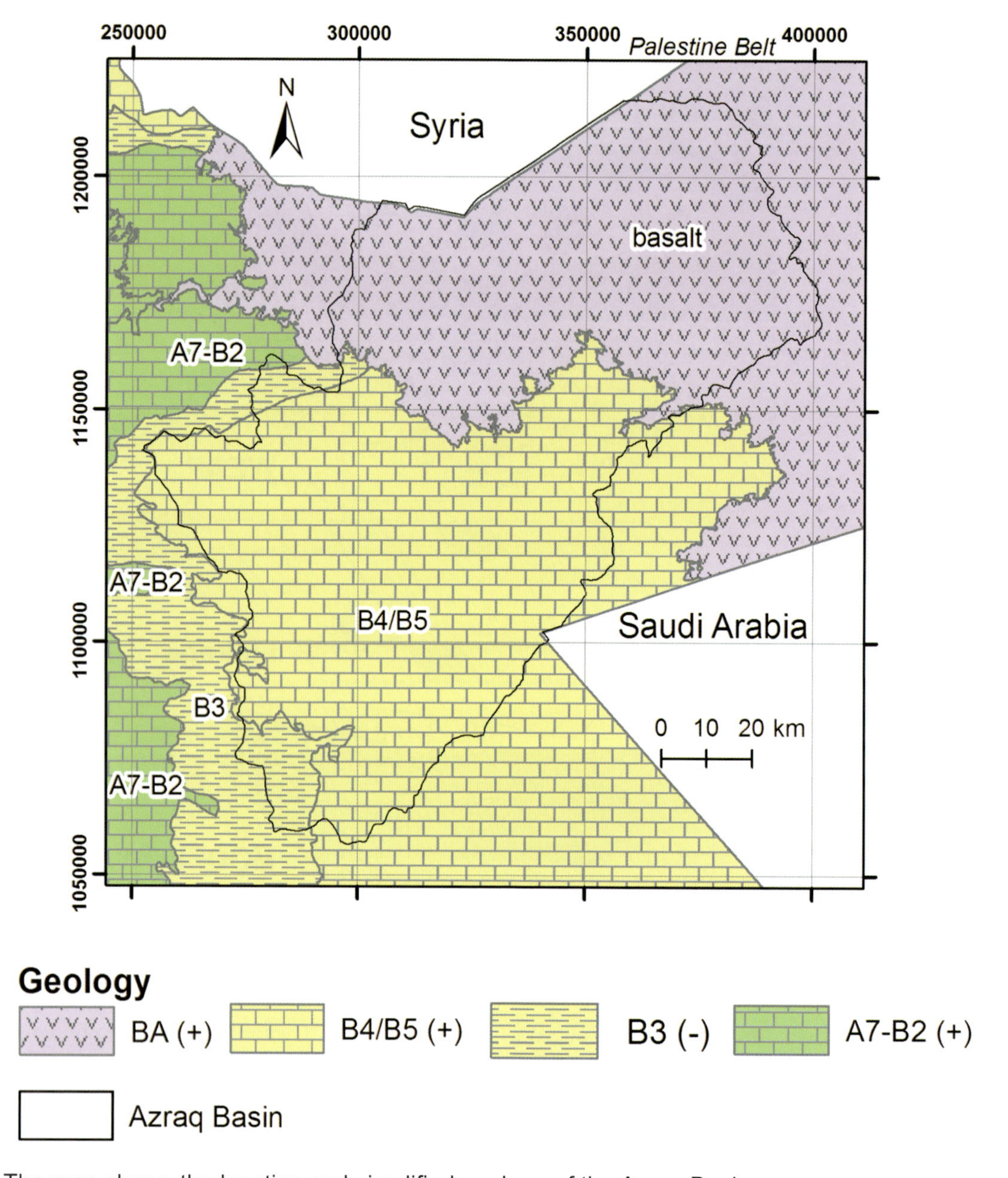

Fig. 2: The map shows the location and simplified geology of the Azraq Basin.

"Shallow Aquifer System" (BA and B4/B5) from the underlying A7/B2 aquifer. The latter crops out in the Highlands west of the Azraq Basin. The structural setting of the area is characterised by predominant sinistral, transtentional fault systems associated with the Dead Sea Rift. Vertical displacements along major faults can reach more than 3000 m. Due to progressively downfaulting during Turonian and Maastrichtian times, the thickness of the hydrogeoligical unit A7/B2 can reach more than 3000 m (Hobler et al. 2001).

The hydrogeological conceptual model consists of three major units. These are primarily jointed limestones intercalated with marl and discordantly covered by partly higher permeable basalts. In reality, groundwater flow mainly occurs along joints. Nevertheless, for such a large-scale groundwater flow model with emphasis on water balances, the assumption of an equivalent porous medium is sufficient. The aquifer systems regarded in the MODFLOW-model are commonly termed the Middle Aquifer System (comprising the formations A1 to A6 and A7, B1 and B2) and the Shallow Aquifer System (consisting of B4, B5, basalts and younger alluvial deposits). These aquifer systems are separated by the marly formation B3. Only the upper part (A7/B2) of the Middle Aquifer System is represented in the model (Ibrahim 1996).

The ground water flow of the modelled aquifers is generally directed towards the centre of the Azraq Basin while in the deeper aquifers, which are not represented in the model, groundwater flow is mainly directed to the Dead Sea and Jordan Valley (Hobler et al. 2001).

3.1 WEAP model

The key factor determining the water budget in the Azraq Basin is agriculture and the corresponding irrigation demand. The model design of the WEAP-model therefore focuses on the effects of changes in agricultural land use and makes use of the WEAP-internal „FAO Rainfall Runoff"-model to simulate surface runoff and irrigation demand.

WEAP provides output data for a spatial unit defined as the "WEAP catchment". Here, the WEAP catchments coincide with the governorates intersecting the Azraq Basin and do not represent water catchments in the hydrological sense. This is due to the fact that the relevant stakeholders need to get the model's results aggregated to administrative units rather than natural ones.

A basic land use classification was executed applying remote sensing techniques, most importantly the analysis of NDVI. As a result, two classes were defined: irrigated and non-irrigated areas were irrigated areas represent a mix of fruit trees and olives. Precipitation was interpolated using inverse distance weighted method on the base of recorded rainfall data at meteorological stations in and around Azraq Basin. The obtained values were averaged over each one respective WEAP catchment.

3.2 MODFLOW model

A particular effort was made to properly represent the structural and geological features of the study site (fig. 3 and 4).

The faults occurring in the study area indicate different hydraulic behaviour. While in some areas a vertical leakage along fault planes seems to be probable, other faults form, at least partly, a horizontal barrier to groundwater flow (Hobler et al. 2001).

A stationary approach, calibrated against groundwater contour maps dating to the mid 1990's, was used to simulate the general flow pattern and to obtain adequate material parameters. Upon this basis, a transient model was developed.

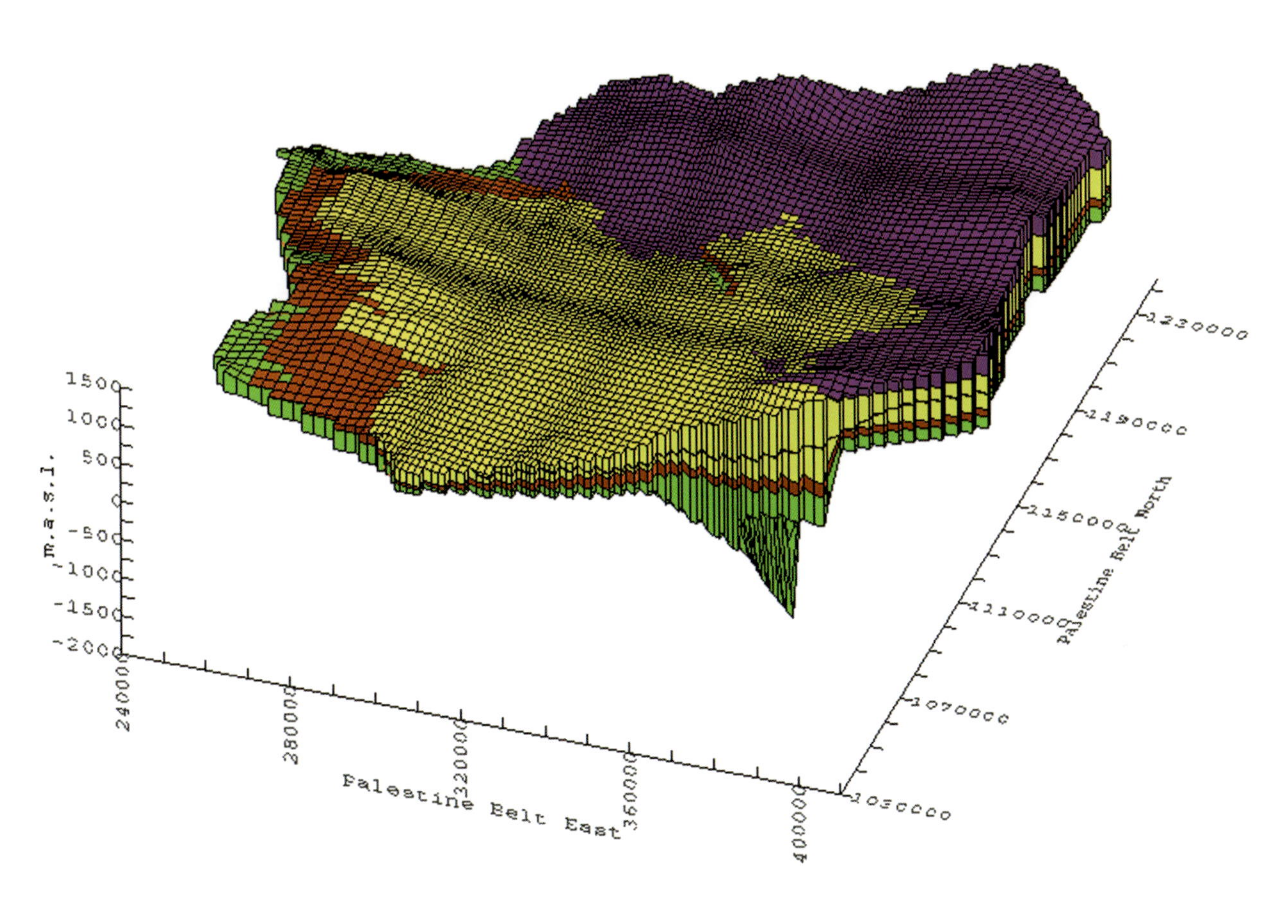

Fig. 3: It is shown the active area of the MODFLOW model and the spatial discretization.

Lacking data concerning actual evapotranspiration, surface run off and secondary infiltration, the input data for the recharge package is primarily based on previous comprehensive studies. Several zones of homogeneous recharge rates were defined. These are confined by isohyetal lines and geological boundaries. A recharge rate as a percentage of rainfall was assigned to each zone. The recharge rate later will be calculated by WEAP.

Along the northern model boundary of the basalt aquifer, an inflow of about 18 million m³ was assumed (Schelkes1997). The representation of moveable water divides (i.e. location depends on actual abstraction rates) along model boundaries presented a particular challenge. Using specified head boundaries for the A7/B2 aquifer (bottom layer) in the stationary approach allowed for the later translation into specified flow (constant or transient) boundaries.

In order to ensure reliability through the prevention of cell drying, it was necassary to allow an upward leakage from the A7/B2 aquifer in the area around Azraq; especially in the area of the AWSA well field. This goes along with previous and recent hydrogeological and hydrochemical studies (Gibbs 1993, Kaudse pers. comm.).

The stationary approach hence provides for the simulation of the general groundwater flow pattern. Due to restrictions imposed by WEAP concerning discretization, The model results remain nevertheless debatable due to the restrictions imposed by WEAP concerning discretization.

4 Results

The calibration of the MODFLOW model substantiates the assumption of an upward leakage from the A7/B2 aquifer into the B4/B5 aquifer in the area around Azraq. In the southern part of the study area, the groundwater flow pattern of the B4/B5 aquifer seem to be at least partly influenced by the deeper aquifer complex A7/B2. The relatively coarse dis-

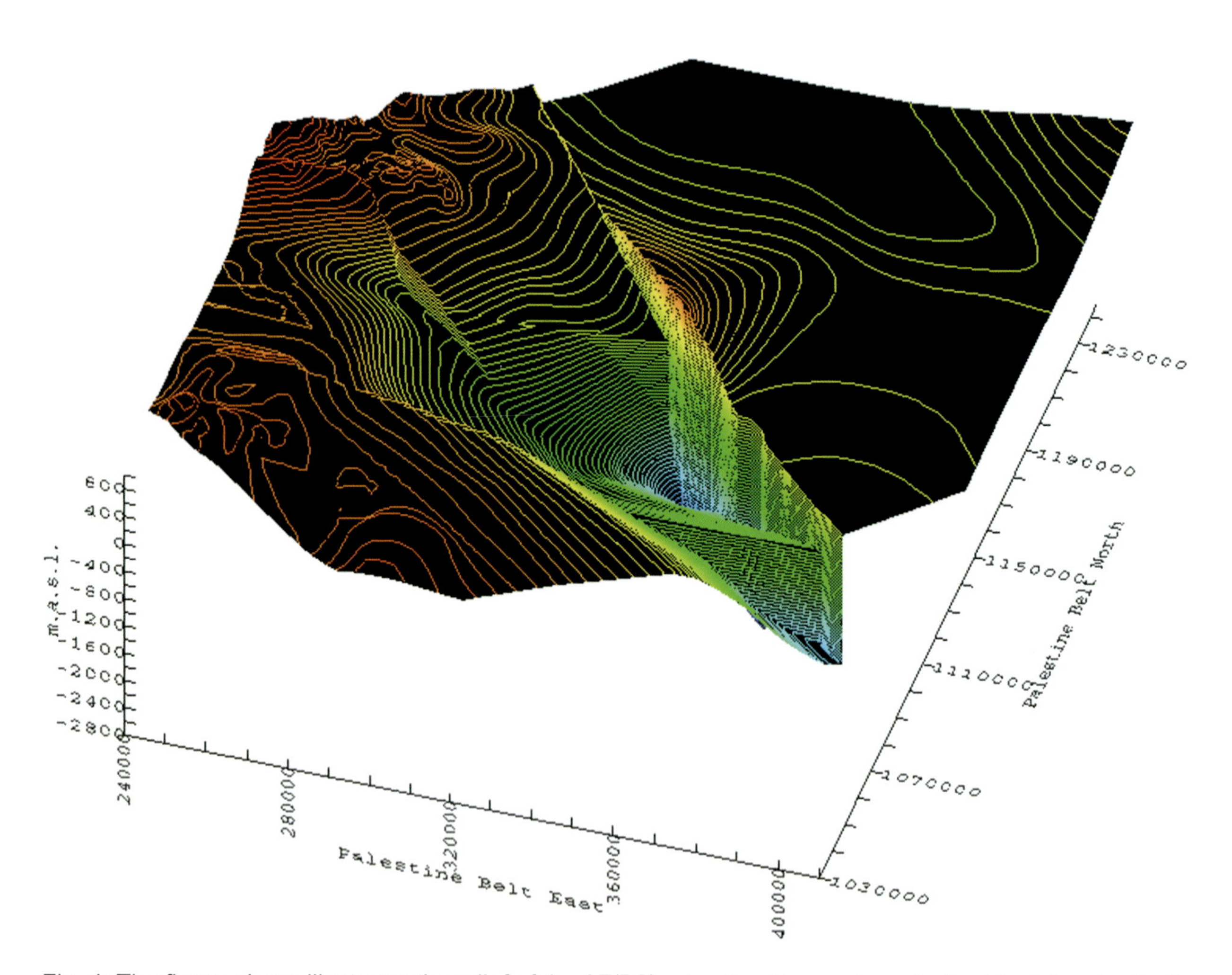

Fig. 4: The figure above illustrates the relief of the A7/B2's structural base characterized by intensive faulting.

cretized finite differences groundwater flow model can hardly represent actual groundwater flow, e.g. perched water tables in the complex strata of the basaltic province or the leakage processes along fault planes. With a relatively high salt content in the A7/B2 around Azraq, further overexploitation of the upper aquifer system may lead to lower quality of pumped water due to an increasing upward leakage.

Since a great effort was made to model the geologic setting of the study site, the present model is a reliable base for future scenario investigations.

The linked WEAP-MODFLOW model will provide a powerful and flexible water resources planning tool. It may act as a Decision Support System for the MWI in order to evaluate water planning scenarios considering a wide range of changes in supply and demand as well as population growth, water policies and climate. Thus, it can be used to investigate the feasibility of rewetting the Azraq Oasis.

Since this contribution is part of a project, which has yet to be concluded, changes concerning the set up of the models and calibration results are reserved.

References

Ministry of Water and Irrigation (2004) : National Water Master Plan (NWMP) Amman.

Al-Sibai, M., Droubi, A., Abdallah, A., Zahra, S., Wolfer, J., Huber, M., Hennings, V. & Schelkes, K., 2008: Incorporate MODFLOW in a Decision Support System for Water Resources Management. Proceedings of Int. Conference: MODFLOW and More 2008, Colorado, USA.

Gibbs, B., 1993: Hydrogeology of the Azraq basin. – Master Thesis, University College of London, London, 198 pp.

Harbaugh, A. W., Banta, E. R., Hill, M. C. & McDonald, M. G., 2000: MODFLOW-2000, the U.S. Geological Survey modular ground-water model – User guide to modularization concepts and the ground-water flow process. USGS Open-File Report 00-92.

Hobler, M., Margane, A., Almomani, M., Subah, A., Hammoudeh, A., Ouran, S., Rayyan, M., Saffarini, I., Hijazi, H. & Kalbouneh, A., 1994: Groundwater Resources of Northern Jordan.- Vol. 3: Structural Features of the Main Hydrogeological Units in Northern Jordan. – Northern Jordan. – Unpublished report prepared by Federal Institute for Geosciences and Natural Resources (BGR) and Water Authority of Jordan (WAJ), Technical Cooperation Project 'Groundwater Resources od Northern Jordan, BGR-archive no. 112 708, Amman and Hannover.

Hobler, M., Margane, A., Subah, A., Almomani, M., Rayyan, M., Khalifeh, N., Al Mahamid, J., Hijazi, H., Zuhdi, Z., Ouran, S. & Hammoudeh, A., 2001: Groundwater Resources of Northern Jordan. – Vol. 4: Hydrogeological features of Northern Jordan. – Unpublished report prepared by Federal Institute for Geosciences and Natural Resources (BGR) and Water Authority of Jordan (WAJ), Technical Cooperation Project 'Groundwater Resources od Northern Jordan, BGR-archive no. 112 708, Amman and Hannover.

Ibrahim, K., 1996: The Regional Geology of Al Azraq area. – Ministry of Energy and Mineral Resources, Natural Resources Authority (NRA), Bulletin No. 36, Amman, 67 pp.

Schelkes, K., 1997: Groundwater Balance of the Jordan-Syrian Basalt aquifer. – Unpublished report prepared by Federal Institute for Geosciences and Natural Resources (BGR) and Economic and Social Commission of Asia (ESCWA), Technical Cooperation Project ‚Advisory Services to ESCWA member states in the field of water resources', BGR-archive no. 116 707, Amman.

University of Jordan, 1996: Water Budget and Hydrogeology of the Azraq Basin. – Technical Bulletin No. **19**, Amman, 39 pp.

Yates, D., Sieber, J., Purkey, D. & Huber-Lee, A., 2005: WEAP21 – A Demand-, Priority-, and Preference-Driven Water Planning Model. Part 1: Model Characteristics. – IWRA, Water International, **30**(4): 487–500.

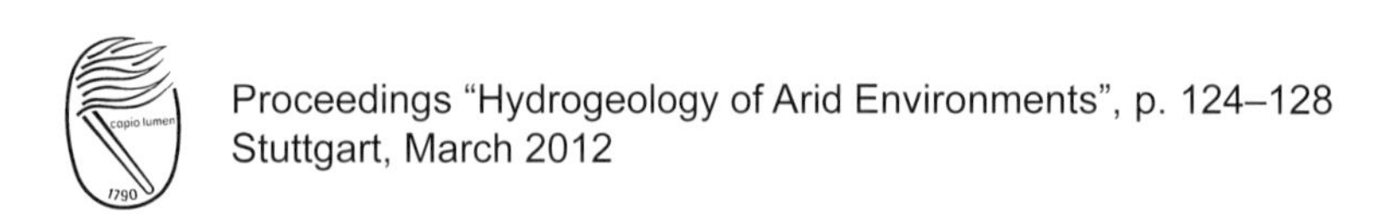
Proceedings "Hydrogeology of Arid Environments", p. 124–128
Stuttgart, March 2012

Extended Abstract

Characterization of Water Storage Dynamics in Arid Areas by Satellite Gravimetry

J. Riegger[1], M. J. Tourian[1, 2]

[1] Institut für Wasser- und Umweltsystemmodellierung, Universität Stuttgart, Pfaffenwaldring 61, 70550 Stuttgart, email: riegger@iws.uni-stuttgart.de
[2] Geodätisches Institut, Universität Stuttgart

Key words: arid areas, water balance, water storage, satellite gravimetry, GRACE

In arid areas there are tremendous difficulties to quantify water balances appropriately as actual evapotranspiration ETa and seapage from surface runoff during precipitation events cannot be measured directly and often river discharge does not occur. Modelling approaches for ETa are not suitable for a quantification as they show variations in a 30% range between the different models (PILPS/AMIP Studies). Thus the knowledge of water storage characteristics and their dynamics is essential for the description of the short and long term system behaviour and the related water balance.

Groundbased observations of the storage components are not available on large spatial scales. Measurements of water storage changes by observations of changes in groundwater levels or soil water saturation (Strassberg et al. 2000) are not reliable due to the inadequate density of monitoring points and the insufficient knowledge of storage coefficients.

Gravity Recovery and Climate Experiment (GRACE) satellite gravimetry provides a direct measure of monthly mass changes over large areas. For land masses monthly mass variations are interpreted as changes in total water storage of the considered area comprising all contributing storage compartments (snow/ice, surface water, soil, (un) saturated underground) in their vertical as well as in their horizontal distribution. A separation of the contributions from different storages is possible, if the related system dynamics i.e. the corresponding time constants differ sufficiently.

The use of GRACE gravity data allows a direct measurement of total continental water storage changes and thus for the first time the closure of the water balance in hydrology and hydro-meteorology on monthly time scales:

$$\frac{dM}{dt} = P - ETa - R$$

hydrologic water balance

$$\frac{dM}{dt} = -\nabla \cdot Q - R$$

hydro-meteorologic water balance

This promotes a number of novel approaches like for the direct determination of actual evapotranspiration and runoff on the basis of satellite gravimetry and atmospheric moisture flux divergence divQ (Rodell & Famiglietti 1999, Riegger & Güntner 2005).

The water balance equation is independent of spatial and temporal scales. On local i.e. on subcatchment scales the runoff is normally not known, as it is depending on terrain, vegetation cover, soil structure, storages etc in a very complex way. Thus simulated runoff is of insufficient accuracy. The only reliable measured runoff R is the aggregated runoff measured by river discharge for the related catchment areas. Hence, the above water balances are most accurate if determined for catchments with aggregated hydrologic and hydro-meteorologic data. For the aggregation topographic catchments are used. Possible groundwater exchanges over catchment borders for non arid areas are considered as negligible compared to runoff generation P-ETa and river discharge.

For (semi)arid areas surface runoff often is intermittent with precipitation events and not measurable. For topographically dischargeless basins the aggregated surface runoff is concentrated in local depressions and is either evaporated or is percolating and recharging groundwater storage.

As for many (semi)arid areas the groundwater outflow is not known, following questions arise:

1. Are there short term changes in storage?
2. Is the system in a transient or in a steady state (long term)?
3. What are the values for groundwater recharge P-ETa and groundwater discharge R?

The components of the water balance show a big range of time scales especially for (semi)arid areas. Precipitation there is typically rare and short. In Saudi Arabia for example only half of the months comprise rain events and these occur within ~4 days per month. The mean precipitation of a rainy day is ~5mm and reaches up to ~125 mm extremes. The time scale of actual evapotranspiration can cover hours to months depending on potential evapotranspiration and water storage.

The reaction time of a groundwater system depends on the hydraulic time constant

$$t = \frac{1}{2.2} \cdot \frac{S}{T} \cdot r^2$$

For catchment areas with river discharge typical distances from the centres of the drained areas to the river network is in the range of several 100 m to km. This leads to typical hydraulic time constants in the range of weeks to months. For arid areas of the considered size distances from the area centre to discharge points typically are >350 km. Thus the minimum reasonable time ($kf = 10^{-3}$–10^{-4} m/s, $S_o = 10^{-3}$–10^{-4}) for a complete disappearance of a recharge event is ~200–20000 yr. This implies, that recharge events comparable to precipitation would increase storage until a water level is reached, which leads to surface runoff and thus shorter time scales.

Observations

Precipitation: Regionalized data from the Global Precipitation Climatology Centre (GPCC) are used for monthly precipitation. GPCC version 5 covers time period 1901–2008. Local daily precipitation measurement for Saudi Arabia are available from 1978–2002.

GRACE: Monthly mass deviations dM with respect to the long term mean are provided by GFZ (GeoForschungsZentrum, Germany) RL4 for the time period of Feb. 2003 to Dec. 2009. The monthly solutions are filtered using the decorrelation filter in combination with a Gaussian 500 km filter. The filtered monthly solutions are converted to equivalent water heights (Swenson & Wahr 2006). Temporal derivatives dM/dt are calculated with a central difference scheme. In order to avoid leakage of large signals from outside of the observed area due to the filter width, gravity data are accumulated on an inside core using an internal buffer zone of ~350 km. This is of specific importance where areas of small signals neighbour areas with high signals like at the southern boundary of the Sahara with the neighbouring tropical regions. This limits the investigated areas to those above 250000 km^2.

Seasonal behaviour

Monthly time series of mass deviation dM, mass derivative dM/dt, Precipitation and actual evapotranspiration ETa calculated from the water balance are shown in Fig. 1a for Saudi Arabia as an example and for the time period 2003–2009. The correlation of mass change rates dM/dt and precipitation time series P is –0,02, even if only rainy months are considered. This means, that gravity data dM and dM/dt are dominated by noise with ~7.7 mm and 4.9 mm/mo RMS respectively (Riegger & Tourian 2011).

Monthly precipitation events of up to 30mm, which clearly exceed the noise level of the gravity data do not show any effect on the storage state dM or the storage change dM/dt. If recharge would contribute a considerable fraction of precipitation, this event would have to appear in the mass change rate and state of mass as the recharge of such a precipitation event cannot be levelled out by groundwater discharge within one month considering the hydraulic time constant of the system. Thus it must be concluded, that precipitation is mainly balanced out by actual evapotranspiration, so that recharge is a small fraction of precipitation. This can be recognized in Fig. 1, where for larger rain events ETa corresponds to precipitation within the error limits.

The use of mean monthly values for the period 2003–2009 (Fig. 1b) reduces the noise in mass change rates to 1,8 +/–0.4mm/mo RMS. The “error bars” in the graphs indicate the RMS of the monthly residual, which comprises climatic variations as well as errors and can serve as upper limits for errors.

For Saudi Arabia the mean monthly behaviour of precipitation shows wet periods during October to May and dry periods from June to September. The mean monthly mass change rates dM/dt are very small even for wet periods. The consequence hereof is, that groundwater recharge (P-ETa) roughly corresponds to groundwater discharge. As the seasonal behaviour of precipitation does neither lead to storage changes on comparable time scales, which would be in contradiction to the hydraulic time scale,

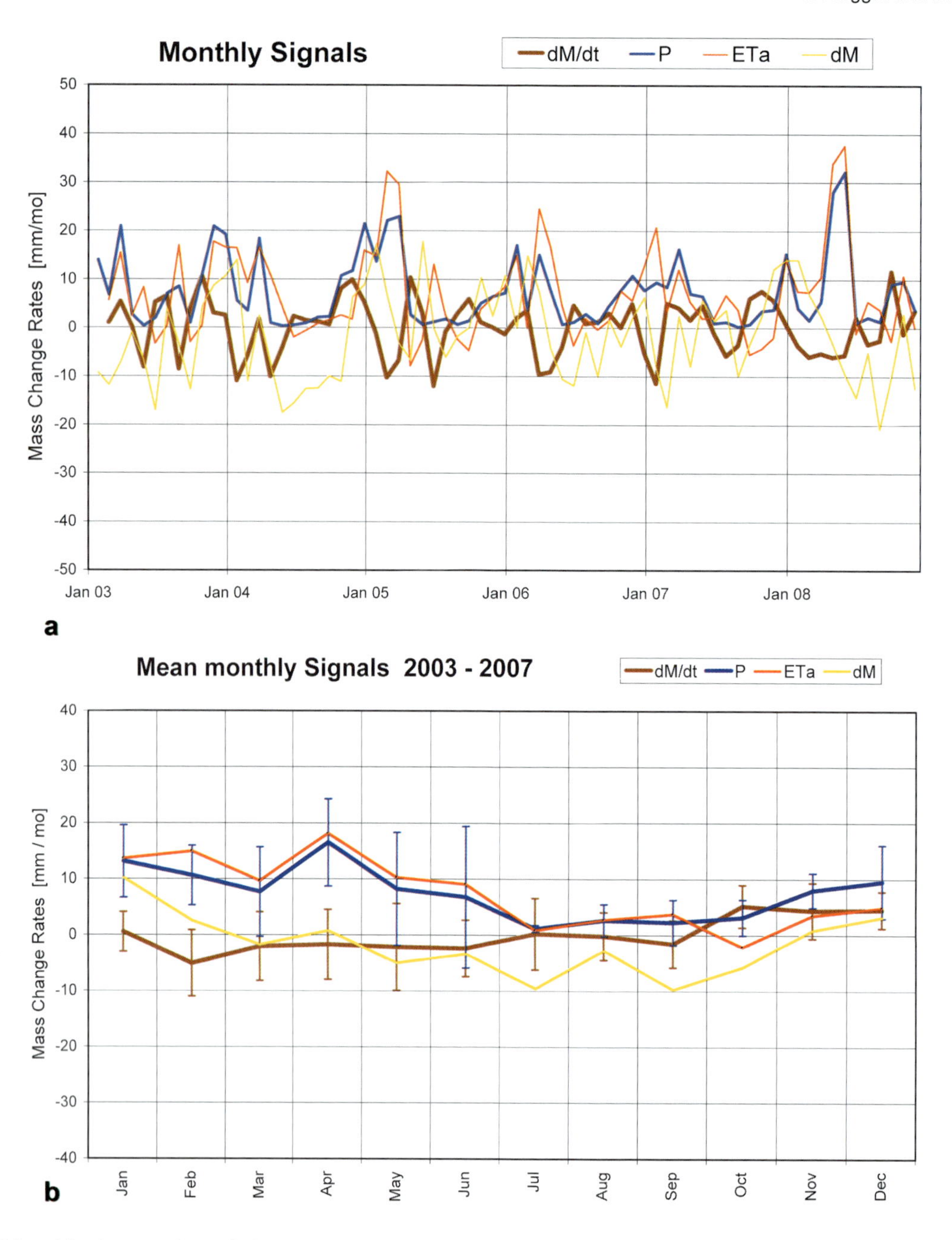

Fig. 1. a. Monthly time series of dM, dM/dt, P and ETa for Saudi Arabia. b. Mean monthly time series of dM, dM/dt, P and ETa for Saudi Arabia.

nor to step functions in storage, it must be concluded, that groundwater recharge is very small. Thus the mean monthly evapotranspiration ETa roughly corresponds to precipitation.

This is confirmed by local hydrogeological investigations over Saudi-Arabia (Kalbus et al. 2011) with estimations of the mean groundwater recharge (P-ETa) of 0.06–0.13 mm/mo on the basis of a groundwater model. This means, that the main part of precipitation is evaporated nearly instantaneously, which is consistent with the fact that, that monthly potential evapotranspiration ~135 mm/mo by far exceeds precipitation events ~20 mm/mo even during rainy months. Indirect recharge via percolation from intermediate surface water during a rain event also plays a role in recharge, but its relative contribution compared to direct recharge cannot be distinguished with this method.

Areas of the central Sahara – all climatically classified by Köppen-Geiger class BWh (hot desert) – show a similar behaviour as for Saudi Arabia, however with even less seasonal precipitation.

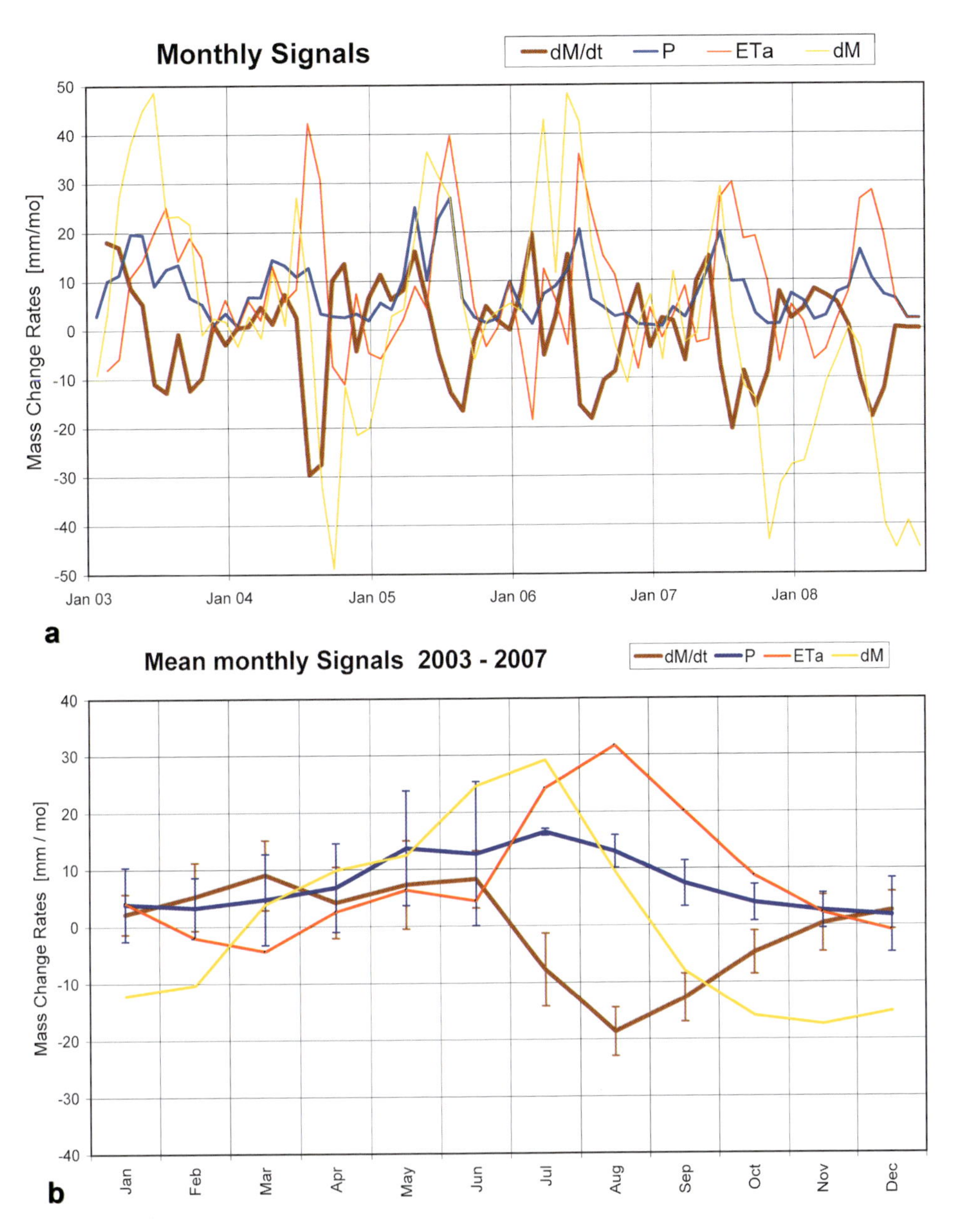

Fig. 2. a. Monthly time series of dM, dM/dt, P and ETa for the Tarim Basin. b. Mean monthly time series of dM, dM/dt, P and ETa for the Tarim Basin.

A different behaviour is seen for the Tarim basin (Fig. 2a) classified as a cold desert (BWk). Even though the mean monthly precipitation (fig. 2b) is not higher than that of Saudi Arabia, there is an accumulation of precipitation and possible discharge from the surrounding mountains, which can be clearly recognized in the increasing storage dM until July. The actual evapotranspiration is not instantaneously balancing precipitation, yet it starts once positive temperatures are reached. Thus a delayed response occurs in evapotranspiration. Opposite to Saudi Arabia and the Sahara a decrease in storage on a monthly time scale is observed. This cannot be explained by a groundwater discharge due to the related time constants, but rather by a groundwater storage, for which the groundwater level at the end of an accumulation phase is shallow enough for an evaporation by capillary rise or possibly by open water bodies.

Further examples like for the Gobi, the Australian deserts etc. show, that satellite gravimetry with present accuracy allows at least a qualitative characterization of the hydrologic behaviour and the monthly dynamics of related water fluxes in (semi) arid areas.

Long term behaviour

While for the Tarim basin the monthly storage change corresponding to input and evapotranspiration is large enough to quantify the seasonal behaviour of water fluxes (Fig. 2a, b), for Saudi Arabia and the Sahara it is dominated by noise. Yet small imbalances between groundwater recharge and groundwater discharge should accumulate and thus appear in long term trends.

For Saudi Arabia the trend determined from GRACE mass deviations dM over 72 months is +0.3 mm/yr. Hydrogeological observations in Saudi-Arabia (Kalbus et al. 2011) quantify the regional water budget with a mean groundwater recharge (P-ETa) of 0.8–1.6 mm/yr, a natural subsurface out flow of 1.3 mm/yr and a groundwater abstraction of 6.5–7.8 mm/yr. With the consideration of a crude oil abstraction of 0.2 mm/yr this corresponds to a total mean mass change rate of –7,2 to –7,7 mm/yr, which should be clearly recognizable over 6 years of observation even for a noise level of ~ 8 mm RMS in the mass deviation dM. Trend calculations for a stochastic data set with the same noise level and measurement period lead to a symmetric distribution with a mean value of $-2.3\ 10^{-3}$ mm/yr and a RMS of 0.35 mm/yr.

This means, that the ground measured trend and the GRACE trend in dM are significantly different for Saudi Arabia and the Sahara.

There are two possible explanations for the differing trends:

1. The hydrogeological investigations underestimate recharge or overestimate outflow.
2. There is a geophysical reason for a mass trend.

As the gravity signal is the superposition of all transport processes from the inner earth mantel via the earths crust up to mass distribution at the surface, in principle the possibility exists that geophysical processes interfere. One of these processes is Glacial Isostatic Adjustment (GIA), which is the viscoelastic response of the crust with respect to load changes at the surface. This is for example observed at high latitudes where the melting of ice masses 10–15000 yrs ago still leads to a rise in elevation and mass. For Saudi Arabia the loss in water storage from close to surface water levels down to the present deep water levels over the last 6000 yrs could lead to a load rebound i.e. to a viscoelastic compensation. This is being checked at the moment with the assumed water storage over the last 6000 yrs.

The Tarim basin shows a significant trend in the state of mass of –6,6 mm/yr with a corresponding RMS of 0.66 mm/yr due to noise. The cause of the mass trend is not clear yet. The time period of GRACE measurement is quite short so that reliable climatic trends are questionable. A reliable climatic trend however is a prerequisite for the separation of long term hydrologic signal and geophysical contributions.

Summary

Our investigations of the largest semi(arid) areas worldwide have shown, that the measurements of water storage dynamics by satellite gravimetry significantly enhance the understanding of their hydrologic characteristics and the quantification of the involved volume flows. GRACE data allow a quantification of the regional water balance and thus the determination of groundwater recharge or actual evapotranspiration on monthly time scales within an error bound of ~5 mm/mo RMS.

Long term trends of the gravity signal in principle provide information on long term water balances with enhanced accuracy, yet are determined by the noise level and the length of the measurement period. However hydrologic trends might be overwritten by the simultaneous occurrence of geophysical signals on long time scales, which cannot be distinguished from climatic contributions at present.

References

Kalbus, E., Oswald, S., Wang, W., Kolditz, O., Engelhardt, I., Al Saud, M. I., Rausch, R., 2011: Large-scale modeling of groundwater resources on the Arabian platform. – International Journal of Water Resources and Arid Environment **1**(1): 38–47.

Riegger, J. & Güntner, A., 2009: Time variation in hydrology and gravity. – Earth, Moon, and Planets **94**: 41–55.

Riegger, J. & Tourian, M.J., 2011: Analysis of GRACE uncertainties by hydrological and hydro-meteorological observations. – J. Geodynamics, submitted.

Rodell, M. & Famigietti, J., 1999: Detectability of variations in continental water storage from satellite observation of time dependent gravity field. – Water Resour. **35**: 2705–2723.

Swenson, S. & Wahr, J., 2006: Post-processing removal of correlated errors in GRACE data. – Geophysical Research Letters **33**: L08402

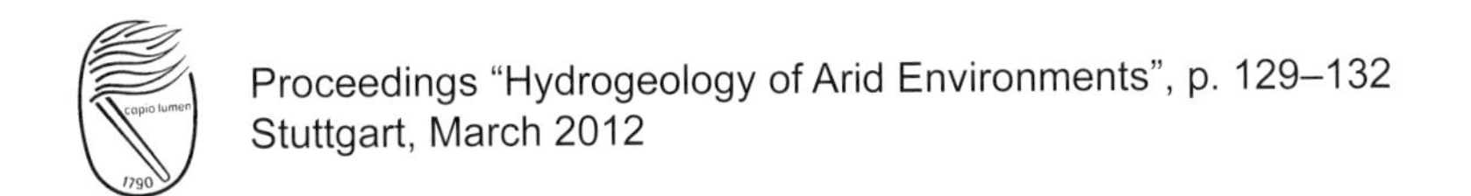

Proceedings "Hydrogeology of Arid Environments", p. 129–132
Stuttgart, March 2012

Extended Abstract

Identification of Potential Groundwater Recharge Using 3D-Spatial Soil Moisture Observations in the Ad-Dahna Desert, Kingdom of Saudi Arabia

Tino Roediger[1], Andreas Meier[1], Christian Siebert[1], Franz Königer[2], Andreas Kallioras[3, 4], Peter Forestier[5], Tobias Fuest[5], Christoph Schüth[3], Randolf Rausch[5], Mohammed Al-Saud[6], Peter Dietrich[7]

[1] Helmholtz-Centre for Environmental Research UFZ, Department Catchment Hydrology, Theodor-Lieser-Str. 4, 06120 Halle (Saale), Germany
[2] Karlsruhe Institute of Technology KIT, Competence Center for Material Moisture (CMM), Institute of Functional Interfaces, Hermann-von-Helmholtz-Platz 1, 76344 Eggenstein-Leopoldshafen, Germany
[3] Technical University of Darmstadt, Institute of Applied Geosciences, Schnittspahnstraße 9, 64287 Darmstadt, Germany
[4] National Technical University of Athens, School of Mining & Metallurgical Engineering, Laboratory of Engineering Geology & Hydrogeology, Heroon Polytechniou Str. 9, 15780 Zografou, Athens, Greece
[5] GIZ-International Services, Riyadh Office, P.O. Box 2730, Riyadh 11461, Kingdom of Saudi Arabia
[6] Ministry of Water & Electricity, Airport Road, Riyadh 11195, Kingdom of Saudi Arabia

Key words: vadose zone, infiltration, groundwater recharge, TDR, tensiometer

Abstract

The Ad-Dahna desert is one of the driest places on the Arabian Peninsula, receiving 90-mm/a precipitations with approximately 3,000 mm/a of potential evaporation. The hydrological processes – such as pore water fluxes within the unsaturated zone –, which take place in desert regions, are considered key issues for understanding and quantifying groundwater recharge. For the purpose of directly measuring and monitoring the water content within the unsaturated column over space and time in a sand dunes area, an experimental field site with a tailored scientific infrastructure was installed in November 2010. The test field was equipped with Time Domain Reflectometry (TDR) sensors, temperature probes, suction cups and climate station. First results provide evidences that the seasonal fluctuations of moisture content profile within the unsaturated zone are well correlated with the local climatic conditions. More specifically, results from one TDR-sensor show that precipitation higher than 6 mm/d can potentially lead to relatively fast downward water infiltration.

Introduction

In arid regions, the available water resources are strongly limited by the extreme climatic conditions. The region of the Middle East is characterized by an extreme growth of population in combination with growing groundwater consumption for agricultural, industrial and domestic use. All aforementioned conditions lead to overexploitation of the already limited groundwater resources. Especially in Saudi Arabia most of the resources are summarized as mega-aquifer systems, which were recharged during wetter conditions some thousand years ago (Hoetzel & Zoetl 1978). Groundwater recharge is the only inflow source of these systems. Therefore, a precise quantification of the groundwater recharge is essential for the management of the limited groundwater resources in these regions. The process of groundwater recharge is based on the local morphological and lithological parameters of the catchment area, which mainly control the infiltration process within the unsaturated zone. In arid regions, different approaches such as those of Dincer (1974)

www.borntraeger-cramer.de
2012/0129/$ 1.00

and Dahan (2006) were used for the investigation of this phenomenon. Therefore a combination of direct push technology and diagonal borehole instrumentation together with TDR (time domain reflectometry) technology were linked in a sand dunes area in Saudi Arabia. During drilling campaign soil cores were retrieved to extract interstitial water for analyzing pore water content, concentration of inert elements and isotope signatures.

Monitoring Site

The monitoring field is located about 70 km SW of Riyadh in the outcrop area of the Wasia-Biyadh aquifer. This aquifer is the lower most section of the Upper Mega Aquifer system (consisting of Biyadh-, Wasia-, Umm Er Radhuma- and Dammam Formation), which builds up the major part of the Arabian Peninsula. In the area of the observation site Wasia-Biyadh aquifer is covered by more than 20m thick Ad-Dahna sand dunes, consisting of medium to fine grained sand. During field works in November 2010 a 20×10m experimental field site with tailored scientific infrastructure was constructed, It consist of 7 vertical direct push boreholes (down to 11 m) and 1 diagonal borehole (total length of 22 m), reaching a depth of 15 m below surface. The direct push boreholes contain TDR sensors, specially developed for the purpose of this research by the TU Darmstadt (Schüth et al. 2011). In the diagonal borehole a flexible ribbon cable is used as TDR sensor, developed by the Karlsruhe Institute of Technology. The TDR technology is complemented with the installation of 6 suction caps in a logarithmic vertical distribution from 0.5 to 8 m depths are installed to collect and analyze the infiltrated pore water. The experimental field site was additionally equipped with a climate station and an artificial rain system; the latter used to conduct and control artificial rainfall events for the investigation of water fluxes under fully controlled rainfall conditions. Field works concerning hydrochemistry, laboratory experiments and remote sensing studies complete the first steps of this research. These methods are used to identify the initial conditions of the monitoring field.

Methods

The TDR method is based on measuring electromagnetic wave travel times and was originally developed to detect breaks or shorts in electrical conductors. Since the early 1980s the method was successively developed to collect sufficient data, which were used for the estimation of soil moisture (Topp et al. 1984). TDR is an indirect geophysical technique. The principle based on the propagation velocities of an electromagnetic impulse along a coaxial cable and enters the TDR probe, which is influenced by the dielectric constant (ε) of a soil mixture (soil matrix, soil water and air) surrounding the sensors. The dielectric constant is a dimensionless ratio of a material dielectric permittivity to the permittivity of space (Stacheder et al. 2009). The TDR system measures the bulk dielectric properties of the soil mixture and infers the volumetric water content (soil moisture) from the ratio between the high dielectric constant of the water (~80), the low dielectric constant of most sediment materials (~2–7) and the very low dielectric constant of air (~1). A summary on TDR-method can be found in Dirksen (1999), Robinson et al. (2003) and Hübner et al. (2007).

Results

Meier (2011) correlates the presented data collected from one of the TDR sensors, logged between March and July 2011 with different natural precipitation events (Fig. 1) ranged between 0 and 6 mm/d (observed mostly during spring). Spring season is characterized by high variations of daily temperatures, ranging from 15 °C to 35 °C, while daily temperature increases of more than 40 °C were recorded during summer season. The calculated volumetric water content (VWC) along the 6 m TDR sensors is shown in Fig. 1.

Appropriate colour variation was used (red – dry ~0% and blue – wet ~30%) representing the different ranges of the calculated volumetric water content. The electromagnetic signals along the TDR sensors were calibrated and converted in VWC (Meier 2011) by a modified Topp et al. (1984) equation (Eq. 1), in which Θ represents the volumetric water content and K_a the apparent dielectric constant:

Eq. 1: $$\Theta_{vol}\% = -3.2823K_a^2 + 26{,}589K_a - 30{,}253$$

The results of the soil moisture profile readings show a distinct fluctuation of low soil moisture (<10%) down to 1 m depths within the measured time period. Below that point, the soil moisture increases continuously and reaches volumetric water contents between 18% and 25% until a depth of approximately 6 m. These results confirm the measurements of Dincer (1974) and prove that sand dunes in arid regions are not dry in depths below 1 m. Furthermore it is observable that the soil moisture generally increases during the period from March to May 2011. In the beginning of June 2011, the trend changes and the soil moisture decreases, reflecting the initiation of strong evaporation effects within the upper

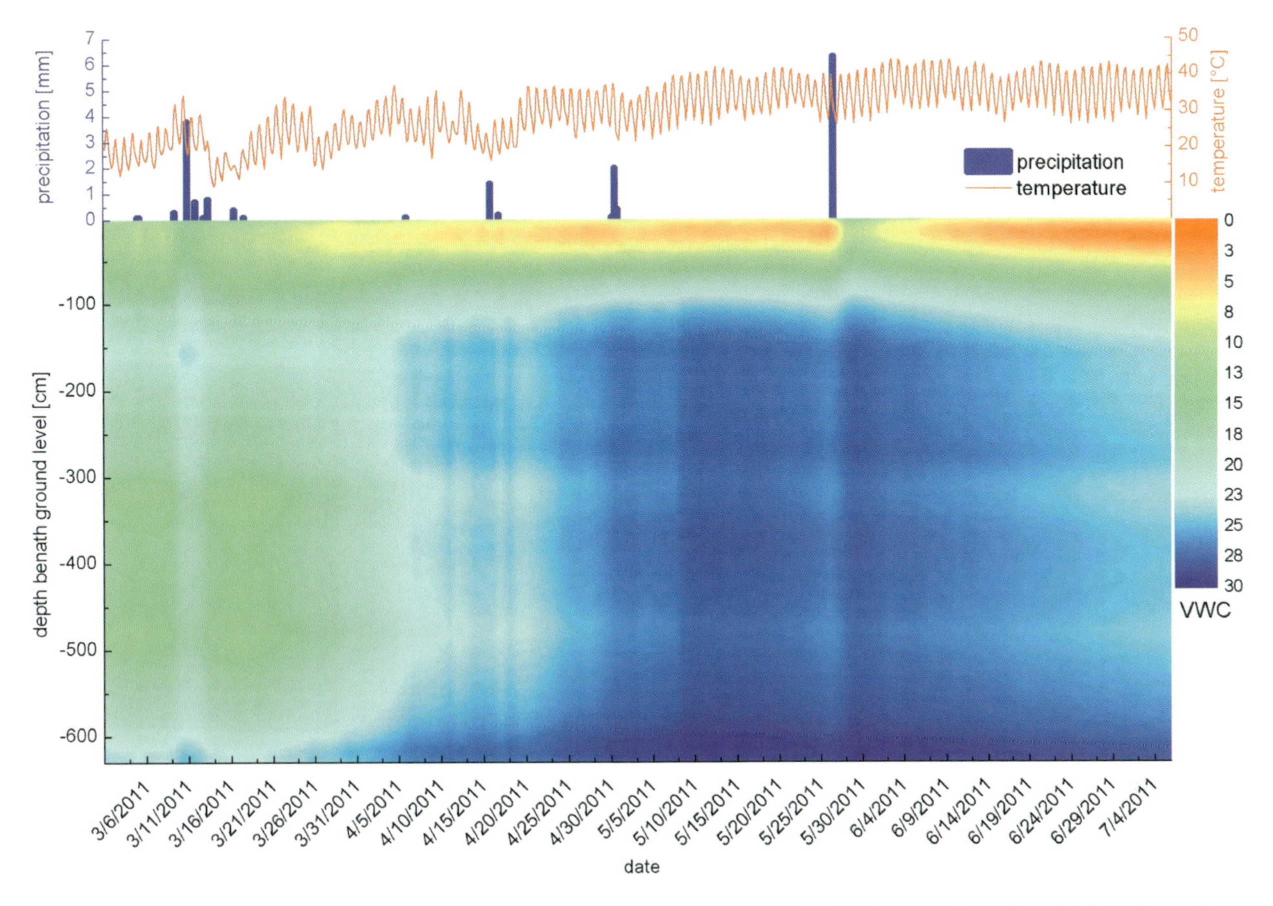

Figure 1: First results – climate data and sand moisture content of the observation field in the Ad-Dahna sand dunes, Saudi Arabia from March until July 2011.

parts of the unsaturated zone. Preliminary correlations between precipitation and VWC evidence that rain events of less 4 mm/d can potentially lead to a slight increase of the moisture content within the first 0.5m of soil. Precipitation rates higher than 6 mm/d may result in relatively fast downward water infiltration within the unsaturated column. Specific rainfall events, which occurred on the 12th of March and 30th of May led to distinct increase in moisture content over the total sensor length of 6m. Continuous analysis of all TDR sensors will contribute to the understanding of these processes.

It is envisaged that the combination of all used methods will allow a continuous and high resolution monitoring of infiltration processes within the unsaturated zone, aiming to the precise determination of groundwater recharge in arid regions.

Acknowledgements

The authors would like to acknowledge the cooperation between Helmholtz-Centre for Environmental Research-UFZ (Leipzig-Halle, Germany); Technical University of Darmstadt (Germany); GIZ-IS/Dornier Consulting (Riyadh Office, Kingdom of Saudi Arabia) and the Ministry of Water & Electricity (Kingdom of Saudi Arabia); within the framework of the German Federal Ministry of Education and Research (BMBF) funded research program IWAS (www.iwas-sachsen.ufz.de).

References

Dahan, O., Rimon, Y., Tatarsky, B. & Talby, R., 2006: Deep Vadose Zone Monitoring System. – Proc. TDR 2006, Purdue University, West Lafayette, USA, Sept. 2006, Paper ID 46, 15 p.

Dincer, T., Al-Mugrin, A. & Zimmermann, U., 1974: Study of the infiltration and recharge through the sand dunes in arid zones with special reference to the stable isotopes and thermonuclear tritium. – Journal of Hydrology **23**: 79–109.

Dirksen, C., 1999: Soil physics measurements. – Catena, Reiskirchen, Germany, 154 pp.

Hübner, C., Schlaeger, S. & Kupfer, K., 2007: Ortsauflösende Feuchtemessung mit Time-Domain-Reflektometrie. – Technisches Messen. **5**: 316–325.

Hoetzl, H. & Zoetl, J.G., 1978: Climatic changes during the Quaternary period. – In: Al-Sayari, S. S. & Zoetl, J. G. (eds.): Quaternary Period in Saudi Arabia. Springer-Verlag, Vienna vol. 2, 301–311.

Meier, A., 2011: Estimation of water infiltration into dune sands of Saudi Arabia by using large scale TDR Sensors. –- Technical University of Dresden, Institute of Geography, Diploma thesis (unpublished), 73 p.

Topp, G. C., Davis, J. L. & Annan, A. P., 1980: Electromagnetic determination of soil watercontent: measurements in coaxial transmission lines. – Water Resources Research **16**: 574–582.

Robinson, D. A., Jones, S. B., Wraith, J. M., Or, D. & Friedman, S. P., 2003: A review of advances in dielectric and electrical conductivity measurements in soils using time domain reflectometry. – Vadose Zone Journal **2**: 444–475.

Schüth, C., Kallioras, A., Piepenbrink, M., Pfletschinger, H., Al-Ajmi H., Engelhardt I., Rausch, R. & Al-Saud, M., 2011: New Approaches to quantify groundwater recharge in arid areas. – International Journal of Water Resources and Arid Environments, **1(1)**: 33–37.

Stacheder, M., Koeniger, F. & Schuhmann, R., 2009: New dielectric sensors and sensing techniques for soil and snow moisture measurements. – Sensors **9**: 2951–2967.

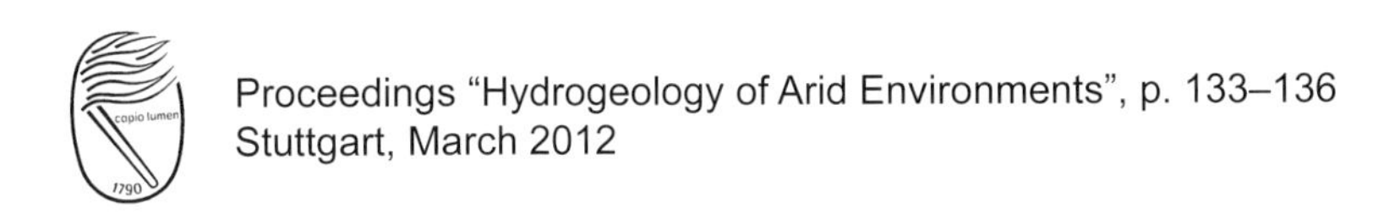
Proceedings "Hydrogeology of Arid Environments", p. 133–136
Stutgart, March 2012

Extended Abstract

Groundwater Modeling in Al Malih Basin, Jordan Valley

Raghid Sabri[1], Marwan Ghanem[2], Maher Abu Madi[3], Broder Merkel[1]

[1] Hydrogeology Department, TU Freiberg, Gustav Zeuner Str.12, D-09596 Freiberg, Germany, email: sabri@student.tu-freiberg.de, merkel@geo.tu-freiberg.de
[2] Geography Department Birzeit University, P.O. Box 14, Birzeit, Palestine, mghanem@birzeit.edu
[3] Institute of Environmental and Water Studies, Birzeit University, P.O. Box 14, Birzeit, Palestine, abumadi@birzeit.edu

Key words: Al Malih basin, groundwater modeling, eastern aquifer, Palestine

Introduction

Most of the available water resources in Palestine are groundwater. The groundwater consists mainly of three aquifers, the western aquifer, the northeastern aquifer and the eastern aquifer. The safe yield from the eastern aquifer is estimated from recharge calculations to be 55–65 million cubic meter, whereas the real extraction is 165 million cubic meter (Palestinian Water Authority 2009). Most of the groundwater aquifers in the Jordan valley are fully or over-exploited (Haddad 2009). The western and north eastern aquifers have two basins, both are considered to be transboundary aquifers in a semi arid region.

On the other hand, the eastern aquifer has six basins including Al Malih basin. The discharges from these basins are towards the Jordan River (Toll et al. 2009). This aquifer occurs in the Upper Cretaceous karstic limestone and dolomite formation (Anker et al. 2009).

This study intended to quantify the potentiality of Al Malih basin, and it aimed as well to test the effect of the groundwater basin on the Jordan River: The basin hypothetically is contributing water to the Jordan River and the model shows how much the utilization of water will affect the discharge to the Jordan River.

Material and Methods

Study Area

Al Malih basin is located in the northeastern of the West Bank. It lies within the eastern aquifer with an area estimated to be 140 km^2. The eastern boundary of the basin is the Jordan River, the northwestern boundary is the groundwater divide between the northeastern aquifer and the eastern aquifer. Al Malih basin lies within the Jordan Valley and exhibits altitudes between 600 masl in the west and –412 masl in the east.

The average annual temperature is between 18.1 °C and 19.4 °C and the humidity ranges between 58.2–58.6%. The average wind speeds is 5 km/hr, the annual evaporation is 973 mm (PCBS 2006), and the average rainfall is 400 mm/year decreasing significantly from west to east. The geological formation of the area is shown in Fig. 1.

Ground Water Modeling

All groundwater modeling parameters, such as transmissitvity, hydraulic conductivity, and aquifer depths were obtained from the literature. The input data are shown in table 1.

These data along with the topography map were used to build the model based on a uniform rectangular grid with 50*50 m cell size. Two layers were assigned with a thickness of 200 m for the first layer and 300 m for the second layer. The boundary conditions of the basin were defined by adding a river boundary and constant head boundaries.

Results and Discussion

Visual Modflow

Running the steady state model over 30 years showed that the recharge (7.9 million m^3/year) is

www.borntraeger-cramer.de
2012/0133/$ 1.00

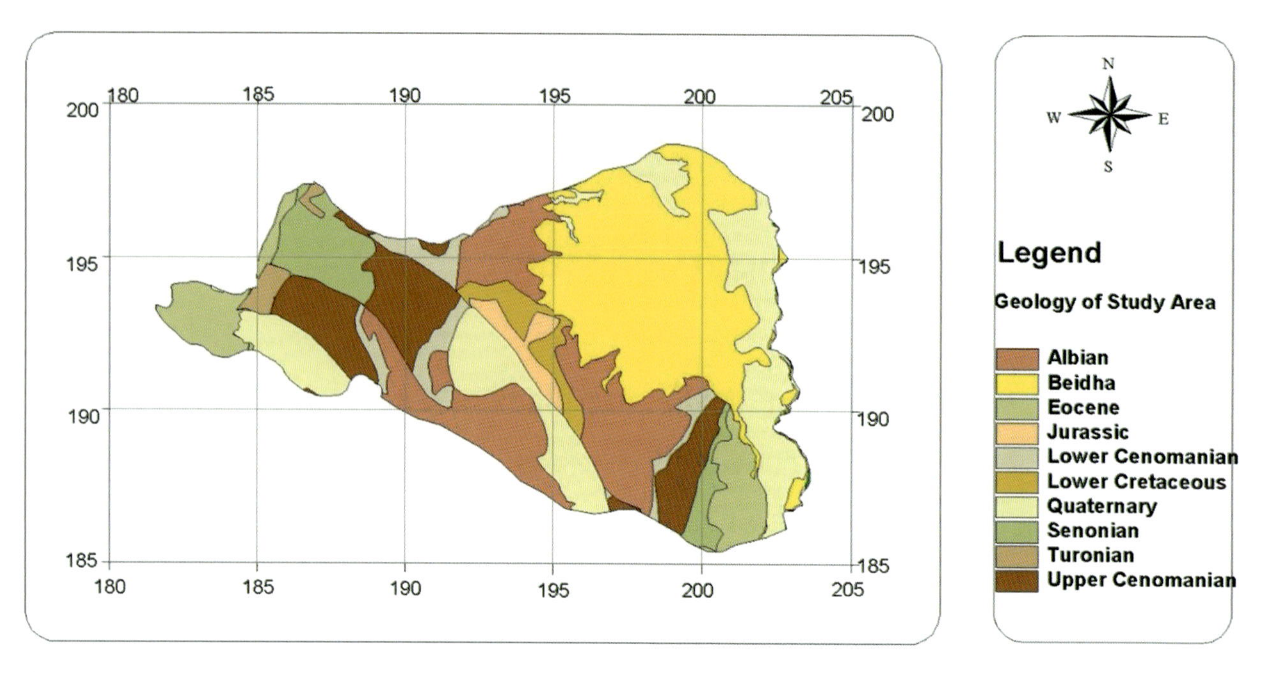

Figure 1: Geological formation of the study area.

much higher than discharge from the springs and pumping wells (2.8 million m³/year) regardless the leakage towards the Jordan River. This showed that the Jordan River is not affected by the discharge of pumping wells and springs.

The output from the Modflow model showed that the highest water head in the basin is in the west side of it and reaches 544 m. The value tends to decrease towards the east and reaches the lowest value along the Jordan River with 100 m. The flow direction of the groundwater in the basin is toward the river from the west to the east as shown in Fig. 2. The maximum iterations number was 50, and the resulted maximum residual was –17.2.

Table 1: Input Data.

Input data	Value
Transmisitivy	910 m²/day (Hurwitz, Stanislavsky et al. 2000)
Kv	1.7 m/year (SUSMAQ 2001)
kH	5 m/year (SUSMAQ 2001)
Kv	1.3 m/year (Hurwitz, Stanislavsky et al. 2000)
kH	40 m/year (Hurwitz, Stanislavsky et al. 2000)
Recharge	0.2 mm/day (Hughes A. & Mansour M. 2005)
Spring discharge	2.698 million m³/year (PWA 2006)
Well abstraction	1.7958 million m³/year (PWA 2006)
Recharge Wadi Al Malih	15 million m³/year

Water Quality

Comparing the spring's quality data in the basin with rainfall data showed a correlation between the rainfall amount and chloride concentration. The Pearson correlation p = 0.4 had a significance level of 0.01. This result match the quick response of the studied springs to precipitation change in terms of quantity (Guttman 2009). The increase of chloride concentration with the increase of precipitation is due to the increase of sea spray from the Mediterranean Sea with the increase of rainfall (Rosenthal et al. 2009).

On the other hand, there is no correlation between rainfall and sulfate. The sulfate concentration in the basin changes with respect to geological formations with the highest sulfate concentration found in Hammam Malih spring within the Jurassic formation Fig. 3.

Conclusions and Recommendations

The numerical results show that the Jordan River is not affected by both the pumpage in the wells and discharge of the springs, which is plausible, because the river leakage is much higher than the discharge value. On the other hand; the river is highly affected by the quality of the groundwater in Al Malih basin.

Jordan River flow rate is at a very low level and the discharge from the springs will not affect the groundwater leakage into the river. So the flow of the Jordan River can only be rebuilt by lowering

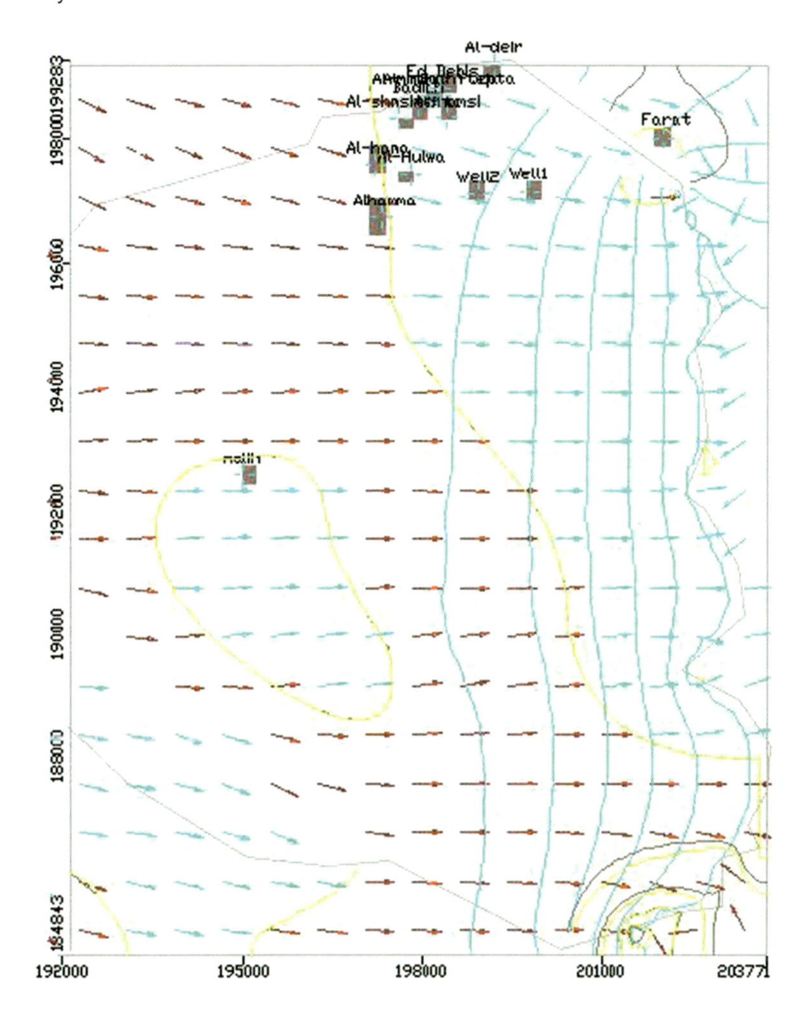

Figure 2: Flow direction.

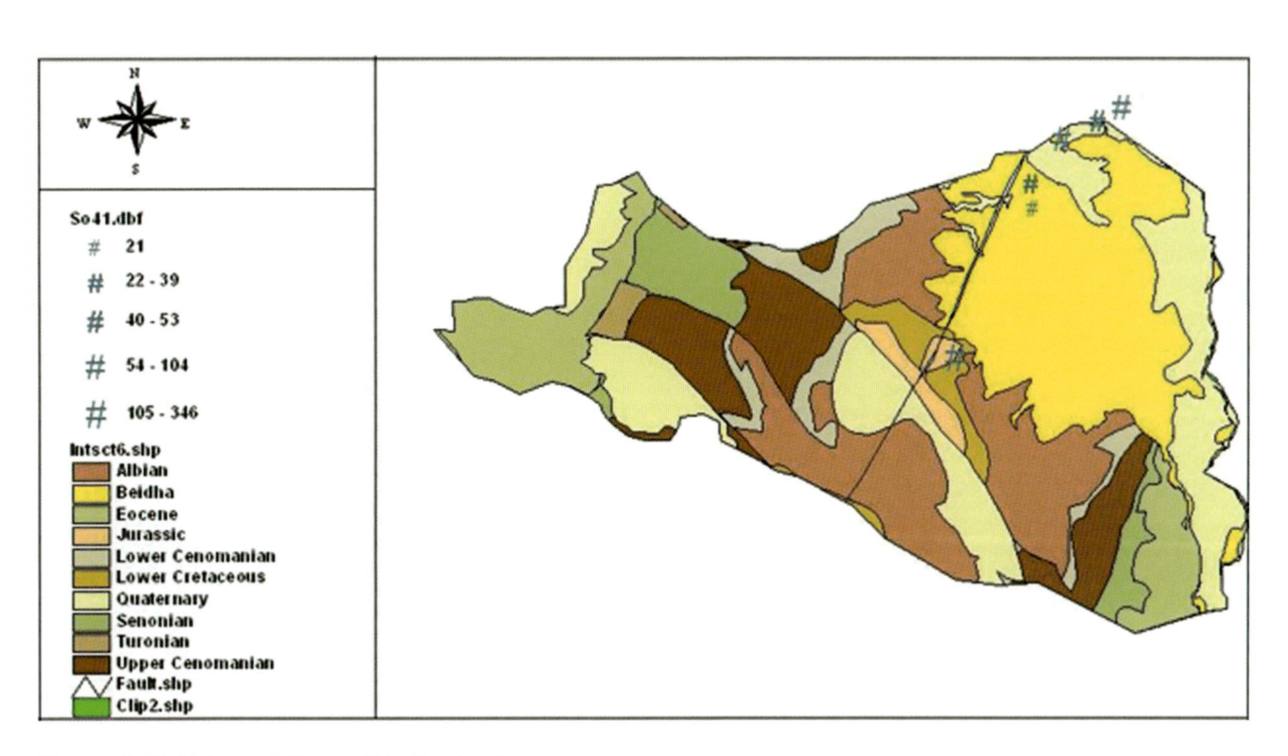

Figure 3: Sulfate variation within the basin.

diverting water from the Lake Tiberias and Yarmok River. Artificial recharge is not recommended in this case; because of the groundwater leakage into the Jordan River.

References

Anker, Y., Shulman, H., et al., 2009: Specific regional hydrological aspects. – In: Hötzl, H., Möller, P. & Rosenthal, E.: The Water of the Jordan Valley, Springer Berlin Heidelberg, 181–198.

Guttman, J., 2009: Hydrogeology. – In: H. Hötzl, P. Möller and E. Rosenthal The Water of the Jordan Valley, Springer, Berlin, Heidelberg, 55-74.

Haddad, M., 2009: Groundwater Management. – In: Hötzl, H., Möller, P. & Rosenthal, E.: The Water of the Jordan Valley, Springer, Berlin, Heidelberg, 465–472.

Palestinian Water Authority, 2009: Water Budget for the West Bank. – Palestine 2009.

PCBS, 2006: Meteorological Conditions in the Palestinian Territory Annual Report 2005. – Palestinian Central Bureau of Statistics. Ramallah, Palestine.

Rosenthal, E., Flexer, A., et al., 2009: The hydrochemical history of the Rift. – In: Hötzl, H., Möller, P. & Rosenthal, E.: The Water of the Jordan Valley, Springer, Berlin, Heidelberg, 75–82.

Toll, M., Messerschmid C., et al., 2009: Aquifers in the western Jordan Valley. – Hötzl, H., Möller, P. & Rosenthal, E.: The Water of the Jordan Valley, Springer, Berlin, Heidelberg, 265–286.

Hughes A. & Mansour M., 2005: Recharge modeling for the West Bank aquifer. Nottingham, British Geological Survey: 90.

Hurwitz, S., E. Stanislavsky, et al., 2000: Transient groundwater-lake interactions in a continental rift: Sea of Galilee, Israel. Bulletin of the Geological Society of America **112**(11): 1694–1702.

PWA (2006). Springs and Well Discharge. Ramallah.

SUSMAQ (2001). West Bank Aquifer conceptual Recharge Estimation, Water resource planning and development, House of Water and Environment.

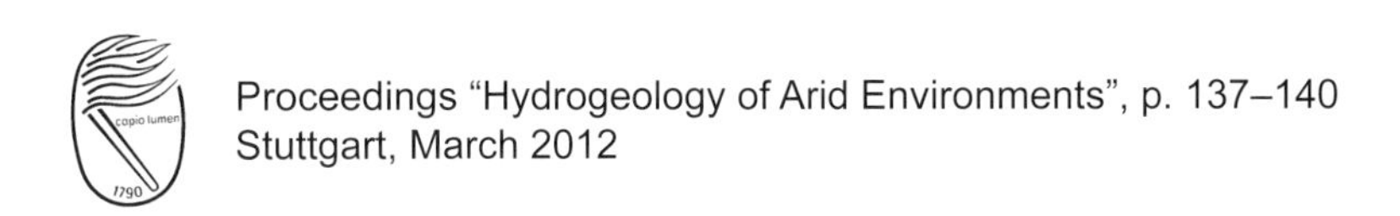

Proceedings "Hydrogeology of Arid Environments", p. 137–140
Stuttgart, March 2012

Extended Abstract

Modeling of Seawater Intrusion Due to Climate Change Impacts in North Gaza Coastal Aquifer Using SEAWAT

Reem Fathi Sarsak[1], Mohammad N. Almasri[2]

[1] Environmental Health Department, UNRWA, Nablus, Palestine, P.O. Box 592, email: r.najjar2@unrwa.org, reem2525@yahoo.com
[2] College of Engineering, Department of Civil Engineering, An-Najah National University, P.O Box 7, Nablus, Palestine

Key words: Chloride, Groundwater, Contamination, Intrusion, SEAWAT, Palestine

1. Introduction

Seawater intrusion is a common contamination problem in coastal aquifers. It takes place in dense population and urban development areas that require intense exploitation of groundwater. The Mediterranean coast is a good example of such a problem (Cameo 2006).

The development and management of fresh groundwater resources in coastal aquifers are seriously constrained by the presence of seawater intrusion. Over the years, many models have been developed to simulate and study the problems related to seawater intrusion. Numerical models provide effective tool to understand groundwater problems. This research presents simulation of seawater intrusion in North Gaza coastal aquifer due to climate change impacts using SEAWAT.

There are many studies of groundwater flow models to help understand and predict the behavior of fresh and saline groundwater under a certain type of exploitation. These studies were important to the management of groundwater. Seawater intrusion problems have been solved by using different methods, ranging from the basic Ghyben-Herzberg principle with the sharp interface models to the more sophisticated theories with the solute transport models such as SEAWAT, which takes into account variable densities (Thuan 2004).

This research deals with the groundwater resources assessment and future forecasting under various development scenarios of North Gaza Coastal Aquifer.

2. SEAWAT Model Development

The basic groundwater flow model for this study was developed based on MODFLOW. The model domain originally consists of 78 rows, 88 columns and one layer. The horizontal grid is divided into uniform cells with size of 200x200 m.

For modeling purposes, the hydrologic year consists of a winter season from October to March and a summer season from April to September.

The SEAWAT model was developed based on the basic groundwater flow model and later used to simulate the seawater intrusion.

2.1 Recharge Components

The net recharge for the study area is comprised of recharge from rain, irrigation return flow, water networks losses, wastewater leakages, existing treatment plants and recharges basins, and recharge from treated wastewater irrigation in the Israeli side of the model domain. A total of 27 recharge zones were considered for the MODFLOW input. Each zone carries a different value based on the annual and seasonal recharge values.

2.2 Abstraction Wells

Within the model area, there are 1,076 wells. This includes 52 domestic wells while the remainders are agricultural wells. 22 wells were selected as head observation wells for the model regional calibration. The selection was based on the availability of good hydrographs for these wells.

www.borntraeger-cramer.de
2012/0137/$ 1.00

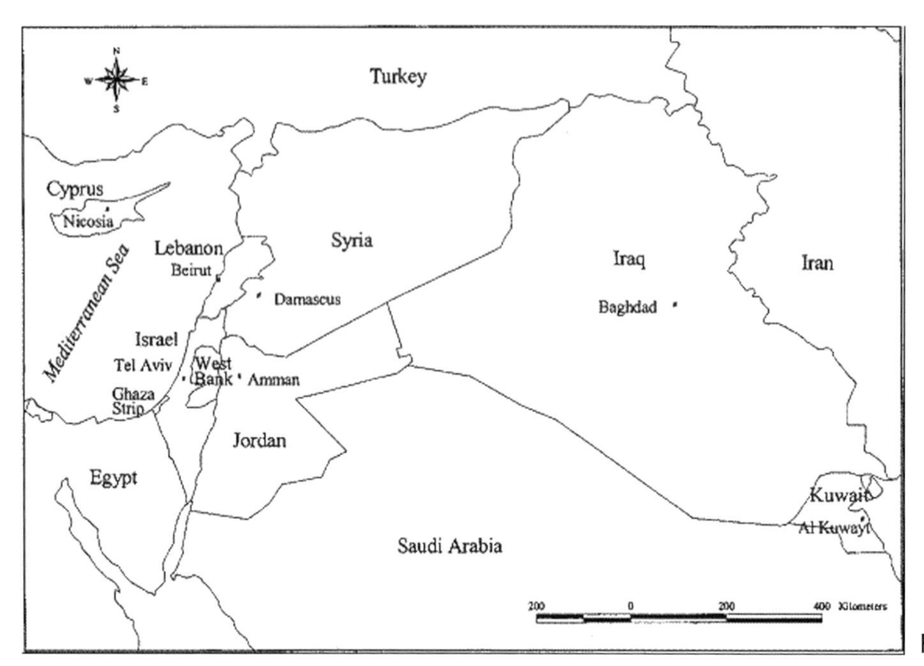

Figure 1: Map of Gaza Strip.

2.3 Simulation Period for SEAWAT Model

In order to evaluate the effects of climate change and pumping on seawater intrusion, the simulation period for SEAWAT model covers 35 years (from 2000 until 2035).

2.4 Boundary Conditions

Constant head cells were assigned along the sea line to the west and the deepest two layers to the east. Initial concentrations were assigned to the model according to the following assumptions:

- Constant concentration (western boundary) is 35,000 mg/l
- Constant concentration (eastern boundary, layers 9 and 10) is 700 mg/l
- Initial chloride concentration =102.5 mg/l (equivalent to 0.1 g/kg) everywhere except at the specific locations of domestic wells which are exceeding this value.
- Specific chloride concentrations assigned to each domestic well location to reflect the actual conditions.
- Recharge concentration was neglected and considered 0 mg/l since the main scope of the work concentrates on salinity from seawater intrusion.

3. Model Application

The study area was determined based on the MODFLOW model which was prepared in 2006 by PWA (Palestinian Water Authority) and EMCC (Engineering and Management Consulting Center) to study the Environmental Assessment for North Gaza Emergency Sewage Treatment Plant Project. This model was used in developing the SEAWAT model since preparing a new model again will be time consuming and is beyond the scope of this research.

3.1 Description of the study area

Gaza Strip is located between longitudes 31 and 25 N, and latitudes between 34 and 20 E. It is a coastal area located along the eastern Mediterranean Sea (see Figure 1). Because of its geographical location, Gaza Strip forms a transitional zone between the semi-humid coastal zone in the north, the semi-arid loess plains in the east, and the arid Sinai Desert in Egypt (PWA 2001).

The study area is the most populated area in Gaza Strip. Presently there are about 840,000 people living over 109 km^2 (about 7,700 capita/ km^2). The population in the study area is expected to increase to more than 1.7 million by 2035 based on PCBS expected population growth rate 3%.

As in Gaza Strip the summers are dry with a short mild rainy season. The mean temperature varies between 12–14 °C in January to 26–28 °C in June. The average annual rainfall is about 250 mm/yr, so the recharge volume is about 27 MCM/yr based on 2009/2010 rainfall data. The average annual potential evaporation is about 1400 mm/yr (SWIMED 2002).

3.2 A Brief Overview of Water Quality in the Study Area

The major water quality problems in the study area are the high salinity and high nitrate concentrations. There are many reasons for this deterioration in water quality, such as increase in population and urban expansion. In addition to improper hydrological and environmental management conditions that represented by high density of wells with high rates of abstraction, intensive agriculture activities, untreated wastewater return flow from septic tanks and networks leakage and inappropriate design of wastewater treatment plant. All these conditions beside the lateral inflow of brackish groundwater from the east, leads to accelerate salinization of this coastal aquifer with chloride. The WHO drinking chloride limit is 250 mg/l, observed concentrations shows that about one half of the wells at the study area exceeded the maximum limit.

3.3 Studied Scenarios

The reference scenario considered herein is Scenario 1. The studied scenarios are in general based on the IPCC (Intergovernmental Panel on Climate Change) projections (Table 1) for the Mediterranean coast for the next 25 years along with the PWA recommendations and projections (especially for scenario 6).

The reference scenario is based on the following assumptions:

- Current pumping rate (Q) for year 2010 is 91.7 MCM (28.9 MCM for agriculture, and 62.8 MCM for domestic).
- Current annual recharge rate (R) is 27.7 MCM
- Current sea level rise (S) is zero

In order to study the sensitivity of the aquifer for these three impacts, specialized scenarios were simulated for each impact.

Table 1: Climate change projections for Gaza Strip (Source, IPCC, 2007)

Indicator	Description	Magnitude
Temperature	Increase	4° to 6 °C
Precipitation	Decrease	–10% to –30%
Evapotranspiration	Increase	10%
Winter rains	Delay	--
Rain intensity	Increase	--
Rainy season	Shortened	--
Seasonal temperature variability	Greater	--
Sea level rise	Increase	1.8–5.9 mm/yr

Scenario 1: Existing conditions: **(reference scenario)**: continue pumping at the current rate with no consideration of climate change in Q, R and S.

Scenario 2: Sensitivity to pumping: take a range for changing pumping rates between -30% and +30% with no consideration of climate change.

Scenario 3: Impact of sea level rise: take the maximum increase in sea level with the assumption that there is no change in both recharge and pumping rates.

Scenario 4: Sensitivity to recharge: take a range for changing recharge rates between -30% and+30% with no consideration of climate change.

Scenario 5: Extreme impacts of climate change: take the maximum rate of sea level rise and the minimum rate of recharge. No change in pumping rate is considered.

Scenario 6: Same as Sc. 4 but with decreasing pumping: this is due to the reuse of treated wastewater and desalination to cover agricultural and municipal abstraction, respectively.

4. Results and Analysis

4.1 Calibration of the SEAWAT Model

The SEAWAT model was calibrated for chloride concentration by entering the available chloride concentrations that were recorded for specific wells by PWA at year 2000, as the initial concentrations, and then the model was run several times to get the simulated concentrations at year 2009 which was compared with PWA observed concentrations for the same year.

The dispersion coefficient was modified everywhere as needed to reflect the actual situation for salinity inside the model.

As there is no available data for the current sea level rise, this simulation was performed using a present day sea level of zero meters. During the simulations, a maximum sea level rise of 59 cm/100 yr was considered as appropriate (IPCC 2007).

4.2 Analysis outcomes

Table 2 indicates the model analysis results as compared to the reference scenario (Scenario 1):

As a result, seawater intrusion in the study area is very sensitive to recharge decrease as compared to pumping rates increase. As such, the most critical impact on seawater intrusion for the study area was found to be recharge variability which in essence depends on climate change. Therefore, it is recommended to search for new resources such as desalination of seawater and brackish water in ad-

Table 2: Summary of scenarios results by the end of simulation period (year 2035).

Indicator	Sc. 1	Sc. 2 (variable pumping rates)		Sc. 3	Sc. 4 (variable recharge rates)		Sc. 5	Sc. 6
		–30%	+30%		–30%	+30%		
[Cl] extent (m)	4,200	4,000	4,300	4,300	4,500	3,900	4,300	2,900
Seawater intrusion (m/yr)	65	60	70	70	80	50	70	35
[Cl] (±%) at wells compared to Sc. 1	--	–20% to –43%	7% to 24%	0.2% to 0.5%	8% to 20%	–17% to –30%	3% to 8%	–81% to –99%

dition to the reuse of treated wastewater in order to reduce the gap in both domestic and agricultural sectors respectively in case of recharge decrease due to climate change.

5. Summary and Conclusion

We found that the chloride concentrations due to seawater intrusion decrease by increasing the distance from sea shoreline with the involvement of two factors; the first is the location of the well from the sea shoreline, and the second is the depth of the well from the ground surface.

The outcome of Scenario 6 was interesting, since it confirms the potency of PWA management plan, which aims to improve the quantity and quality for groundwater aquifer at Gaza strip by proposing many alternatives to reduce the pumping rates for both municipal and agricultural sectors, and this reduces the in-land seawater intrusion. Seawater intrusion is very sensitive to recharge decrease as compared to the increase in pumping rates. In-land intrusion rate was found to be 80 m/yr as recharge decrease by 30% and 70 m/yr as pumping rate increases by 30%.

As a broad conclusion, PWA must go ahead in implementing the strategic plan for desalination plants for both brackish groundwater and saline seawater to cover the future water demand. Existing wastewater treatment plants must be developed to increase their capacity and efficiency in order to reduce the reliance on the aquifer.

Improving the municipalities' water network system (system efficiency) is a key factor in reducing losses and thus reduces pumping requirements in the end.

Regarding the agricultural sector, it should be managed through efficient use of water, adopting new crop patterns and utilization of alternative water resources (for instance low water quality and treated wastewater).

References

Cameo, E., 2006: Seawater Intrusion in Complex Geological Environments. – Department of Geotechnical Engineering and Geo-Sciences (ETCG), Technical University of Catalonia.

IPCC, 2007: The Fourth Assessment Report (AR4) was released in 2007, Intergovernmental Panel On Climate Change.

PWA, 2001: Hydrogeological data book for the Gaza Strip, Technical report, Palestinian Water Authority, Gaza Strip.

PWA, 2007: Rainfall Data in Gaza Strip, Strategic Planning Directorate, Water information bank department.

PWA, EMCC, 2006: Environmental Assessment, North Gaza Emergency Sewage Treatment Plant Project, Final Report.

SWIMED project, 2002: Sustainable Water Management in Mediterranean coastal aquifers:

Thuan Tran, 2004., Multi-Objective Management of Seawater Intrusion in Groundwater: Optimization under Uncertainty, Wageningen Agricultural University geboren te Can Tho, Viet Nam.

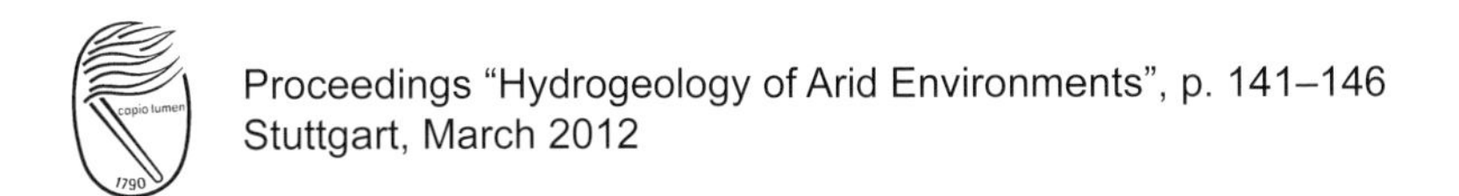
Proceedings "Hydrogeology of Arid Environments", p. 141–146
Stuttgart, March 2012

Extended Abstract

Water Balance for the Aleppo Basin, Syria – Implications of Land Use on Simulated Groundwater Abstraction and Recharge

A. Schlote[1], V. Hennings[2], U. Schäffer[3]

[1] Groundwater Department; [2] Soil Department, [3] Remote Sensing Department, Federal Institute for Geosciences and Natural Resources, Stilleweg 2, 30655 Hannover, Germany, email: Alexander.Schlote@bgr.de

Key words: Aleppo Basin, groundwater, hydrological model, Syria, water balance, WEAP

Introduction

Within the framework of the Syrian – German Technical Cooperation Project "Advisory Service for the Ministry of Irrigation in the Geo-Environmental Sector" Phase II, between the Ministry of Irrigation (MoI, Damascus/Syria) and the Federal Institute for Geosciences and Natural Resources (BGR, Hannover/Germany), a first quantitative water balance estimation for the project area "Aleppo Basin" in northern Syria for the hydrological year 2009/2010 was developed utilizing the Water Evaluation and Planning System Software (WEAP, SEI 2011). The WEAP system includes various options to determine percolation or groundwater recharge rates. For this investigation the simulation module MABIA, based on simplified initial and boundary conditions of the soil water balance was chosen, based on the FAO 56 approach and the dual crop coefficient concept.

Daily climatic data, sufficient with regard to spatial resolution, information on soil hydrologic properties and contemporary and spatially comprehensive land use data including information on irrigation schemes could be obtained from a variety of governmental as well as freely accessible sources. Different verification methods, such as statistical comparisons and correlation and linear regression analyses as well as data processing methods such as Kriging interpolation, linear regionalization and weighting were applied in order to gain reliable input data. Since land use is a crucial factor for groundwater recharge, a special focus was put on the development of a comprehensive land use classification with LANDSAT TM/ETM data on the basis of a time series, which includes information on all-seasonal land use sequences. The land use classification was specially processed with respect to the hydrological year 2009/2010.

The accuracy of the FAO 56 approach in (semi) arid environments has been validated in several studies throughout the world. In some cases, this has even been done on the basis of local lysimeter measurements. In general, its results are showing quite a good model performance. Data from field plots, e.g. from the International Center for Agricultural Research in Dry Areas (ICARDA) near Aleppo where continuously measured soil water contents allow to calculate water flows from soil to groundwater, can serve as a data source for validation. Additionally local WEAP results can be compared to estimates obtained by pedotransfer functions (PTFs) that were developed within a cooperation project between BGR and the Arab Center for the Studies of Arid Zones and Dry Lands (ACSAD) in Damascus.

Looking at the modeling results, the relation between groundwater recharge and groundwater abstraction is leading to a constant depletion of groundwater resources in the area of the Aleppo Basin, a process for which the analysis of available monitoring logger data adds further indications.

Study Area, Data and Methodologies

The project area "Aleppo Basin" covers about 12.371 km². It divides into the two interior surface water catchments of Quaik River (≈7.104 km^2) and Gold River (≈5.267 km^2), draining into terminal

www.borntraeger-cramer.de
2012/0141/$ 1.50

lakes (Sabkhas). Watershed Delineation was conducted on SRTM 90m resolution Digital Elevation Model. Average annual rainfall amounts range from ~500 mm/yr in the north-western area to less than 100 mm/yr in the south-eastern area of the Basin, whereas potential evapotranspiration ranges from ≈1.470 mm/yr in the north-west to ≈1.810 mm/yr in the south-east. Thus, the climatic water balance is strongly negative. Rainfall is mainly concentrated in the winter rain season forming the major fraction of annual precipitation. Consequently, surface runoff mainly occurs during the winter rain season too and rivers fall dry in summer. Dominant land use types are rainfed crops as well as olives, irrigated cotton and vegetables. Irrigation water in the Gold River catchment is supplied from Lake Assad (Euphrates) in the east, transferred through a major freshwater channel that extends up to Aleppo city. In the Quaik River catchment, southwest of Aleppo city, irrigation is supplied from treated wastewater and groundwater. Domestic and industrial sectors represent other major water users in the basin. An extension of the freshwater channel into Southern Aleppo Lands irrigation scheme is currently under construction (Fig. 1).

Hydrogeology in the Aleppo Basin is dominated by a system which consists of an upper aquifer (Helvetian, Neogene; Extends mainly in the west of the Quaik river catchment, groundwater resources are depleted), a middle aquifer (Eocene/Oligocene Limestone, Paleogene) and a deep aquifer (upper Cretaceous Limestone) (Brew et al. 2001). While the Paleogene aquifer yields most of the available fresh water of good quality, water from the Cretaceous aquifer is of very limited use due to its high salinity. Measurements of water levels from the past three years are showing a constant drop of the water table in Paleogene and Cretaceous aquifers.

For parametrization of the WEAP-model using the MABIA-module, preparation of specific datasets was necessary. Those are outlined in Table 1. For exploiting all data at hand and in order to account for intra-basin spatial heterogeneities of hydrological properties, such as climate, soil, land use and irrigation, a sophisticated model structure was developed. The two existing catchments were divided into subunits representing different-dominant subsystems of the catchment's hydrological cycle. Since land use and irrigation are determining hydrological properties which are subject to planning and management decisions, the 21 intra-basin subunits were delineated primarily on the basis of an all-seasonal, generalized land use classification derived from combined LANDSAT TM5/ETM7 data (summer + winter scene) including supplementary information from other sources (crop calendar, irrigation

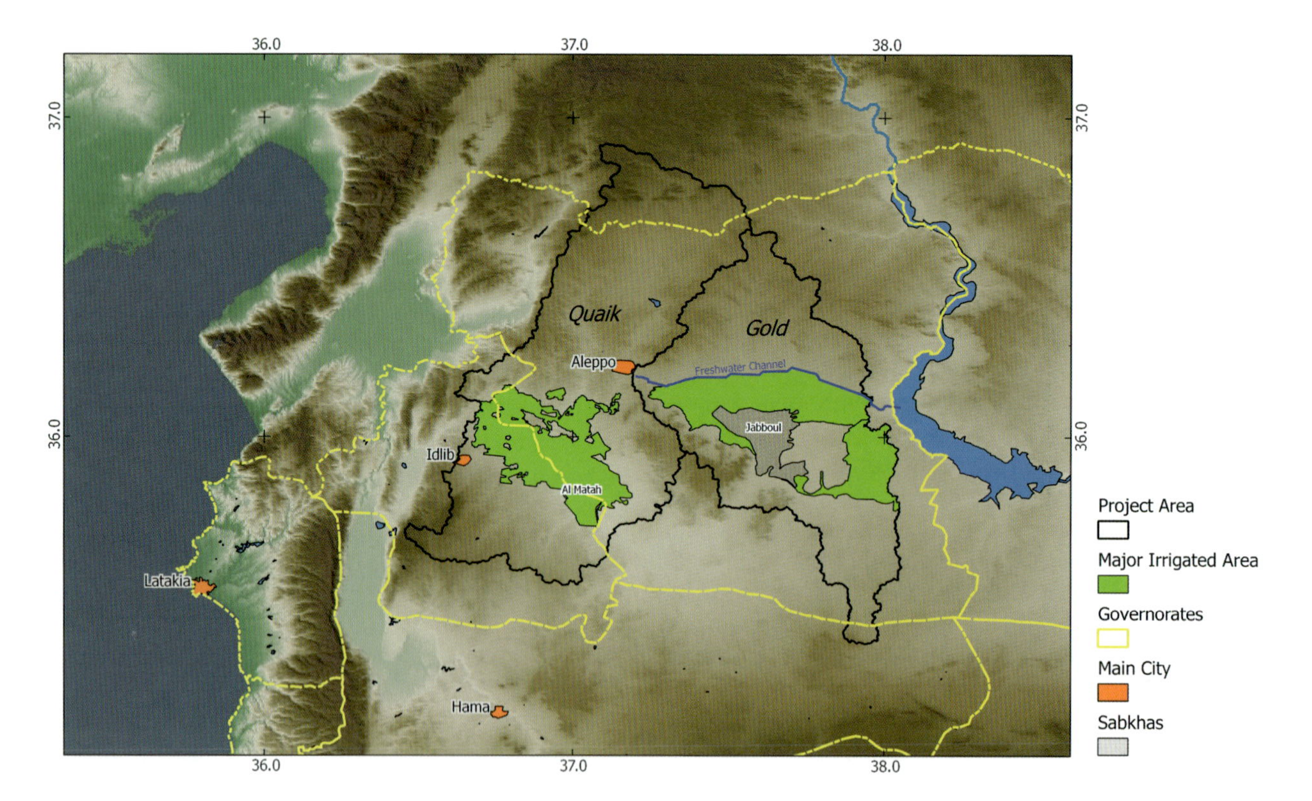

Figure 1: Project Area Aleppo Basin, Irrigation schemes.

Table 1: Datasets for WEAP-Component of Aleppo Basin.

Data category	Source	Preparation Method	Result
Rainfall	– TRMM 3B42 remote sensing product	Verification with available ground measurements	Daily data, 0.25° resolution grid
Potential Evapotranspiration	– Climate station (ICARDA) – CLIMWAT (FAO)	Construction of annual isolines, regionalization of daily station data	Daily regionalized station data
Soil	– Soil maps – ICARDA report on soil resources	Digitalization of soil maps, derivation of AWC* values	AWC values for each soil unit
Land Use	– LANDSAT satellite images – Land use data (MAAR)	Interpretation of satellite images, generalization	Generalized land use classification (raster dataset)
Irrigation	– Land use map (MAAR 2003) – Irrigation area map (MoI 2002)	Delineation of irrigated areas	Shapefiles

*Available water holding capacity

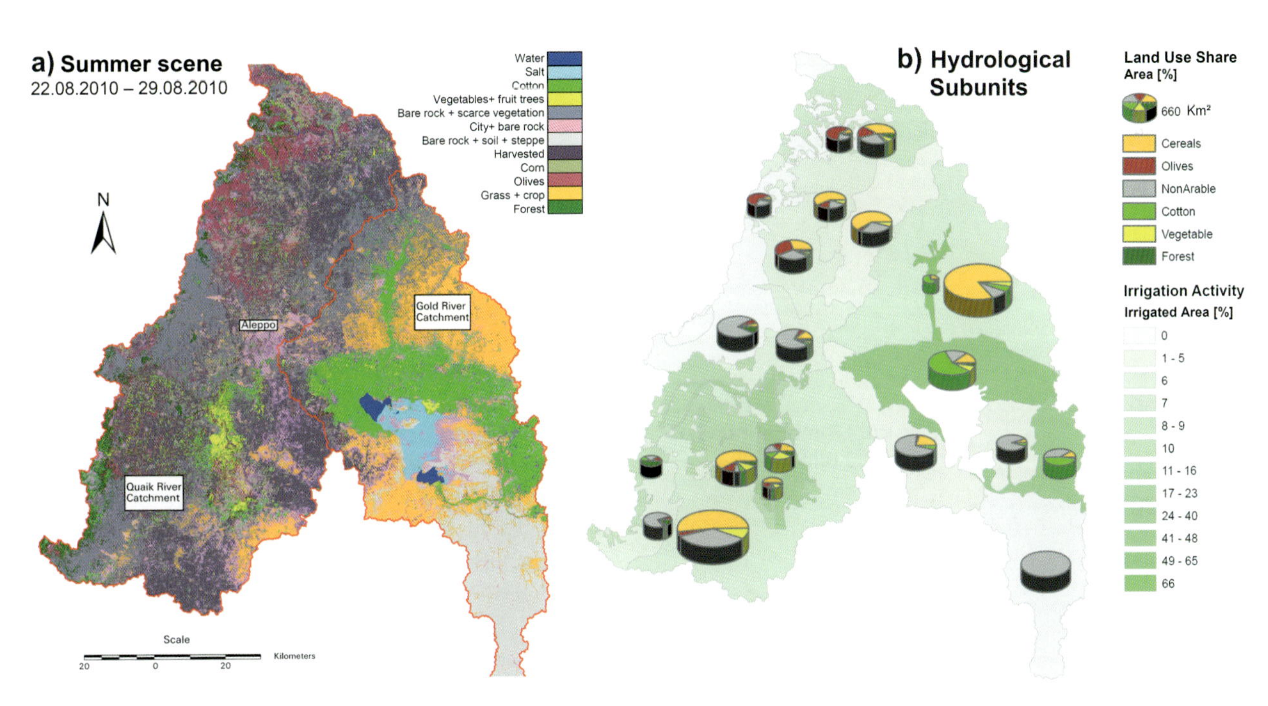

Figure 2: Land Use Classification. a) LANDSAT-TM5 Mosaic (30 x 30m pixels, summer scene example) of the Aleppo Quaik River and Gold River Catchment areas (Schäffer 2011). b) Hydrological subunits based on generalized land use and irrigation criteria.

data). Hydrological properties of land use classes were configured utilizing MABIA's crop library and crop scheduling wizard. For single data values, area weighted averages were calculated (climate, soils). Maximum infiltration rate (water depth that can infiltrate into the soil over a 24 hour period) values for each subunit were taken from WEAP-MABIA Tutorial (Jabloun & Sahli 2011), based on hydrological properties of soil textural classes. Values for available water holding capacity (AWC) were preprocessed and then entered. Water transfer/abstraction takes places through linear WEAP-elements such as transmission links, diversions and return flows.

Results and Discussion

Table 2 shows the annual figures for demand site inflows and outflows calculated by WEAP. Interba-

Table 2: Aleppo Basin Demand Site Inflows and Outflows, Hydrological Year 2009/2010.

Demand Site Inflows and Outflows	Water Volume [Billion m³]
Inflow from Euphrates	1.27
Other Inflow	0.09
Total Interbasin Water Inflow	**1.36**
Inflow from GW_Paleogene	1.10
Inflow from GW_Cretaceous	0.08
Total Groundwater Abstraction	**1.18**
Inflow from Quaik River	0.06
Inflow from Treatment Plant Aleppo[2]	0.03
Total Accessible Internal Water Resources	**0.09**
Outflow to Quaik River	0.50
Outflow to Gold River	0.13
Total Surface Water Runoff	**0.63**
Outflow to Treatment Plant_Aleppo	0.11
Consumption	0.17
Outflow to GW_Paleogene from Catchments	0.06
Outflow to GW_Paleogene from River Channels (Quaik + Gold)	0.37
Total Groundwater Recharge	**0.43**

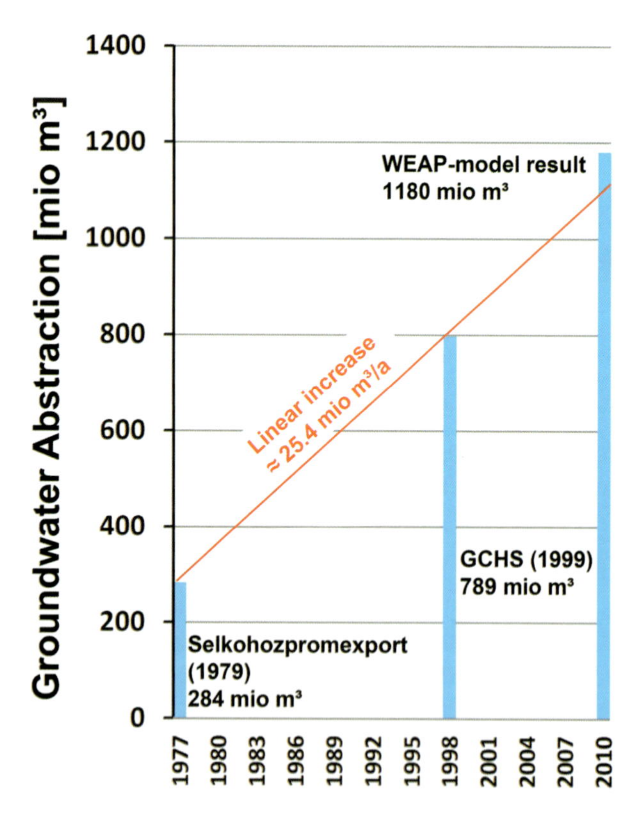

Figure 3: Increase of groundwater abstraction rates, measured and modeled.

sin water transfer covers about half of total water demand in the basin, groundwater abstraction accounts for almost the other half of water demand coverage. Bulk water transfer and groundwater abstraction occurs in summer, bulk surface runoff in the rivers occurs in winter.

Information on total annual groundwater abstraction from historical studies (Selkhozpromexport 1979, GCHS 1999) can be compared to the results of the WEAP-Model, since they were applied to an area which approximately matches the borders of the project area.

Extrapolating the trend forward with an annual increase of 24.5 million m^3 /y, starting in 1998 would lead to a total groundwater abstraction of ≈1.08 Billion m^3 in 2010 (modeled: ≈1.18 Billion m^3, Table 2). Since total water demand and total groundwater abstraction can be considered as relative accurate through verification with different independent sources, magnitudes of total interbasin water inflow and total accessible internal water resources can be considered as relative accurate too on the demand side. Some uncertainties however remain on the supply (-and resources) –side, especially with surface water runoff and groundwater recharge.

WEAP-MABIA modeling results for groundwater recharge (GWR) can be compared to estimates

Table 3: Comparison of annual WEAP-MABIA – GWR results for Quaik River Catchment (Aleppo Basin) and PTF - GWR-estimates, hydrological year 2009/2010.

WEAP- Intra-Basin- Subunit	Precipitation [mm/a]	Potential Evapotranspiration [mm/a]	AWC per soil profile [% of volume]	AWC per soil profile [mm/10 dm]	GWR - Wheat		GWR - Olives	
					WEAP-MABIA [mm/a]	PTF [mm/a]	WEAP-MABIA [mm/a]	PTF [mm/a]
Q_OlivesTurkey	263	1551	6.50	78.00	0	0	0	0
Q_MixedTurkey	258	1558	7.80	93.60	0	0	0	0
Q_Olives	414	1616	7.50	90.00	47	54	55	58
Q_Mixed	235	1601	8.25	99.00	0	0	0	0
Q_RockySteppe2	239	1621	8.60	103.20	0	0	0	0
Q_Rainfed1	282	1628	8.25	99.00	0	0	0	0
Q_RockySteppe1	293	1638	8.55	102.60	0	0	0	0
Q_Urban	144	1694	9.83	117.96	0	0	0	0
Q_IrrigationGW	248	1682	9.83	117.96	0	0	0	0
Q_IrrigationWW	202	1735	9.83	117.96	0	0	0	0
Q_IrrigationMIX	225	1705	9.93	119.16	0	0	0	0
Q_Forest	382	1631	9.20	110.40	8	0	15	0
Q_RockySteppe3	351	1650	8.80	105.60	5	0	12	0
Q_Rainfed2	214	1731	9.00	108.00	0	0	0	0

calculated by pedotransfer functions, which were developed within the technical cooperation project "Management, Protection and Sustainable Use of Groundwater and Soil Resources" between the Arab Center for the Studies of Arid Zones and Dry Lands (ACSAD) in Damascus and BGR. In order to predict mean annual GWR-rates by PTFs, linear models based on precipitation, potential evapotranspiration and available water capacities are adequate. Based on correlation coefficients, mean annual percolation rates can be predicted by PTFs very precisely. The PTFs were developed utilizing data from 44 meteorological stations in Syria and they comprise six land use scenarios: Three crops under rainfed agriculture (wheat, barley, olives) and three crops under irrigation (wheat, citrus trees, peas as an example for small vegetables). A comparison of annual GWR-rates calculated by WEAP-MABIA and PTFs respectively is displayed in Table 3 for the rainfed land use classes wheat (in WEAP equivalent to cereals) and olives. A high level of conformity is evident in both cases.

Due to lack of runoff data for 2009/2010, modeled surface runoff in Quaik river was calibrated to long term average discharge values from different studies. According to ICARDA 1994, a substantial fraction of groundwater recharge occurs primarily inside and along the Quaik river channel and in a relatively small area south of Aleppo City due to predominating highly fractured and weathered limestone with no soil cover or with soil confined to pockets between bare rocks (ICARDA 1994). Thus, simple runoff calibration was conducted by estimating a monthly outflow factor of 15 % along the entire river channel as transmission losses into Paleogene aquifer. For Gold River, it was assumed that 5 % of channel flow infiltrate in the riverbed and percolate into Paleogene aquifer.

As a result, the annual decrease in Paleogene groundwater storage is 0.67 Billion m³ (= Inflow from GW_Paleogene – Total Groundwater Recharge, Table 2). This water volume accounts for a decline of the groundwater table for about 2 m, according to estimated Paleogene aquifer properties, which is also backed up by 2010 measurements from 3 different monitoring wells. It confirms the applied calibration method. The results illustrate a non-sustainable use of groundwater in 2009/2010. Long-term measurements with groundwater hydrographs in most observation wells in the area are showing a constant decline of groundwater tables too.

For estimating the impacts of specific possible interventions or modifications to the basin's hydrological system, future scenarios were developed. One modification to the system will be the extension of the irrigation channel which will be in use from 2012

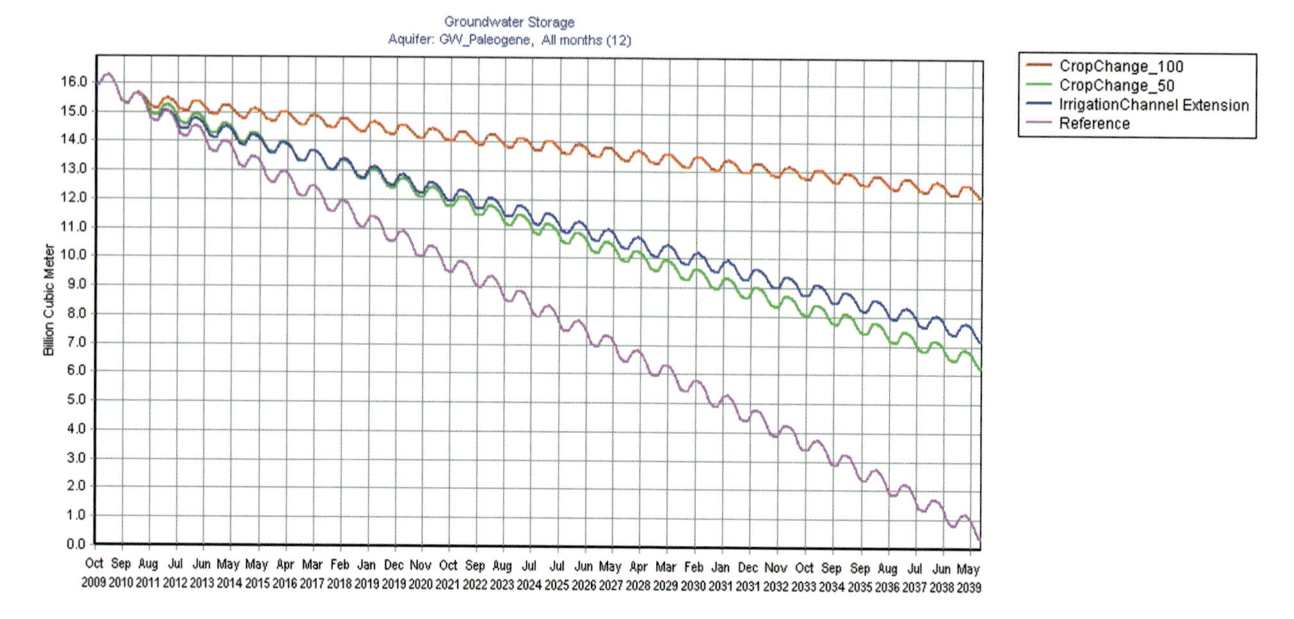

Figure 4: Change in groundwater storage of Paleogene aquifer, different scenarios.

onwards. The extension will enhance interbasin water transfer from Euphrates river and thereby ease the demand for groundwater. Other possible modifications could be land use changes. For instance, cotton has a much higher water demand than most of the other crops. Two crop change scenarios were set up. In CropChange_50 scenario, cotton is reduced by 50% and replaced with irrigated cereals. In CropChange_100 scenario, cotton is completely replaced with irrigated cereals. The impacts of those modifications on groundwater storage of the Paleogene aquifer become visible in Figure 4. According to the reference scenario, groundwater use under current conditions might lead to the complete depletion of Paleogene aquifer at the end of 2039. Regarding the other scenarios, it becomes clear that crop changes could have a very strong mitigating impact on groundwater demand. A complete readjustment from cotton to irrigated cereals could prolong Paleogene groundwater availability substantially. A combined scenario (Irrigation Channel Extension + CropChange_100, not included in Figure 4) would lead to a positive groundwater balance and therefore sustainable groundwater use.

Conclusion

The WEAP-MABIA Water Balance takes spatial heterogeneities of input data, irrigation processes, specific crop-properties, sequences and patterns, interbasin water transfers and other processes into account. Water demand is modeled as a function of land use, allowing for comparative land use change experiments. Results for total groundwater abstraction could be validated with independent data. Magnitudes of areal groundwater recharge rates could be validated by comparison with pedotransfer functions. Uncertainties remain with local and regional groundwater recharge rates, especially in the river channels, due to current lack of runoff data.

References

Brew, G., Barazangi, M., Al-Maleh, A. K. & Sawaf, T. (2001): Tectonic and Geologic Evolution of Syria. – GeoArabia, **6**, 4: 573–616, 19 fig.; Bahrain – ISSN 1025-6059.

ICARDA (1994): Farm Resource Management Program.- Annual Report for 1993, chapter 2: Special Focus: Groundwater and Supplemental Irrigation, pp. 8–59; Aleppo.

Jabloun, M. & Sahli, A. (2011): WEAP-MABIA Tutorial -A collection of stand-alone chapters to aid in learning the WEAP-MABIA module. – 93 p.

SEI – Stockholm Environment Institute (2011a): Water Evaluation and Planning System. – http://www.weap21.org

Proceedings "Hydrogeology of Arid Environments", p.147–150
Stuttgart, March 2012

Extended Abstract

Design and Setup of a High Resolution Hydrometric Monitoring Network in a Semi-Arid Karst Environment – West Bank

Sebastian Schmidt[1], Steffen Fischer[1], Mathias Toll[1], Fabian Ries[2], Omar Zayed[3], Joseph Guttman[4], Amer Marei[5], Menachem Weiss[6], Tobias Geyer[1], Martin Sauter[1]

[1] Georg-August-Universität Göttingen, Geowissenschaftliches Zentrum Göttingen, Angewandte Geologie, Goldschmidtstraße 3, 37077 Göttingen, Germany, email: sschmid1@gwdg.de
[2] Albert-Ludwigs-Universität Freiburg, Institut für Hydrologie, Fahnenbergplatz, 79098 Freiburg, Germany
[3] Palestinian Water Authority, P.O.B. 2174, Ramallah, West Bank
[4] Mekorot, Israel National Water Company, 9 Lincoln St., Tel Aviv 61201, Israel
[5] Al-Quds University, Department of Earth and Environmental Studies, P.O.B. 20002 Jerusalem, West Bank
[6] Hydrological Service of Israel, 234 Yafo St., P.O.B. 36118, Jerusalem 91360, Israel

Key words: Catchment instrumentation, spring monitoring, Jordan Valley, West Bank

1. Introduction

Semi-arid to arid environments are characterised by a general shortage of water resources. Surface water flow is often ephemeral and occurs only after strong precipitation events, especially in karstified landscapes with predominant subsurface drainage. Groundwater recharge displays a high spatial and temporal variability because of preferential flow pathways in the subsurface. A succession of several years with less than average precipitation can lead to a long-term decline of water resources. Compared to humid environments, often less hydrologic and hydrogeologic data are available. However, these data are a prerequisite for the system analysis, i.e. to quantify available water resources and ensure a sustainable use of those resources. To meet the specific requirements of (semi-)arid environments regarding hydrometric data assessment, an adapted design of the measuring network is demanded.

2. Study area and hydrogeologic challenges

The study area is located on the western graben shoulder of the Jordan Rift Valley northwest of the Dead Sea (West Bank). Groundwater is the main water resource in the region. The regional aquifer system is composed of karstified Cretaceous carbonate rocks with a thickness of about 800 meters. Ground level ranges from 1,000 m above sea level in the mountain range in the west to about 400 m below sea level at the Jordan River and Dead Sea in the southeast (Fig. 1).

The steep topographic gradient leads to an overall steep gradient in the hydraulic head distribution, however with large local differences. The mean annual precipitation ranges from about 600 mm/y in the mountainous area in the western part to about 100 mm/y near the Dead Sea. The recharge area of the carbonate aquifers corresponds to the outcrop area of the permeable carbonate strata and is located in the western mountainous part of the study area. Two main aquifer horizons can be distinguished. The upper aquifer discharges in the arid Jordan Valley via several springs which are the reason for oases like e.g. Jericho. The spring water is used for irrigated agriculture and domestic water supply. However, due to complex hydrogeological conditions, the catchments of the springs, the flow system of the aquifers, as well as the spatial and temporal distribution of groundwater recharge, are not yet very well described and understood. Therefore a monitoring network was designed and imple-

www.borntraeger-cramer.de
2012/0147/$ 1.00

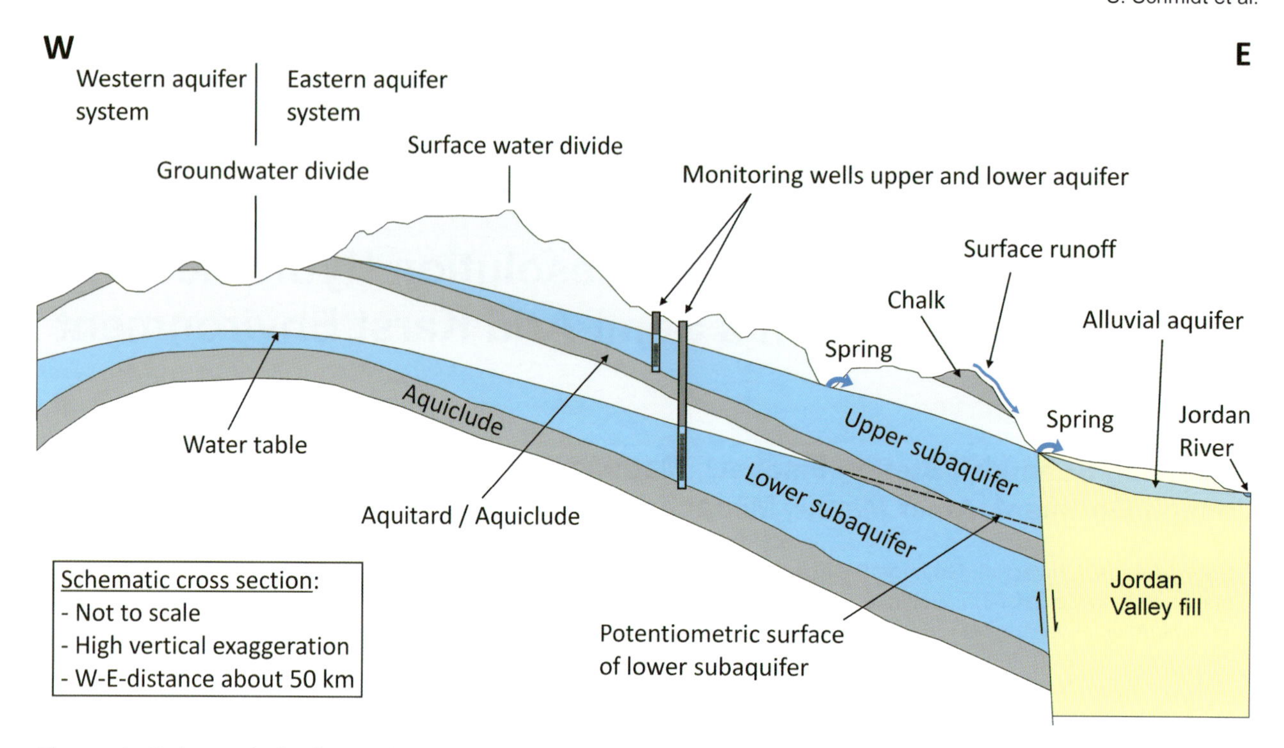

Figure 1: Schematic hydrogeologic cross section through the study area. The upper and lower subaquifers are mainly composed of fractured and karstified Albian to Turonian limestones and dolostones, the aquitards and aquicludes are chiefly composed of marl and chalk. However, lateral facies changes occur, enabling local aquifer connection (ANTEA 1998).

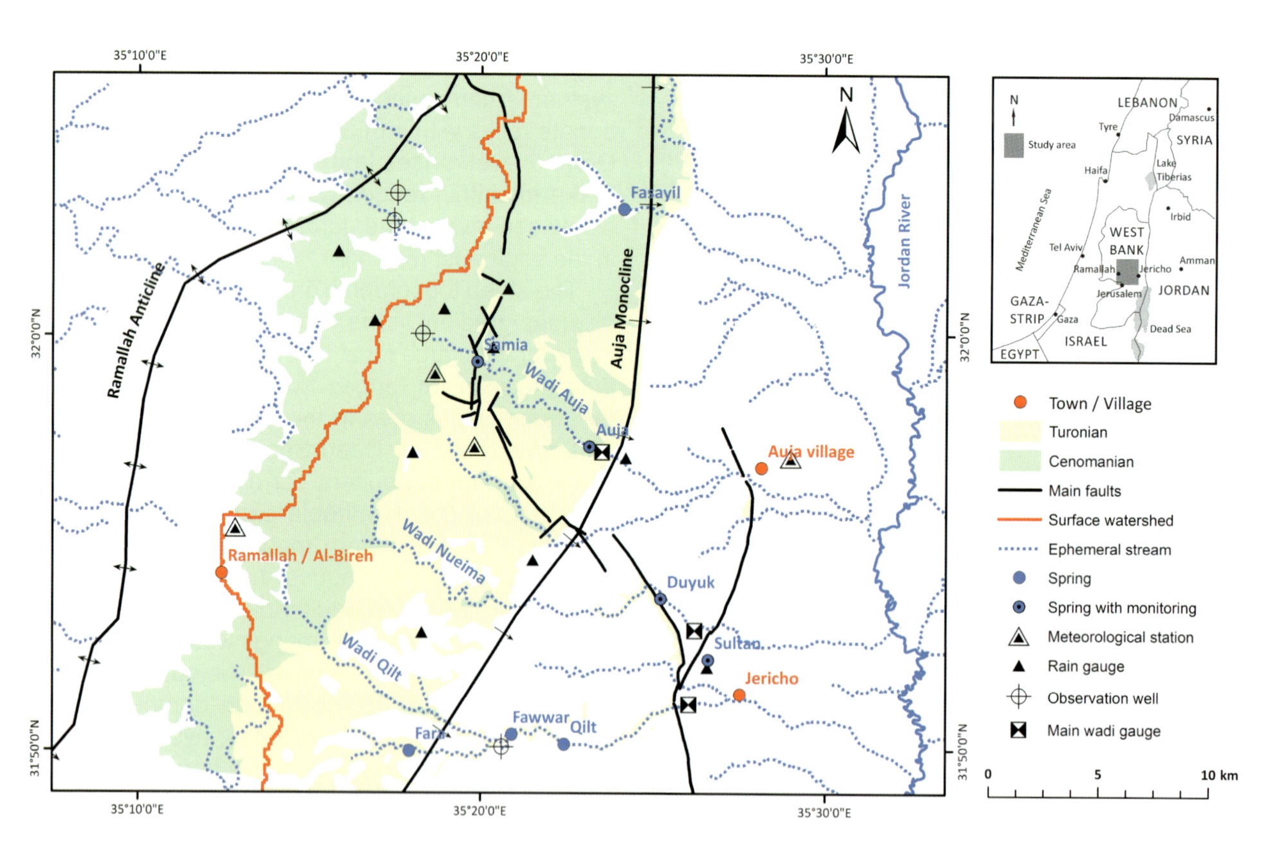

Figure 2: Map of the study area with the main components of the monitoring network. The outcrop of the Cretaceous formations (= potential recharge area of springs) was digitized from Sneh et al. (1998).

mented in the area north of Ramallah and west of Jericho (Fig. 2).

3. Setup of the monitoring network

Flow and transport processes in the aquifer systems shall be characterised. Because of the large heterogeneity of karst systems, data with a high spatial and temporal resolution are required. Numerical models for prediction of spring discharge require time series of precipitation, spring discharge, groundwater levels, surface runoff and meteorological parameters to calculate potential evapotranspiration, and soil moisture. Physicochemical spring parameters might be used to estimate aquifer characteristics.

Spring gauging stations and monitoring of spring physicochemical parameters

Emphasis is placed on the measurement of karst spring responses to precipitation. The time series of spring discharge and physicochemical parameters provide integral information about the whole groundwater catchment. For accurate long term discharge measurements, solid gauging stations had to be constructed. This was accomplished at two main springs in the study area. Figure 3 displays one example. For each gauging station, a stage-discharge relation was established with discharge measurements during different flow stages. To assess time series of spring water electrical conductivity, temperature, turbidity, and pH, multiparameter-probes were installed at four springs.

Monitoring stations for groundwater level

Groundwater level monitoring provides data for the assessment of groundwater flow and recharge processes, the assessment of the water percolation through the unsaturated zone and groundwater flow modeling. Four, deep groundwater wells in the carbonate aquifers were equipped with pressure transducers. At three of the wells, vandalism safe caps were installed. Well depth ranges from 150–575 m, depth to water table from 100–440 m. Therefore absolute pressure transducers attached to polypropylene ropes were installed in the wells. Close to the land surface barometric pressure transducers are installed in the wells to monitor changes in barometric pressure.

Meteorological stations, rain gauges and precipitation sampling

Meteorological instruments need to be installed at exposed but safe places. Therefore comparatively few sites are suitable. Schools, other public buildings and water supply infrastructure, like pump houses or elevated reservoirs turned out to be suitable sites (Fig. 3). In total, four meteorological stations and 11 rain gauges were installed. The meteorological stations measure time series of temperature, relative humidity, global radiation, wind speed, and precipitation. Those parameters are necessary to calculate potential evapotranspiration rate, i.e. by the Penman-Monteith method (Allen et al. 1998). The meteorological stations are aligned in a transect perpendicular to the topography, to cover

Figure 3: Examples of installations: Left: spring gauging and hydrochemical parameter monitoring station at the Auja spring. The weir was constructed for the research. The spring tapping structure in the background was profoundly rehabilitated. From the structure, a underground pipe leads to the weir. In the structure, the multiparameter-probe is installed (middle). Flow at time of picture was about 500 l/s. Right: Meteorological station on top of a water reservoir.

the strong gradient in meteorological parameters from the mountain range down to the Jordan Valley. The rain gauge network is denser around Wadi Auja to conduct high resolution rainfall-runoff research. Precipitation sampling for major ions and stable isotopes is conducted manually at three locations along the meteorological gradient by water authority staff, teachers and waterworks personnel.

Monitoring stations for flood runoff

Flash floods are very difficult to gauge, because of the high flow velocity and sediment load. Pre-calibrated structures like weirs and flumes (Bos 1989) are preferable. Two concrete weir structures at main wadis in the region were already available, but had to be rehabilitated (one station was rehabilitated by Helmholtz Centre for Environmental Research – UFZ, Halle) and equipped with instruments for stage measurement and record. A third main wadi is gauged at a large road culvert. The main study site for rainfall-runoff processes is Wadi Auja, where six additional (sub-)catchment gauging stations were installed. The sediment load of the ephemeral flow is a major problem for stations with an insufficient sediment passing capacity. Therefore, at those sites, continuous maintenance and removal of accumulated sediment is required.

Monitoring stations for soil moisture profiles

At four places soil moisture sensors, arranged as depth profiles, were installed in combination with a data logger. Those data provide high resolution data regarding infiltration, soil water movement, soil water balance and estimates for deeper percolation, hence the water that will become groundwater recharge at the water table.

4. Selected results

The Auja spring shows a very fast reaction after precipitation events. For example, after a very intense precipitation event during early 2010, the spring discharge started to raise just 14 hours after the onset of the precipitation in the catchment. Discharge increased 8-fold from about 60 l/s to 480 l/s within 110 hours. Currently a period of less than average precipitation prevails. Therefore, Auja spring stops flowing during the summer. During the last three years the spring was flowing for only 30 percent of the time.

Despite the very stable discharge of the Sultan spring, and the very large storage of the aquifer, the spring nevertheless displays characteristics of a karst spring. Following intense precipitation events in the recharge area, a strong response in the physicochemical paramters can be observed. Within a few days after the onset of precipitation, a strong increase in turbitity followed by a strong decrease in electrical conductivity can be noted. This is an indicator, that newly recharged event water is discharged at the spring with only a short lag time.

Acknowledgements

The work is conducted within the framework of the multi-lateral research project “SMART – Sustainable Management of Available Water Resources with Innovative Technologies” funded by BMBF (German Federal Ministry of Education and Research), references No. 02WM0802 and 02WM1081. The people and institutions contributing to the setup of the monitoring network are too numerous to be acknowledged individually. We are thankful for the help of everyone.

References

Allen, R. G., Pereira, L. S., Raes, D. & Smith, M., 1998: Crop evapotranspiration (guidelines for computing crop water requirements). – FAO Irrigation and drainage paper no. 56, Rome, 300 pp.

ANTEA, 1998: Well development study of the Eastern Aquifer Basin, Northern Districts of Palestine, Volume 1: Interim Report, Conceptual Model. – Unpublished Antea report No. A11903, 119 pp.

Bos, M. G. (Ed.), 1989: Discharge Measurement Structures, 3rd ed. – ILRI Publication 20, International Institute for Land Reclamation and Improvement, Wageningen, 401 pp.

Sneh, A., Bartov, Y. & Rosensaft, M., 1998: Geological Map of Israel 1:200000, sheet 2. – Ministry of National Infrastructures, Geological Survey of Israel, Jerusalem.

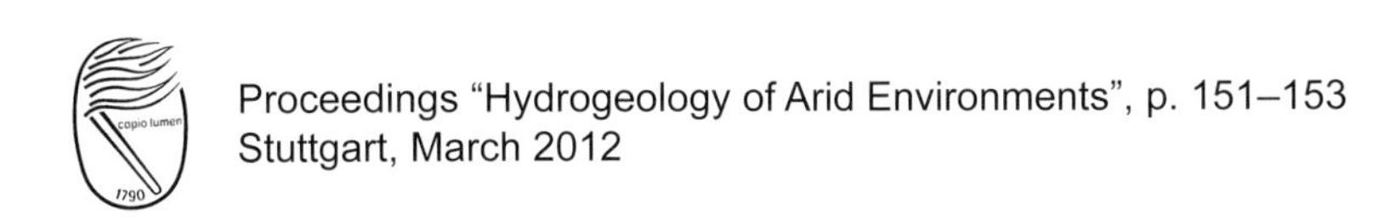
Proceedings "Hydrogeology of Arid Environments", p. 151–153
Stuttgart, March 2012

Extended Abstract

Fluoride Problem in Semi-Arid Region – a Case Study from India

S. K. Sharma

Dr. S. K. Sharma, Head, Environmental Education Department, Carman Residential and Day School, Dehradun 248007, India, E-mail: SKS105@rediffmail.com

Key words: dilution, fluoride, flurosis, osteoporosis, tooth mottling

Introduction

India is among those nations around the world, where health problems occur due to the consumption of fluoride contaminated water. Fluoride problems are wide spread in India and has been reported from many parts of the country. The semi-arid region of Rajsthan is one of them. Over 70% irrigation and 95% drinking water supply schemes are based on ground water resources which has resulted in a rapid depletion in ground water level and deterioration in water quality. The withdrawal of ground water is more than the recharge due to over exploitation. The geo-chemical factors have resulted in increase of hydrochemical parameters viz. salinity, nitrate, and fluoride in ground waters and have adversely affected the lives of inhabitants. A systematic study has been carried out to understand the behavior of fluoride in natural water and its affect on human health in the southern part of Rajsthan in Bhilwara which is facing the acute problem of fluoride, in terms of geological setting, hydrological and climatic conditions and agricultural practices.

Materials and Methods

A total of 160 ground water samples from dug wells, shallow hand pumps and deep tube wells were collected during May (pre-monsoon) and were analyzed at the chemical lab of Central Ground Water Board, as per APHA (1995) procedure for fluoride (F), calcium (Ca), Sodium (Na), Bicarbonate (HCO_3) concentration together with the determination of temperature and pH .

Results

The analyses of the water samples collected have been summarized in Table 1 which indicate the fluoride concentration up to 11.5 mg/l. Nearly, 90% of the samples show excess of fluoride against the World Health Organization's safe permissible limit of 1.5 mg/l.

Discussion

Fluoride incidence in groundwater is mainly a natural phenomenon influenced basically by the local and regional geological setting and hydrogeological conditions. Main rock formations of the water sampling locations are Schist, Gneiss, Phyllite, Granite, Quartzite and Alluvial formations. Most of the sampling locations are from aquifers in the unconfined/ semi confined conditions, in general, in weathered and fractured zones. It seems more appropriate that rocks rich in fluoride minerals have contributed to the enriched fluoride content of groundwater during the course of weathering of rock types. Areas with semi-arid climate, crystalline rocks and alkaline soils are mostly affected (Handa 1975). The major fluoride bearing minerals present in the rocks are Fluorapatite, Fluorite, Sepiolite, Palygorskite, Cryolite, Muscovite, Biotite, Hornblende, Asbestos etc. Among these, the most important being the fluorite, sepiolite as well as palygorskite and the leaching of fluoride from the metamorphic rocks like granite gneiss and schist of Proterozoic age. Fluoride gives statistically significant positive correlation with HCO_3 as well as Na and negative correlation with Ca in groundwater samples of the study area. Coupled

www.borntraeger-cramer.de
2012/0151/$ 0.75

Table 1: Summary of physico-chemical properties of groundwater samples.

Sampling site / State / number of samples	Average pH	Average Temperature (°C)	Fluoride Range (mg/l) / Calcium Range (mg/l)	HCO_3 Range (mg/l)	Sodium Range (mg/l)
Bhilwara / Rajsthan / 160	8.2	27	2.1 to 11.5 / 26.0 to 60.0	270 to 1390	450 to 900

with high Na content, these correlations reveal that weathering is the major source of fluoride in groundwater. In general, relatively high pH conditions have a tendency to displace fluoride ions from the mineral surface. From the above, it is obvious that relatively high alkalinity favors high concentration of fluoride in groundwater of the study area. The arid climate with low rainfall and high evapotranspiration and insignificant natural recharge favors salinisation of groundwater resulting in the precipitation of calcite. Soils become more alkaline with very high pH limit the solubility of calcite. These conditions allow fluoride to concentrate in the groundwater environment (Vikas et al. 2005). Fluoride concentration in excess of permissible limit in drinking water causes dental and skeletal fluorosis in the study area.

Optimal concentration of Fluoride around 1 mg/l has been found to have a significant mitigating effect against dental caries and it is accepted that some fluoride presence in drinking water is beneficial. However, chronic ingestion of concentrations much greater than 1.5 mg/l (WHO guideline value) is linked with development of dental fluorosis and, in extreme cases, skeletal fluorosis. High doses have also been linked to cancer. Health impacts from long-term use of fluoride-bearing water have been summarized (Dissanayake 1991) as: <0.5 mg/l: dental caries; 0.5–1.5 mg/l promotes dental health; 1.5–4 mg/l dental fluorosis; >4 mg/l dental, skeletal fluorosis and >10 mg/l crippling fluorosis. Dental fluorosis is by far the most common manifestation of chronic use of high-fluoride water. As it has greatest impact on growing teeth, children under age 7 are particularly vulnerable. Fluorosis has no known treatment other than early detection and limiting the amount of fluoride ingested.

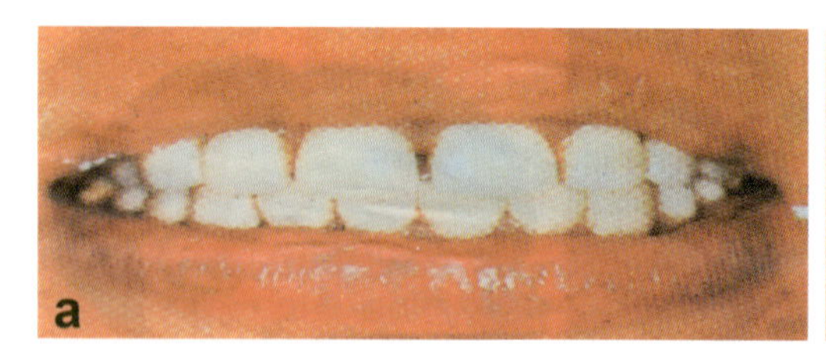

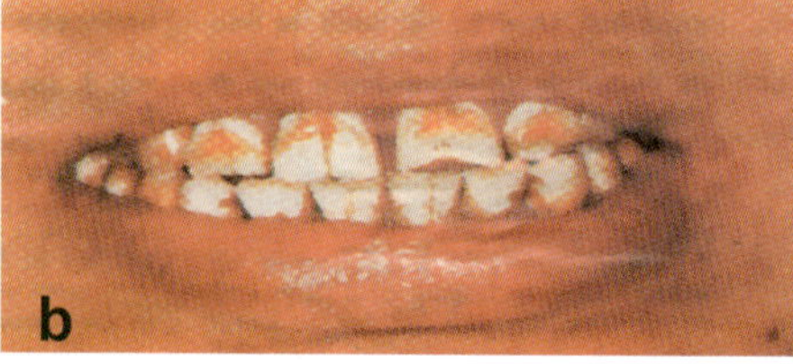

Figure 1:
(a) Normal teeth.
(b) Mottled teeth.

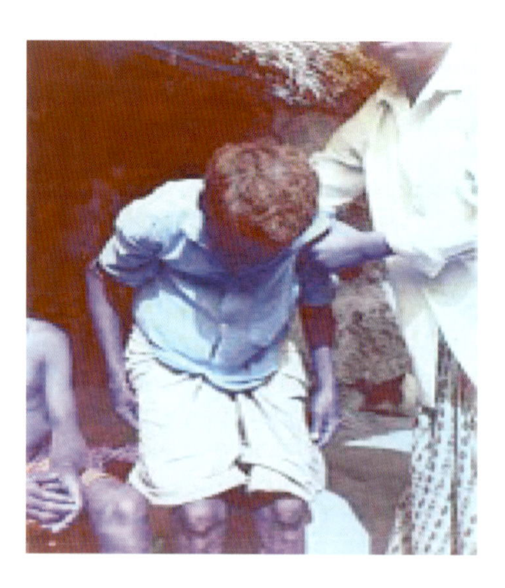

Figure 2:

Reported case of dental fluorosis:
In the following Fig. 1, the left hand side picture (a) shows a person with Normal Teeth whereas (b) the right one is suffering from the dental fluorosis having brownish yellow Mottled Teeth, a common feature in high fluoride (more than 1.5 mg/l) region.

Reported case of Skeletal fluorosis:
The following Fig. 2 shows a case of skeletal fluorosis when the fluoride is readily incorporated into the Crystalline structure of bone, and accumulates over time and causes skeletal fluorosis which increases risks of bone fracture. Skeletal fluorosis is a bone and joint condition associated with prolonged exposure to high concentrations of fluoride (>10 mg/l). Fluoride increases bone density and causes changes in the bone that lead to joint stiffness and pain.

Remediation techniques

One very simple, cheap and inexpensive way to reduce the fluoride level in groundwater is to artificially recharge the aquifers with freshwater. Alternately, recharging the aquifer with the rainwater appears to be more practical. Therefore, it is in this context that the rainwater harvesting in the affected areas need to be encouraged and practiced. Probably the most well-known and established method is the Nalgonda technique, commonly used in India, where a combination of alum (or aluminium chloride) and lime (or sodium aluminate), together with bleaching powder, are added to high-fluoride water, stirred and left to settle. Fluoride is subsequently removed by flocculation, sedimentation and filtration. The method can be used at domestic scale (in buckets) or community scale (fill-and-draw type defluoridation plants; Nawlakhe & Bulusu 1989). It has moderate costs and uses materials which are usually easily available. Other precipitation methods include the use of dolomite, calcium chloride or chalk. Chalk is available locally everywhere in India and it has the advantage to increase the pH and the total concentration of dissolved calcium in the water. These methods have been tested locally and are capable in principle of reducing fluoride in treated water to below the WHO guideline value. Gypsum treatment is the classical method of alleviating the soil alkalinity. It has advantages in being cheap as gypsum is abundant in India, even in the hard rock areas. The only problem with the treatment is that it will give a harder water. the permissible SO_4^{2-} content would be exceeded and the accepting such water by the population might not be possible. But this may be an advantage as the higher intake of Ca++ mitigate the effect of fluoride uptake in the intestine (Teotia & Teotia 1988). The locally available chalk can be used in water treatment, which appears to be a good material to decrease the fluoride levels in the waters. It has the advantage to increase the pH and the total concentration of dissolved calcium thus, bringing in an appreciable decrease in fluoride concentration in the water. Efficient irrigation / agricultural practices also help in decreasing the alkalinity of soil thus reducing the concentration of fluoride in groundwater.

References

APHA 1995: Standard methods for the examination of water and wastewater, 19th edition, American Public Health Association, Washington, D.C.

Dissanayake, C. B., 1991: The fluoride problem in the groundwater of Sri Lanka – environmental management and health. – Intl. J. Environ. Studies **19**: 195–203.

Handa, B. K., 1975: Geochemistry and genesis of fluoride-containing ground waters in India. – Ground Water **13**: 275–281.

Nawlakhe, W. G. & Bulusu, K. R., 1989: Nalgonda technique – a process for removal of excess fluoride from water. – Water Quality Bulletin **14**: 218–220.

Teotia, S. P. S., Teotia, M. & Singh, R. K., 1981: Hydrogeochemical aspects of endemic skeletal fluorosis in India – an epidemiological study. – Fluoride **14**: 69–74.

Vikas, C., Kushwaha R. K. & Pandit, M. K., 2009: Hydrochemical status of groundwater in district Ajmer (NW India) with reference to fluoride distribution. – Jour. Geol. Society of India, Vol. **73**, June 2009, pp 773–784.

WHO 1984: Guidelines for drinking water quality. Values 3; Drinking water quality control in small community supplies. – WHO, Geneva, p. 212.

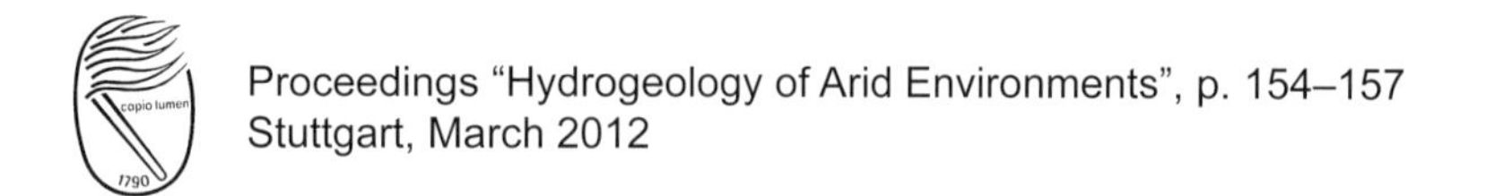
Proceedings "Hydrogeology of Arid Environments", p. 154–157
Stuttgart, March 2012

Extended Abstract

The Upper Mega Aquifer System on the Arabian Peninsula – Delineation of Sub-Aquifer Interaction Using Hydrochemical and REE+Y Patterns

C. Siebert[1], T. Rödiger[1], R. Rausch[2], J. Döhler[2], N. Michelsen[3], M. Al-Saud[4]

[1] Helmholtz-Centre for Environmental Research UFZ Halle, Germany;
[2] GIZ/Dornier Consulting Riyadh, Saudi Arabia;
[3] TU Darmstadt, Institute for Applied Geosciences, Darmstadt, Germany;
[4] Ministry of Water & Electricity Riyadh MoWE, Saudi Arabia.

Key words: multi aquifer system, rare earth elements, groundwater development

During a field campaign in November 2009, about 30 water samples from different formations of the Saudi Arabian Upper Mega Aquifer system were collected. The aim was to define specific rare earth element and Yttrium (REY) patterns as well as major dissolved ions characteristics for each of the sub-aquifers. Especially the observed REY patterns reflect the lithological differences of the respective formations and can be used to determine the origin of the waters. Also evidence for mixing of waters due to leakage between aquifers, either structurally or antropogenically induced, can be found based on rare earth element patterns.

Introduction

The Upper Mega Aquifer system (UMA) of the Arabian Peninsula is one of the world's large groundwater reservoirs (2×10^6 km^2). It consists of more than 2,500 m of sedimentary rocks ranging from the lower Cretaceous Biyadh Formation to the Cenozoic Dammam Formation. Determined ages of the respective groundwaters of up to 10.000 years (Hoetzl & Zoetl 1978) indicate slow groundwater flow velocities and long residence times, resulting in some cases in high salinity values.

Along tectonic features, such as the Ghawar anticline, hydraulic windows between the formations of the UMA are assumed that may result in cross formation flow. This could be further intensified by extensive groundwater abstraction, e.g. within the Umm Er Radhuma (UER) formation, that led to dropping groundwater levels resulting in the alteration of regional flow fields.

Rare earth elements and yttrium (REY) patterns in groundwater samples are a powerful tool to determine the lithological origin of the waters due to water-rock interactions. Based on distinctively different REY patterns for different formations also mixing due to leakage between different aquifers can be investigated (e.g. Johannesson et al. 1997, Möller et al. 2003, Siebert et al. 2009, 2011). In addition, analyses of the major dissolved ions in the waters can provide further evidence for mixing processes.

The present study was therefore initiated to analyze (i) the general REY patterns and major ion composition of the groundwaters of the UMA in oder to (ii) detect possible cross formation flow. Moreover (iii), results will serve as additional information to be implemented into a numerical groundwater flow model of the UMA. For this, 30 groundwater samples were taken along a transect from the Tuwaiq Mountains (Riyadh) to the Gulf coast (Dammam), from proposed recharge- to the well-known discharge- and exploitation (well fields) areas (Fig. 1) and analyzed.

Results

Based on information from the MoWE database, wells were selected for sampling that could be clearly correlated to sub-aquifers close to their respective outcrop areas. Waters could be classified as follows: (i) Wasia- and Biyadh: Ca-HCO_3-SO_4 and Ca-

www.borntraeger-cramer.de
2012/0154/$ 1.00

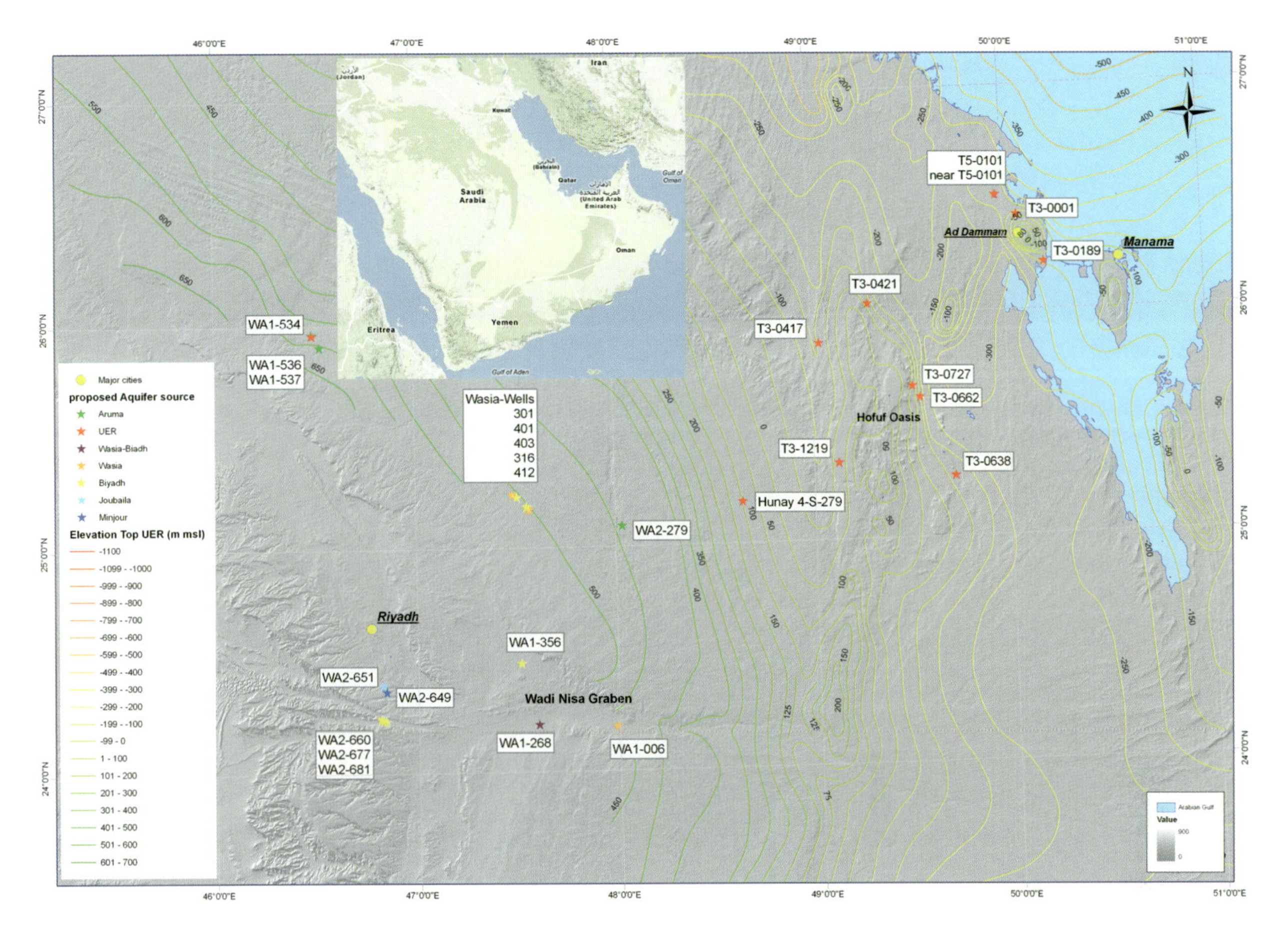

Figure 1: Map of the sampling locations along the proposed flow path from Riyadh NE-wards to the Arabian Gulf. Insert: Overview map of Arabian Peninsula (google).

SO_4-HCO_3 type, (ii) Aruma: Ca-HCO_3 type, and (iii) UER: Ca-SO_4-Cl type. Hence, already close to the recharge areas, groundwaters from the sub-aquifers of the UMA show distinct differences in their ionic composition. Total salinity was highly variable, e.g. in the same area groundwaters from Aruma are distinct less saline than those from UER. Waters from the Biyadh aquifer at Wadi Nisa are chemically similar to waters from the Wasia aquifer (Wasia Well field) (Fig. 2), however, waters from the Biyadh aquifer collected in the Wasia Well field and in the northern shoulder of Wadi Sahba are much more saline.

REY patterns of the analyzed groundwaters indicate 3 major types (Fig. 3). **Group 1** waters decrease from La to Lu, with variable negative Eu-, mostly strong positive Y- and often small positive Gd-anomalies. Such patterns are typical for groundwaters controlled by carbonate dissolution. Waters of **Group 2** are characterised by convex shaped patterns with declining abundancies of REY towards Lu. They show variable Ce- and distinct negative Eu- and positive Gd- and Y-anomalies. Waters of **Group 3** are characterised by increasing trends from La to Lu with negative Eu- and positive Y-anomalies.

Conclusions

By dissolving sulfatic minerals along the flow paths, waters in the Wasia and Biyadh sandstone and in the Aruma aquifer develop towards a Ca-SO_4-HCO_3 type (Fig. 2). In UER-waters, two distinct trends are observable. Starting from a Ca-Mg-SO_4-HCO_3 type, waters develop along their flow-path either to a Ca-Na-SO_4-Cl type in Hofuf region or to Na-Ca-Cl-SO_4 type in Hofuf and Dammam regions. In the Hofuf area, a tectonical destruction of the Rus aquitard can be assumed, allowing water to flush out gypsum and transport it into the UER aquifer below, which is locally phreatic due to groundwater abstraction. Due to intense irrigation and enhanced actual evapotranspiration agricultural return-flow is enriched in Cl. Additional continuous dissolution of halite and a possible influence of Gulf-water further increase the observed high NaCl-contents in nearshore UER-waters.

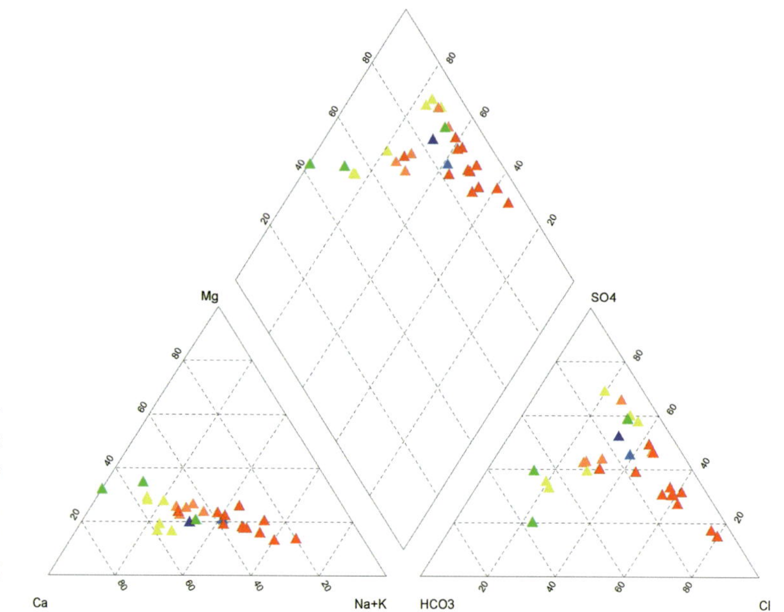

Figure 2: Piper plot of all sampled waters. According to the well database waters are coloured according to their designated aquifers: Minjur – dark blue, Joubaila – light blue, Biyadh – yellow, Wasia – orange, Aruma – green, UER – red.

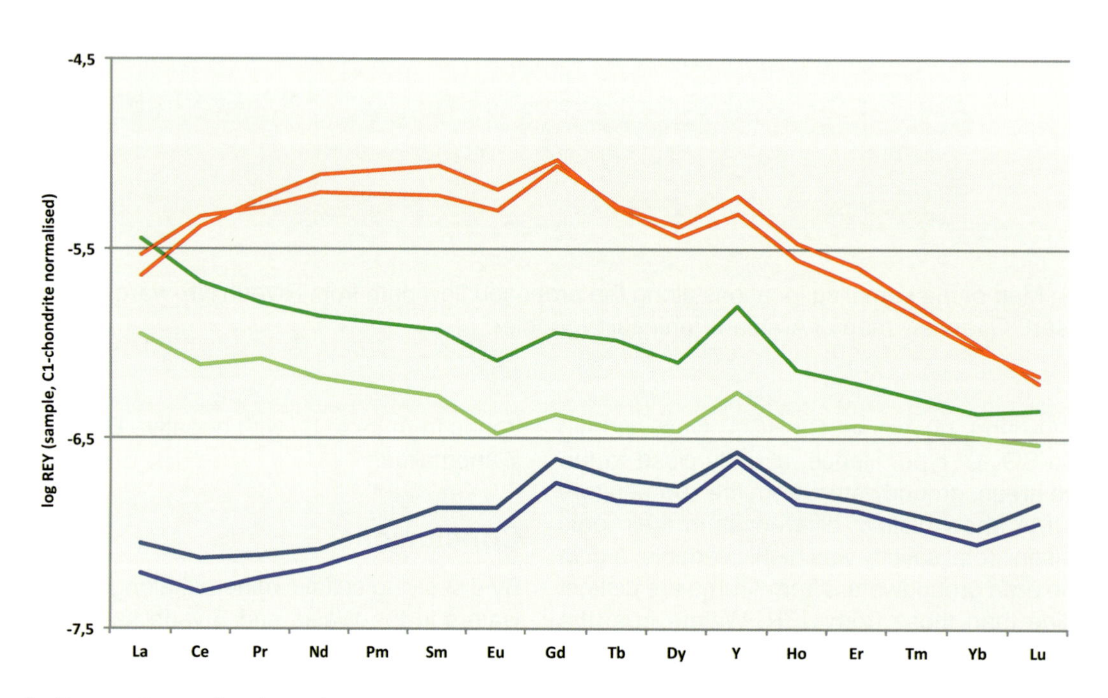

Figure 3: Rey pattern of selected water samples representing REY groups: group 1 (greenish), group 2 (reddish) and group 3 (bluish).

By evaluating REY-patterns, groundwater from the Aruma and the UER formation can be separated from water of the sandstone formations below. Additionally, REY pattern allow a differentiation between waters from Biyadh- and Wasia sandstone aquifers. Based on these findings, it is possible to assign extracted waters to respective aquifers and to identify areas of inter-aquifer flow. Although the number of analyzed samples is low with respect to the length of the transsect, the analysis of REY patterns shows therefore very promising results for the interpretation of the origin of groundwater pumped from the UMA. The method could therefore be used to substitute for missing or incomplete drilling- and well-information, as mis-interpretation of water origin can induce erroneous interpretations concerning the expected water quality and quantity. That in turn can have a strong impact on water management schemes. In conclusion, particularily REY-patterns can be used to assign water from existing groundwater abstraction wells to specific aquifers and respective water sources (Siebert et al. in prep.).

References

Hoetzl, H. & Zoetl, J. G., 1978: Climatic changes during the quaternary period. – In: Al-Sayari, S. S. & Zoetl, J. G.: Quaternary Period in Saudi Arabia. Vol. 2, 301–311p., Springer Vienna.

Möller, P., Rosenthal, E., Dulski, P., Geyer, S. & Guttman, Y., 2003: Rare earths and yttrium hydrostratigraphy along the Lake Kinneret–Dead Sea–Arava transform fault, Israel and adjoining territories. – Applied Geochemistry **18**: 1613–1628.

Johannesson, K. H., Stezenbach, K. J. & Hodge, V. F., 1997: Rare earth elements as geochemical tracers of regional groundwater mixing. – GCA 61 (17): 3605–3618.

Siebert, C., Geyer, S., Möller, P., Berger, D. & Guttman, Y., 2009: Lake Tiberias and its dynamic hydrochemical environment. – In: Möller, P., Rosenthal, E. (eds.): The Water of The Jordan Valley. Berlin-Heidelberg, Springer-Verlag, pp. 219–246.

Siebert, C., Rosenthal,E., Möller, P., Rödiger, T. & Meiler, M., 2011: The hydrochemical identification of groundwater flowing to the Bet She'an-Harod multiaquifer system (Lower Jordan Valley) by rare earth elements, yttrium, stable isotopes (H, O) and Tritium. – Appl. Geochem. DOI: 10.1016/j.apgeochem.2011.11.011

Siebert, C., Rödiger, T., Rausch, R. & Al-Saud, M. (in prep.): Large scale groundwater characterisation by REE+Y in the Cretaceous and Paleogenic Aquifers along the transsect Riyadh-Dammam, Saudi Arabia.

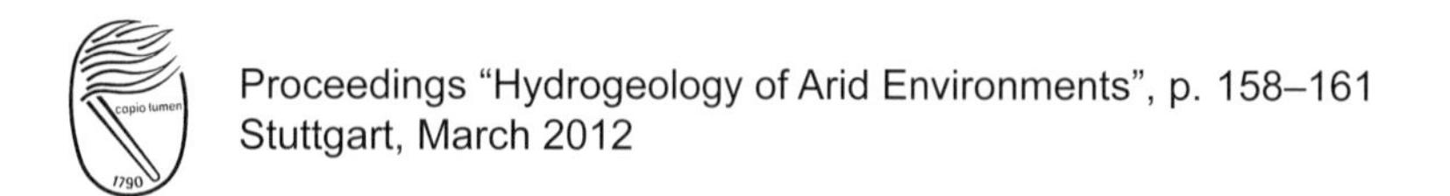
Proceedings "Hydrogeology of Arid Environments", p. 158–161
Stutgart, March 2012

Extended Abstract

Moving Well Solution to Optimal, Multi-Objective Groundwater Use – Method and Application

Tobias Siegfried[1] and Wolfgang Kinzelbach[2]

[1] hydrosolutions Ltd., Technoparkstrasse 1, 8005 Zürich, Switzerland
[2] Institute of Environmental Engineering, HIL G 37.3, Wolfgang-Pauli-Str. 15, 8093 Zürich, Switzerland

Key words: groundwater mining, multi-objective optimization, genetic algorithms, North-West Sahara Aquifer System, non-cooperative water use, tradeoff analysis

Introduction

Especially in semi-arid and arid environments, where irrigation accounts for the bulk of consumptive freshwater use, aquifers are key resources for buffering climate variability. Unfortunately, they are increasingly depleted and degraded in many parts of the world . As a result, adverse consequences from groundwater mining have emerged as key freshwater resources challenges over the last decades.

Here, a new method for optimal multi-stakeholder aquifer management is presented that is based on the coupling of a genetic optimization algorithm with a finite-difference representation of the groundwater system. The coupled simulation-optimization approach allows for the quantification of an arbitrary number of Pareto-optimal solutions and trade-offs between different users and best uses in aquifers of arbitrary complexity. It helps to address the question of how groundwater should be optimally allocated in time and space over a variable number of pumping facilities, with (possibly) competing use, given economic and environmental constraints. We show the usefulness of the methods presented with an application to non-cooperative groundwater mining in the North-West Saharan Aquifer System (NWSAS) in Northern African.

Method

In economic decision-making, stakeholders normally attempt to maximize returns on investment or minimize costs from allocating scarce resources. Planning and resource management in groundwater-irrigated economies is no different as it involves questions related to the optimal operation and existing and distribution of new wells and how pumping should be distributed in time over these facilities, given demand, supply and capital constraints.

Often, multiple stakeholders are present in freshwater management which utilize joint resources non-cooperatively as in the case of transboundary aquifers. Under these circumstances, external effects arise which can impact the overall effectiveness of the allocation regimes. Also, any optimal allocation strategy of one stakeholder depends on activities of the other planners in such settings. An optimizing planner thus has to anticipate the allocation activities of the others, in response to her own strategy, and vice-versa.

In this sense, no single optimal solution for space-time resource use exists in multi-stakeholder allocation problems as there are always tradeoffs involved between any two allocation strategies and a decision-support model should be able to help to identify the set of best compromise solutions, for which no better allocations exist. These solutions are called the set of Pareto-optimal solutions based upon which tradeoff negotiations over allocation strategies can be carried out, given individual stakeholder preferences.

The multi-objective optimization model presented here belongs to the class of location-allocation problems (Daskin 1995) and has the following characteristics:

- The model couples a finite-difference aquifer representation of arbitrary complexity with a multi-objective optimization routine for minimizing (present costs) or maximizing (present benefits) a

www.borntraeger-cramer.de
2012/0158/$ 1.00

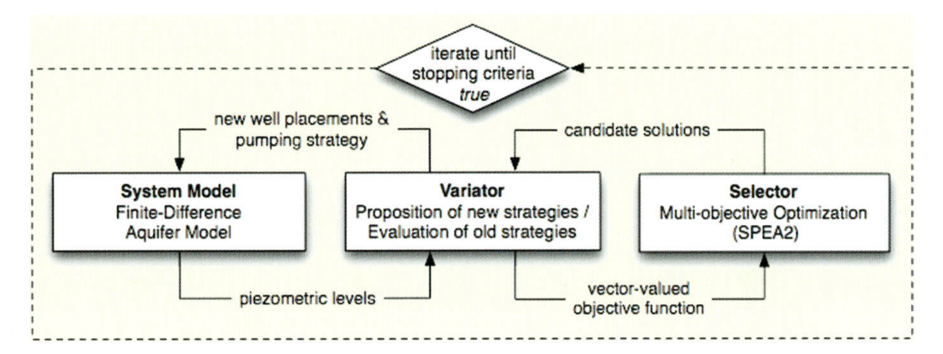

Figure 1: Top-level representation of multi-objective simulation-optimization algorithm developed. For illustration, the solid arrows indicate the individual components whereas the dashed arrows show one evolution cycle of the genetic algorithm optimization routine.

goal function over an arbitrary time horizon (see Figure 1).

- It associates present provision costs with an arbitrarily complex space-time groundwater allocation strategy. Facility costs of wells and pipelines as well as pumping and conveyance costs are fully taken into account in the goal function.
- The goal function is multi-objective with an unrestricted number of objectives, corresponding to the number of stakeholders involved.
- The optimization approach involves discrete (i.e. well locations and the discrete, time-varying control policy) and continues decision variables (i.e. pumping), subject to conditions that feasible solutions must satisfy.

As the optimization is carried out without the inclusion of preference information of individual stakeholder, the resulting optimal set consists of equally important solutions from the stakeholders perspective. Thus, the output can be used in decision-making processes for the quantification of tradeoffs.

An elitist, multi–objective evolutionary algorithm (EA) to approximate the Pareto-set is used. An EA models the evolution of individual space-time allocation strategies through successive generations using probabilistic operators (called Variator, see Figure 1). The Variator encapsulates the representation of solutions, the generation of new solutions, and the calculation of objective function values. It is described in (Siegfried 2004), (Siegfried & Kinzelbach 2004) and (Siegfried et al. 2009) and has been developed specifically for well placement and management problems in groundwater[1].

1 All of the Variator coding has been carried out in MATLAB (Matlab 2011) and is available from the author upon request. A toy application can be downloaded at http://www.tik.ee.ethz.ch/pisa/.

Desirable solutions, i.e. cost-effective or large net benefit strategies, are retained by the Selector which approximates the the set of Pareto-effective strategies. For the Selector, the EA SPEA2 was chosen. In comparison to similar other evolutionary techniques, this algorithm has shown good performance in problems with large and complex search spaces (Zitzler et al. 2002). The coupling of the simulation part, i.e. the system model and the Variator, with SPEA2 is facilitated by the availability of a platform and programming language independent interface (Bleuler et al. 2003). This implementation allows the problem–independent Selector SPEA2 to interact with the problem-specific Variator with minimal overhead (Zitzler et al. 2002) and (Zitzler 2001).

Application

In northern Africa, fossil transboundary groundwater is being mined, mostly for irrigation purposes in Algeria and Tunisia, but also for large-scale domestic supply in Libya. Water is pumped from the two main aquifers in the region, the intercalary continental (IC) and the terminal complex (TC), that together form one of the largest groundwater system in the world, i.e. the north-west Saharan aquifer system (NWSAS). Figure 2 shows a map of the region.

The three riparian states have been allocating the non-renewable common-property resources of the NWSAS over the past fifty years in a non-cooperative way. The main problems associated with groundwater mining are rising energy requirements (costs) due to increasing drawdowns and large conveyance distances respectively and the salinization of groundwater in regions where natural pollution sources exist, such as in the vicinity of the Tunisian and Algerian Chotts (salt lakes).

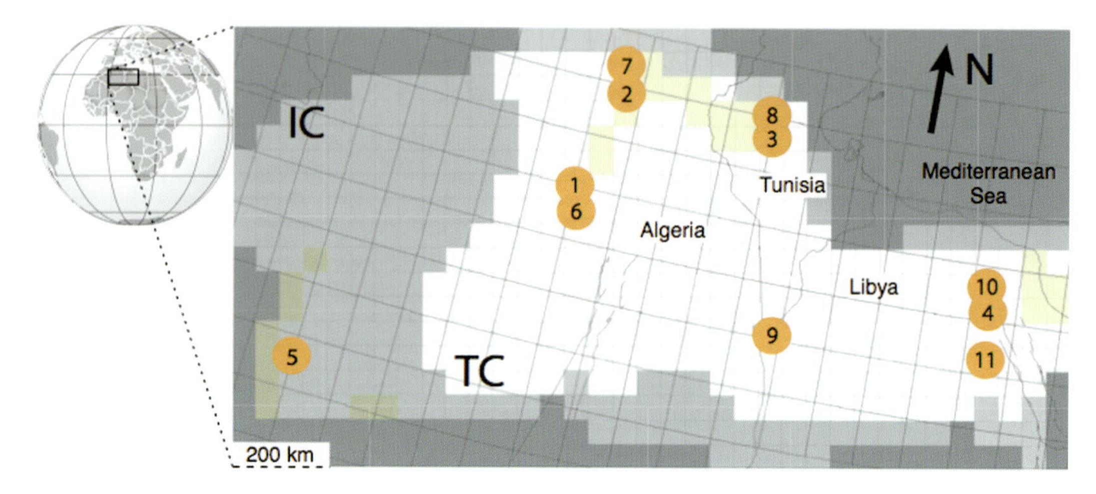

Figure 2: Map of the study region. The two aquifers IC and TC cover an area of 1 million km^2 approximately. Their extent is shown as implemented in the finite-difference MODFLOW groundwater model (Harbaugh & McDonald 1996). A 50 km cell discretization was used. Inactive cells are greyed out, sinks are modeled as drains cells are shown in yellow color. The main demand centers are denoted with numbers and refer to location and aquifers in use[2].

Given an expected six-fold increase of total pumping[3] from the CT and IC until the year 2050, the question is whether gains from pursuing cooperative allocation strategies, given the specific national demands, exist and, if so, what these imply in terms of policy recommendations. For this purpose, cooperative optimal multi-objective strategies are explored through the application of the above-presented model and compared to present and projected non-cooperative status quo strategies. The model was setup as a traditional discounted present cost-minimization exercise with well-field locations and pumping rates being the decision-variables for the three stakeholders (countries) over a 50 year time horizon.

Results

Key optimization results are:

- While per unit groundwater prices will keep rising in the future, rates of increases are 4 times lower of cooperative allocation strategies as compared to the Status Quo, on average. At the same time, per unit groundwater prices (expressed in present costs, averaged over the optimization time horizon) under cooperative allocation scenarios are significantly lower for each of the three stakeholders (77 percent for Algeria, 90 percent for Tunisia and 85 percent for Libya), again as compared to the non-cooperative Status Quo. Thus, significant capital gains can be achieved by each stakeholder from cooperation.
- The common property nature of the groundwater resources implies that optimal pumping from the perspective of each stakeholder is a best response to individual allocation strategies by riparians. In this sense, attaining cooperative benefits from intelligent, adaptive pumping requires coordination across stakeholders.
- Features of optimal solutions include moving wells in space (Figure 3) as well as intermittent pumping at particular well-locations between different aquifers in the basin.

The results suggest that cooperative use is incentivized since stakeholders can jointly and individually benefit from optimal pumping in the common-property aquifer resources. Moving pumping between different locations emerges as an important cost-minimizing feature. The basic underlying idea is intuitive. First, induce drawdown at a particular place until it becomes prohibitively expensive to continue pumping there, then move to a different location or a different aquifer at the same location so as to let the piezometric levels recover at the original location. Once partially recovered, start pumping again at the original location and continue with this kind of intermittent pumping over the optimization horizon.

2 {1,6}: Ghardaia / Zelfana / Ouargla (TC&IC), {2,7}: M'Rhair / El Oued (TC&IC), {3,8}: Nefzawa/Djerid (TC&IC), {4,10}: Libya Coast (TC&IC), {5}: Adrar / Gourara / Turat / In Salah (IC), {11}: Libya Sud (IC).

3 An estimated 80 m^3/s second was pumped in the three countries in the year 2000. National projections indicate a total demand of 550 m^3/s by the year 2050.

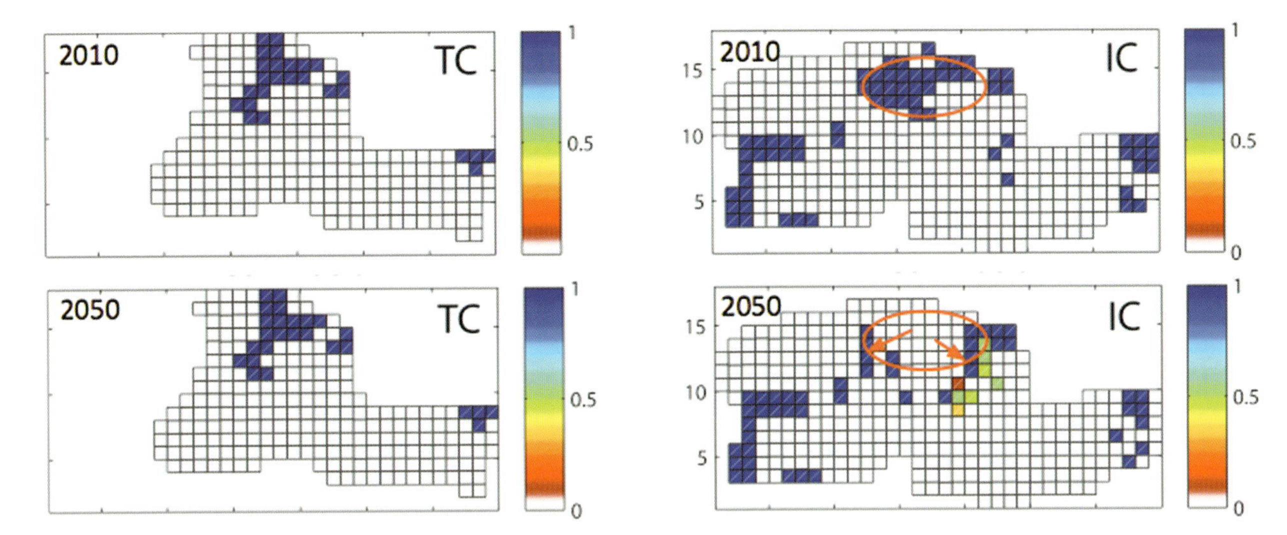

Figure 3: Density plots of optimal well locations in the year 2010 and 2050 as proposed by Pareto-optimal solutions. Optimality requires moving active pumping wells away from the Algerian and Tunisian Chotts because of excessive drawdowns in the confined IC there (red circle and arrows).

The same is true for pumping between different aquifers at fixed locations where multiple groundwater reserves exist, i.e. CT and IC at the indicated locations in Figure 2.

These findings provide valuable input to the recently established trinational consultation mechanism in the region since the simulation-optimization model can be utilized as a decision-support tool (Observatoire to Sahara et Sahel 2003). Because stakeholder mistrust can be an issue in situations like this, abstraction activities could be regularly monitored by remote sensing by measuring irrigated perimeters so as to ensure proper adherence of cooperative strategies. Furthermore, through centralized data collection and management, the groundwater models can be constantly updated and improved in quality and updated optimal strategies thus calculated in response to new data.

References

Daskin, M. S., 1995: Network and discrete location models, algorithms and applications. – New York [etc.]: Wiley.

Bleuler, S., Laumanns, M., Thiele, L. & Zitzler, E., 2003: PISA – a platform and programming language independent interface for search algorithms. – Evolutionary Multi-Criterion Optimization, Proceedings 2632, 494–508.

Harbaugh, A. & McDonald, M., 1996: Programmer's documentation for MODFLOW-96, an update to the U.S. Geological Survey modular finite-difference groundwater flow model. – Tech. Rep. U.S. Geological Survey Open-File Report 96-486.

Observatoire du Sahara et du Sahel OSS, 2003: The north-western Sahara aquifer system: Joint management of a trans-border basin. – Tech. Rep. Observatoire du Sahara et du Sahel, Tunis.

Siegfried, T. & Kinzelbach, W., 2005: A multiobjective discrete stochastic optimization approach to shared aquifer management: Methodology and application. – Water Resour. Res. **42**, W02402, doi:10.1029/2005WR004321.

Siegfried, T., 2005: Optimal utilization of a non-renewable transboundary groundwater resource – Methodology, case study and policy implications. – Schriftenreihe des Instituts für Hydromechanik und Wasserwirtschaft, Zurich, ISSN 1423–7997.

Siegfried, T., Bleuler, S., Laumanns, M., Zitzler, E., Kinzelbach, W., 2009: Multi-Objective Groundwater Management Using Evolutionary Algorithms. – IEEE Transactions on Evolutionary Computation, Vol. **13**, No. 2, April 2009, doi: 10.1109/TEVC.2008.923391.

Zitzler, E., 2001: SPEA2: Improving the strength Pareto evolutionary algorithm. – TIK Report 103. Swiss Federal Institute of Technology, Zurich.

Zitzler, E., Laumanns, M. & Thiele, L., 2002: SPEA2: Improving the strength pareto evolutionary algorithm for multiobjective optimization. – In: Giannakoglou, K., Tsahalis, D., Periaux, J., Papailiou, K. & Fogarty, T. (eds.): Evolutionary Methods for Design, Optimization and Control, CIMNE, Barcelona, Spain.

Proceedings "Hydrogeology of Arid Environments", p. 162–165
Stuttgart, March 2012

Extended Abstract

Wadi System Components under Arid Climate to Estimate Transmission Losses and Groundwater Recharge through Analytic/Numeric Solutions

Ali Unal Sorman[1]

[1] Middle East Technical University, Civil Eng. Dept., Ankara, Turkey. email: sorman@metu.edu.tr

Key words: Transmission loss, recharge, arid climate

Introduction

The **World Water Assessment Programme** (WWAP) is established to develop the tools and skills needed to achieve a better understanding of the basic hydrological processes and their components (rainfall, runoff, evaporation, transmission and recharge); management practices and policies that will help improve the supply and quality of global resources under scarce resources due to arid climate.

Part 1. Wadi System Components

This section will discusses the cycle elements in a summarized form in addition to the estimation of bare soil evaporation which is highly dependent on the availability of soil moisture in the upper surface layer, the temporal and spatial distribution of rainfall and the accumulated rainfall depth and moisture-holding capacity of the soil layers are needed in studies of arid regions.

Quantification of water balance components, under arid conditions, is essential to the development of water management policies. This study demonstrates the application of the mass water balance approach for the assessment of water resources in a typical watershed located in the southwestern region of Saudi Arabia (SA) including process components as rainfall, evapotranspiration, recharge, soil moisture and wadi flow.

Part 2. Estimation of Wadi Recharge from Channel Losses

This part examines the infiltration-recharge processes under arid climatic conditions. Results presented in this research paper are the outcome of long-term field investigations involving rainfall, runoff, infiltration and recharge characteristics at several sites of the experimental basin in the Kingdom of SA.

Transmission loss assessment in ephemeral streams of arid regions is difficult due to the transient nature of the surface and subsurface flow processes. However, simplified procedures based on the mass balance approach provide a reasonable estimate of mean infiltration losses. If T_L represents the cumulative volume of transmission loss, V_{UP} the cumulative upstream inflow volume, V_{DS} the cumulative downstream outflow volume and V_{TR} the tributary runoff contribution to the main channel, then it follows by mass balance expressed by eq. 2.1

$$T_L = V_{UP} - V_{DS} + V_{TR} \qquad \text{Eq 2.1}$$

Large variations in transmission loss occurred due to varying runoff magnitude and spatial variability. Runoff availability was shown to have a dominant influence on the amount of transmission loss and subsequent recharge. Runoff information was therefore classified into four groups based on runoff volumes recorded at the upstream and downstream stations.

The analysis of results indicated that the predominant parameters controlling the magnitude of transmission loss and groundwater recharge are the flood hydrograph and soil characteristics. Hydrograph shape is determined by wadi channel width, depth and duration of inundation. The data analyses showed that the magnitude of transmission losses, and the resulting groundwater recharge expected under arid conditions, can be reasonably predicted using the equations formulated in this study. It is also considered that the equations can be used to estimate recharge in areas with similar hydrological and morphological characteristics.

Part 3. Groundwater Recharge Estimation from Ephemeral Streams

In Part 3, the estimation of groundwater recharge is done, it is important to simulate the behavior of the unconfined aquifer under transient conditions using analytical and numerical models, for the purpose of model calibration and mass balance prediction. A numerical model is formulated to estimate recharge in an arid-zone wadi hydrology and its validity is tested by comparing it with an analytical solution of the derived equations.

The calculated recharge values matched the piezometric levels observed at a well site at the edge of the wadi channel. The total recharge depths found by integration in the time domain provided a good estimate of the transmitted volume of water per unit length of wadi channel. The findings were confirmed by runoff volume measurements at gauging stations located in the basin. Then, a numerical approach is applied which is an extension and numerical form of an integral formulation (Morel-Seytoux & Miracapillo 1988, Morel-Seytoux et al. 1988).

Analytical Solution Used for Comparison

The results from the analytical solution by Ortiz et al., 1978 were presented as dimensionless curves at the centre of the mound, and water table profiles were generated by the solutions and represented by Figure 1a. These profiles may be used to estimate the height of the mound or to determine the aquifer properties from observed data.

The water table profile is approximated below the infiltration basin by a line at a distance Zrf above the initial water table location and by the profile h(x, t) in the region not below the direct infiltration.

The origin for the x-axis is at the limit of the infiltration boundary of the area of width 2B. The rapid change in water table elevation in the region of sharp curvature of the flow is represented by the head drop, ΔH which is a function of time and equal to eq. 3.1. Then, the mathematical solution of the problem is decomposed into two parts.

$$\Delta H(t)=\left[Z_{rf}(t)-h(x=0,t)\right] \qquad \text{Eq 3.1}$$

Numerical formulation for constant and transient recharge rate, I_0

It is easier to solve the integro-differential Eq. 3.1 numerically in discretized form (Morel-Seytoux & Miracapillo 1988). This provides q(t) at discrete time intervals. The numerical solution of the original Eq. 3.2 for q(t) is done using physically measurable parameters; the half-width of the channel (B), saturated effective conductivity (K), saturation deficit, transmissivity (T), drainable porosity and recharge rate (I_0) as shown by Eq. 3.3.

$$KBI_0 t = B(\breve{\theta}-\theta_0)q(t)+K\int_0^t q(\tau)d\tau + \left[\frac{2KB(\breve{\theta}-\theta_0)}{\sqrt{T\pi\varphi}} * \int_0^t \sqrt{t-\tau}\,\frac{\partial q(\tau)}{\partial \tau}d\tau\right] \qquad \text{Eq 3.2}$$

$$q(t)=\frac{KBI_0 t}{\left[A\right]}-\frac{K}{\left[A\right]}\sum_{v=1}^{t-1}q(v)-\frac{2BK(\breve{\theta}-\theta_0)}{\left[A\right]\sqrt{T\varphi\pi}}\left[B\right] \qquad \text{Eq 3.3}$$

where

$$\left[A\right]=\left[(\breve{\theta}-\theta_0)B+K+(2BK(\breve{\theta}-\theta_0)/\sqrt{T\varphi\pi})\Delta(1)\right] \qquad \text{Eq 3.4}$$

$$\left[B\right]=\left[\sum_{v=1}^{t-1}\left[q(v)-q(v-1)*\Delta(m)-q(t-1)\Delta(1)\right]\right] \qquad \text{Eq 3.5}$$

$$\Delta(m)=\frac{2}{3}\left[m^{\frac{3}{2}}-(m-1)^{\frac{3}{2}}\right];\ \mathrm{m}=(\mathrm{t}-\mathrm{v}+1) \qquad \text{Eq 3.6}$$

$$\Delta(\mathrm{I}) = 0.667 \qquad \text{Eq 3.7}$$

Transient Recharge Rate, I_0

Following the steps described earlier in the derivation of Eq. 3.2, the unknown q(t) is expressed similarly in terms of the known transient recharge rate, I_0 (t), as: leading after integration, to

$$KB\int_0^t I_0(\tau)d\tau = B(\breve{\theta}-\theta_0)q(t)+K\int_0^t q(\sigma)d\sigma + \left[\frac{2KB(\breve{\theta}-\theta_0)}{\sqrt{T\varphi\pi}} * \int_0^t \sqrt{t-\sigma}\,\frac{\partial q(\sigma)}{\partial \sigma}d\sigma\right] \qquad \text{Eq 3.8}$$

The integro-differential Eq. 3.8 can be expressed in discrete form as for the steady case and its solution for q(t) is provided by Eq. 3.9 in discretized form, where it can be solved numerically for any input parameters using a computer program coded by the authors:

$$q(t)=\frac{KB}{\left[A\right]}\sum_{v=1}^{t}I_0(v)-\frac{K}{\left[A\right]}\left[\sum_{v=1}^{t-1}q(v)\right]-\frac{2KB(\breve{\theta}-\theta_0)}{\left[A\right]\sqrt{T\varphi\pi}}\left[B\right] \qquad \text{Eq 3.9}$$

where [A], [B] and Δ(m) are the same as in the case of steady recharge.

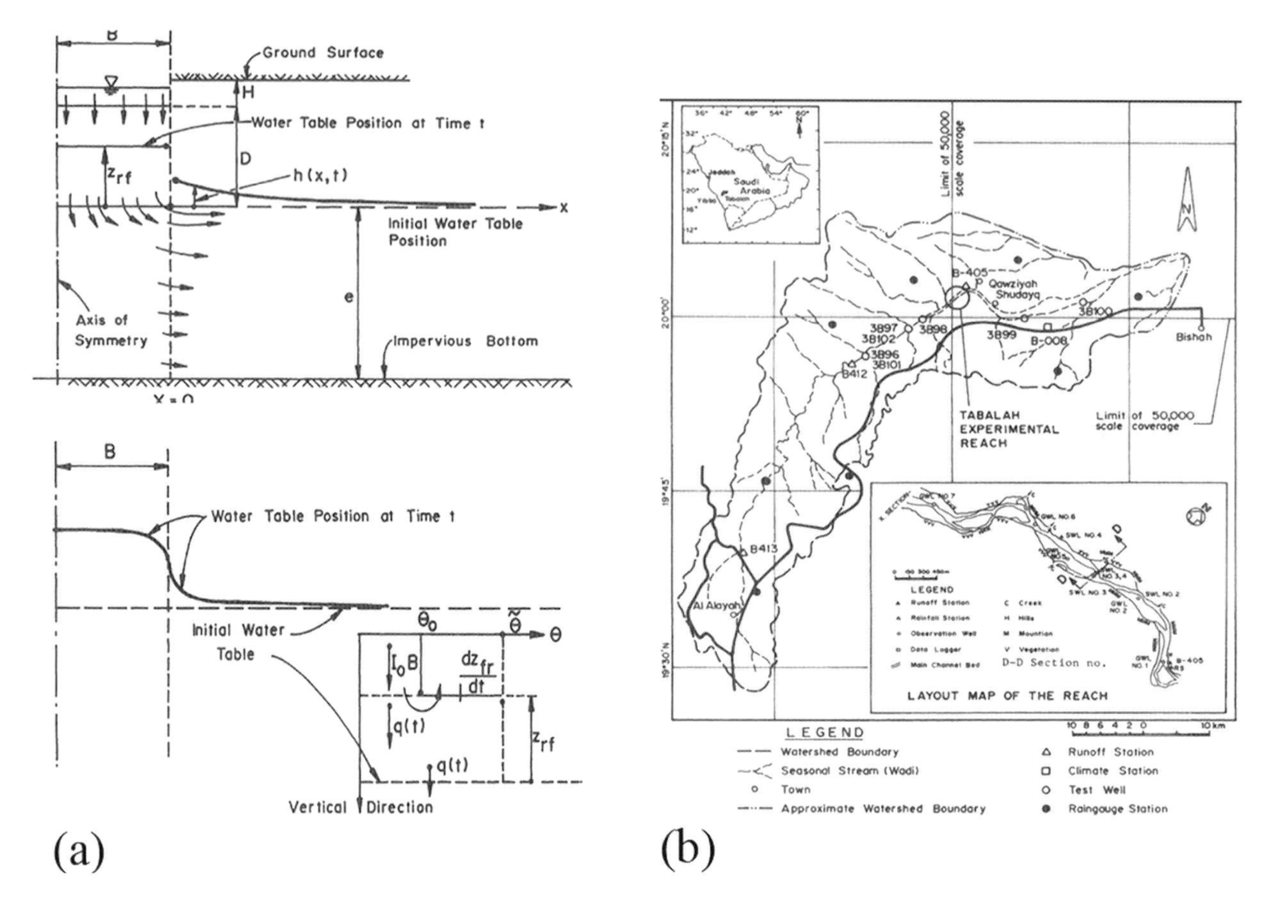

Figure 1. a: Schematic representation of cross-section and water table profile (from Morel-Seytoux 1988). b: Wadi Tabalah and layout map of the reach.

For calibrating the numerical model and comparing it with the field data, a detailed investigation of the mechanism that relates infiltration to recharge is required. An experimental site in Wadi Tabalah, as shown in Figure 1b, was selected along the main channel.

Discussion of results

Comparison of the analytical solution with the numerical technique for constant recharge rate, and the numerical solutions with the transient recharge rates from observed data have both shown that a versatile numerical technique can be used reliably to estimate recharge rates under arid climatic field conditions.

The results indicate that mound heights at the centre of the recharge area, Zrf, and at the edge, h(0, t), are almost identical for the set of hydraulic and soil parameters. Given the proven validity of the numerical model for the parameter values corresponding to field conditions, piezometric level observations in one of the wells (no. 3) at the edge of the recharge zone were used to infer the pattern of variable recharge rates. For that purpose, the observed groundwater table levels at well 3 are tabulated at two-hourly time intervals for 11 May 1987 in Table 1.

Conclusion

Although hydrologists and geologists have been working with wadis for a long time in the Arab Region and other arid and semi-arid regions of the world, much work remains to be done to develop adequate techniques of resource exploitation which recognize the susceptibility of wadi systems to over-exploitation and ensure their sustainable development. Many workers have reviewed the state-of-the-art, and identified future needs (Al-Weshah 2002, Wheater & Al-Weshah 2002).

Major areas of need in the Arab Region include improved data management systems and data collection networks in wadi basins. There is a need for improved measurement techniques for wadi flow, wadi sediment load and groundwater response. In

Table 1: Groundwater table levels, recharge rates at the center and edge.

Symbol	Description	Time (hours)						
		2	4	6	8	10	12	14
$I_0(t)$	(m/h) recharge rate	0.022	0.055	0.045	0.040	0.035	0.032	0.028
$Z_{rf}(t)$	(m) centre	0.277	1.294	1.483	1.673	1.740	1.752	1.770
h(t)	(m) edge	0.204	0.590	0.835	1.025	1.145	1.217	1.250
q(t)	(m^2/h) specific discharge	0.073	0.704	0.648	0.648	0.596	0.536	0.494
GW table rise	(m) observed at well 3	0.210	0.600	0.860	1.000	1.140	1.200	1.250

addition to conventional measurements there is a need to measure other aspects of wadi hydrology such as watershed characteristics including vegetation cover, topography, soil characteristics, geology and land use.

High quality research is needed to investigate processes such as spatial rainfall, infiltration and groundwater recharge. Both detailed research and regional analyses are required for a better understanding of wadi hydrology.

Modeling of wadi aquifer systems is an effective tool for the sustainable management of wadi systems. Development of rainfall-runoff models appropriate for wadi catchments is a basic need. However, there are still many uncontrolled or unknown elements that hinder understanding of physical systems.

References

Al-Weshah R. A. (Ed.), 2002: Water Resources of Wadi Systems in the Arab World: Case Studies. – IHP, UNESCO Cairo Office, IHP, NO. **12**, Cairo

Morel-Seytoux, H. J., 1988: Soil-aquifer-streams interaction – a reductionist attempt toward physical-stochastic integration. – J. Hydrol. **102**: 355–379.

Morel-Seytoux, H. J. & Miracapillo, C., 1988: Prediction of infiltration, mound development and aquifer recharge from a spreading basin or an intermittent stream. – Hydrowar Report No. **88.3.** Hydrology Days Publications, Atherton, CA. 96 pp.

Morel-Seytoux, H. J., Miracapillo, C. & Abdulrazzak, M. J., 1988: A reductionist physical approach to unsaturated aquifer recharg. – In: Dahlblom, P. & Lindl, G. (Eds.). – Proc. Int. Symp. on Interactions Between Groundwater and Surface Water. IAHR, Ystad, Sweden, 181–187. Vol. 1.

Ortiz, N. V., McWhorter, D. B., Sunada, D. K. & Duke, H. R., 1978: Growth of groundwater mounds affected by in-transit water. – Wat. Resour. Res. **14**: 1084–1088.

Wheater H. & Al-Weshah, R. A. (Eds.), 2002: Hydrology of Wadi Systems. – IHP Regional Network on Wadi Hydrology in the Arab Region, IHP-V, Technical Documents in Hydrology, No. **55**, UNESCO, Paris.

Proceedings "Hydrogeology of Arid Environments", p. 166–170
Stuttgart, March 2012

Extended Abstract

Mitigating the Current and Future Challenges for the Drinking Water Supply of Damascus, Syria

Mathias Toll[1*], Khaled Shalak[2], May Al-Safadi[2], Ahmad Hadaya[2], Ahmad Abdullah[1], Refaat Rajab[1], Georg Houben[1], Thomas Himmelsbach[1]

[1] BGR Federal Institute for Geosciences and Natural Resources, Groundwater Resources, Stilleweg 2, 30655 Hannover, Germany, *mathias.toll@bgr.de
[2] DAWSSA Damascus Water Supply and Sewerage Authority

Key words: climate change, karst groundwater, Figeh spring system, Syria

Introduction

The Figeh spring system is located around 15 km northwest of the Syrian capital of Damascus and accounts for around two-thirds of the water supply of Damascus (Fig. 1). The Figeh spring system consists of the Figeh main spring, Figeh side spring, Haroush and Deir Moukareen wells. The long-term average discharge of the Figeh main spring is 200 Mm^3/a. The discharge of the Figeh main spring is highly variable with fluctuations ranging from 1.5 to 27 m^3/s.

The recharge area lies in the northeast trending part of the Anti-Lebanon mountain chain. Stable isotopes analyses estimated a mean recharge altitude of around 2,100 ± 75m (Al-Charideh 2011), where most of the recharge area lies above 1,800m and reaches up to 2,500 m asl. A detailed description

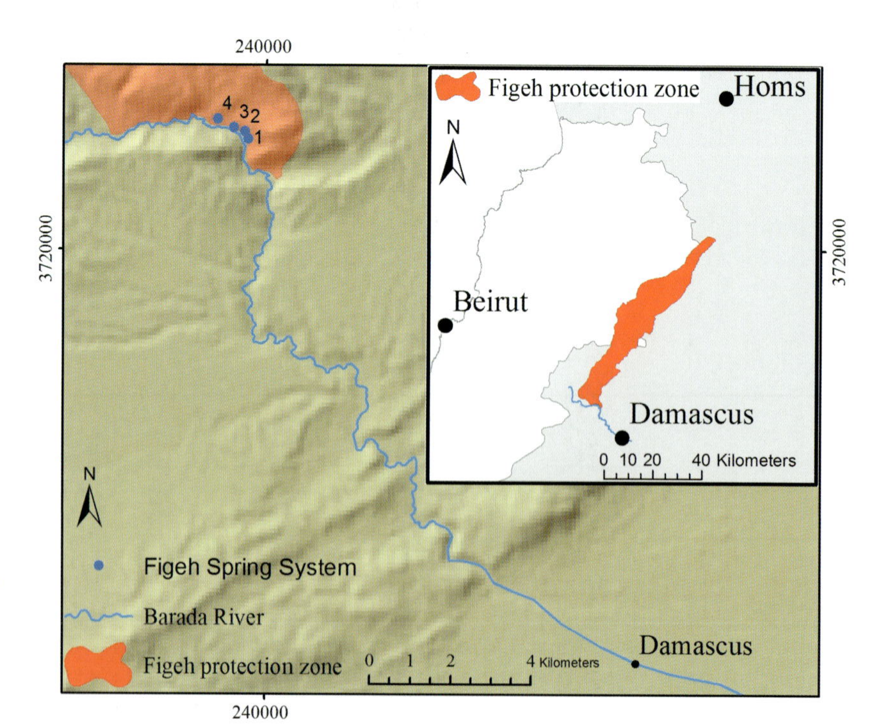

Figure 1: Location of the Figeh spring system with its protection zone (1 Figeh main spring, 2 Figeh side spring, 3 Haroush, and 4 Deir Moukareen wells).

www.borntraeger-cramer.de
2012/0166/$ 1.25

of the hydrogeology of the Figeh spring system can be found at Lamoreaux et al. 1989. The total extent of the catchment was previously estimated to cover a limestone outcrop area of 665 km^2 (Sogreah 1973). The recharge consists solely of infiltrating precipitation, both snow and rain. More than 70% of the precipitation that falls during the wet season is snow. This creates an important water buffer storage in the mountains. A future temperature rise, as predicted by regional and global climatic models, might lead to a significant decrease or a complete loss of this important water buffer storage, which in turn would have a significant influence on the water availability throughout the year (future challenge).

The Damascus Water Supply and Sewerage Authority (DAWSSA) manages the spring water discharge. The water is mostly used for drinking water purposes. Since 1989 the whole catchment area is under groundwater protection by the water law no. 10.

During the flood period the natural outflow of the Figeh spring is enough to supply the capital of Syria with drinking water. At this peak flow period the discharge of the spring is too large to be diverted completely to Damascus and the villages downstream. Excess water is discharged into the adjacent Barada River. Since 2006 aquifer storage and recovery tests are performed at the Quaternary aquifer in the Damascus Basin with excess water from the Figeh spring. Several challenges were faced at the sites. These challenges included e.g. low injection rates and the recovery of acceptable amounts of water and the presence of high concentrations of nitrate in the recovered water (Kattan et al. 2011).

During the dry period additional wells and caissons in the surroundings of the springs pump water from the aquifer. Overexploitation changed the groundwater flow patterns in the vicinity of the spring system and resulted in infiltration of adjacent Barada River water (current challenge). Barada River, especially during summer and autumn, consists mostly of contaminated wastewater from unsewered domestic settlements in the Barada Valley.

Current Groundwater Quality Challenge

The groundwater of the Figeh aquifer is low mineralized. Selkhozpromexport (1986) stated that the salinity ranges from 200 to 600 mg/l in the Cenomanian-Turonian rocks. Groundwater temperature is fairly constant around 13°C and only slightly decreases during the high flow period. Burdon & Safadi (1964) investigated the hydrochemical composition of groundwater from the Upper Cretaceous (8 sites) karst aquifer. The comparison shows, that the Ca-Mg-HCO_3 water type is predominant.

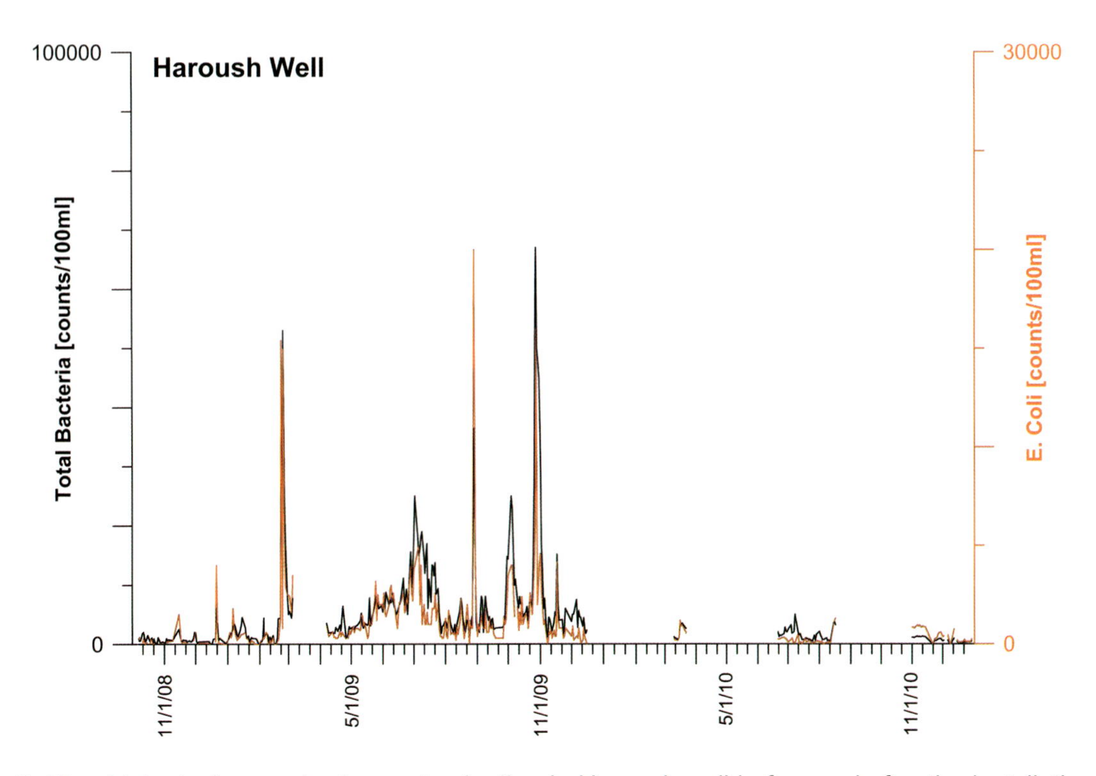

Figure 2: Microbiological groundwater contamination in Haroush well before and after the installation of a wastewater collector in the Barada River streambed.

Kattan (1997) found that the variations of the ion concentrations and correspondingly of the specific electrical conductivity are relatively small during the dry season, however, their values decrease during peak flow season. The Old Haroush Spring and Old Side Spring as well as Barada River show a sharp decrease in calcium concentration and less pronounced decrease in magnesium concentration during high flow season (flood period). This behavior of the two springs may suggest rapid recharge from precipitation or even a hydraulic contact between river and springs. On the contrary, Figeh main spring exhibits a sharp decrease in magnesium concentration accompanied by a slight increase in calcium concentration.

The possible connection between Barada River and Haroush spring is further indicated by micro-contaminants and the significant decrease of microbiological pollution at Haroush well after the completion of a wastewater collector in the Barada River streambed. A significant improvement of water quality took place after the installation of a sewer pipe in a concrete streambed in 2009. Figure 2 shows the bacterial contamination before and after the installation of the wastewater collector. Time series analyses for different micro-contaminants (54 different components) revealed waste-water borne pollutions in Haroush well. The unstable compound Caffeine and its metabolites were mostly encountered, while more persistent micro-contaminants like Sulfamethoxazole and Carbamazepine were always present (Fig. 3). This might indicate that at times fresh, but very low contamination with wastewater exists at Haroush well. The presence of Sulfamethoxazole and Carbamazepine in all samples indicates residues of the old contamination with wastewater. It should be noted, that concentrations of micro-pollutants are extremely low (ng/L) and therefore do not pose a threat to human health. No continuous major pollution indicators were found at Figeh main spring. Additional structurally engineered measures are implemented to cope with the altered hydraulic system to better protect this vital source of drinking water for the capital Damascus.

Future Groundwater Quantity Challenge

Continuous long-term isotope (3H, δ18O and δ2H) records between 1970 and 1990 of the water emerged from the Figeh Spring System and of the precipitation in the Antilebanon Mountains are missing. This corrupts a reliable and precise determination of the isotopic mean residence time (MRT) of the base flow and of the isotopic altitude effect of the recharge area. In spite of this concern the re-assessment of the accessible isotope data reflects

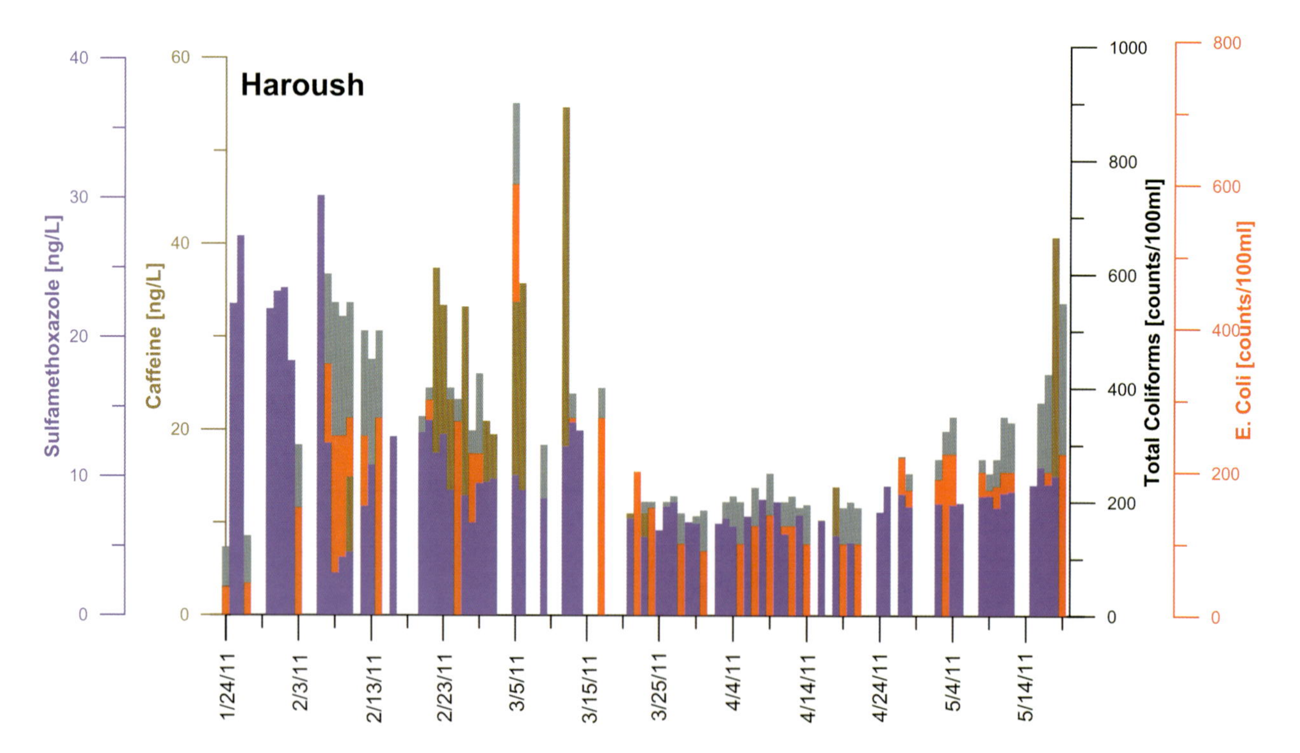

Figure 3: Time series analyses for two different micro-contaminants: Caffeine (brown) Sulfamethoxazole (purple) and total coliforms and Escherichia coli at Haroush well.

on one side the hydraulic disturbance, due to the present groundwater abstraction for the drinking water supply of Damascus, and provides evidence on the other side that the hydraulic situation has been changed since at least 1981, the time, when additional water was pumped from the Figeh aquifer. The published and recently measured data referred to the 3H record of precipitation at Bet Dagan provide evidence that the isotopic MRT of the late base flow of the Figeh Spring System increased since then. Groundwater pumped from the Figeh Spring System contains an admixture of about 25% of old deep groundwater without detectable tritium (Geyh 2010).

Decreasing precipitation trends observed in various parts of the Eastern Mediterranean region together with recent drought periods raise concerns about water availability under changing future climate conditions. A regional climate model developed by the Karlsruhe Institute of Technology (Smiatek & Kunstmann 2011) calculated a precipitation decrease in order of 11 % in winter and 8 % in spring together with an increase in temperatures of up to 1.6 °C in the near future (until 2050) thus challenging the future water supply of the Syrian capital. Especially the rise of temperature will lead to a significant reduction of the snow buffer storage which in turn will have its effect on the discharge behavior of the spring. Figure 4 illustrates the different discharge characteristics of an average cool wet season (2003/04) and an exceptional warm wet season (2009/10). The loss of most of the snow buffer storage can be clearly seen by the direct increase of discharge after rainfall events.

Recent stable isotope studies, e.g. Koeniger et al. 2012, are focused on a better understanding of the snow storage on the discharge behavior of the Figeh main spring. Automatic and manual snow-water equivalent measurements along with continuous measurement of selected groundwater parameters complete the investigation on the effect of the snow buffer storage on the spring discharge.

The enhanced understanding of the system allowed the formulation of proposals to mitigate the current and future challenges which includes structurally engineered modifications, regional adapted meteorological measurement stations, new water management options, and suggestions for new promising areas for managed aquifer recharge (MAR).

Acknowledgements

The work is conducted within the framework of the Syrian-German-Technical-Cooperation Project "Protection of the Figeh Spring System" funded by the German Federal Ministry for Economic Cooperation and Development (BMZ).

References

Al-Chariedeh, A., 2011: Environmental isotope study of groundwater discharge from the large karst springs in West Syria: a case study of Figeh and al-Sin springs. – Environ. Earth Sci. **63**: 1–10.

Burdon, D. & Safadi, C., 1964: The karst groundwaters of Syria; Journal of Hydrology **2**: 324–347.

Geyh, M., 2010: Isotope Hydrological Study on the Figeh Spring System; final unpublished report; technical cooperation project no.: 2009.2088.4; Federal Institute for Geosciences and Natural Resources (BGR); Winsen; 55 p.

Kattan, Z., 1997: Environmental isotope study of the major karst springs in Damascus limestone aquifer sys-

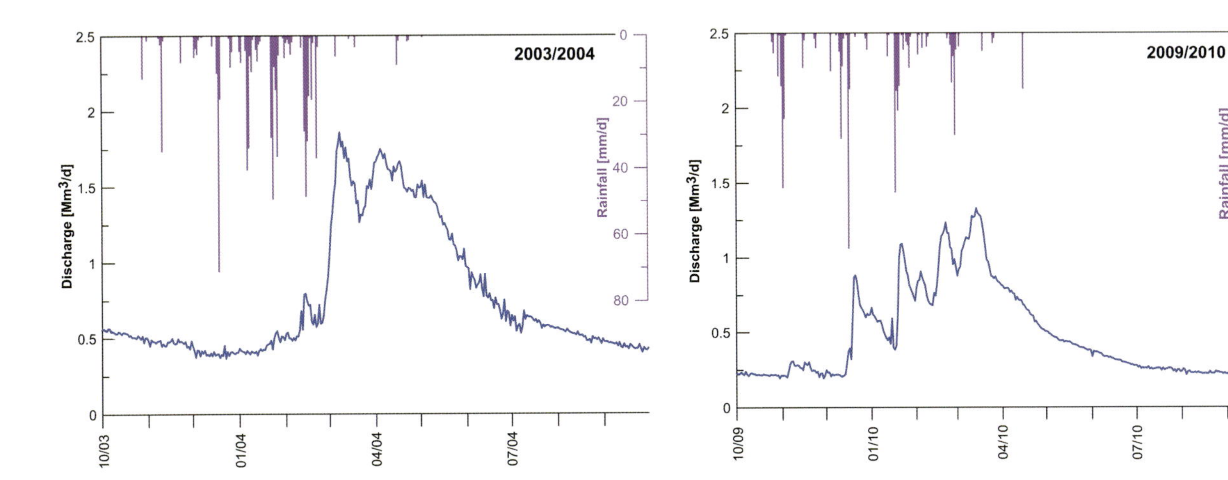

Figure 4: Left: Rainfall- discharge behavior of Figeh main spring during a normal cold wet season (2003/04). Right: Rainfall- discharge behavior of Figeh main spring during a warm wet season (2009/10).

tems: Case of Figeh and Barada springs. – Journal of Hydrology 193: 161–182.

Kattan, Z., Kadkoy, N., Nasser, S., Safadai, M. & Hamed, A., 2011: Isotopes and geochemistry in a managed aquifer recharge scheme: a case study of fresh water injection at the Damascus University Campus, Syria. – Hydrological Processes Vol. 24, 13: 1791–1805.

Koeniger, P., Toll, M., Himmelsbach, T. Shalak, K., Hadaya, A. & Rajab, R., accepted: Stable isotope studies in semiarid, karstic environments reveal information for sustainable management of water resources in Damascus, Syria; Proceedings of International Conference on Hydrogeology of Arid Environments; 14th – 17th of March 2012, Hannover, Germany

Lamoreaux, P. E., Huges, T. H., Memon, B. A. & Lineback, N., 1989: Hydrologic Assessment – Figeh Spring, Damascus, Syria; Environ. Geol. Water Sci. **13**: 73–127.

Selkhozpromexport, 1986: Water resources use in Barada and Auvage basins for irrigation of crops, Syrian Arab Republic. – USSR, Ministry of Land Reclamation and Water Management, Moscow.

Smiatek, G. & Kunstmann, H., 2011: Climate change investigations for the Figeh spring area/ Syria. – Karlsruhe Institute of Technology (unpublished report): 28 p.

Sogreah, 1973: Etude hydrologique et hydrogeologique de la Source Figeh. – Rapport final (unpublished), R 11442; Societe Grenobloise d'Etudes et d'Applications Hydrauliques (SOGREAH), Grenoble.

Proceedings "Hydrogeology of Arid Environments", p. 171–176
Stuttgart, March 2012

Extended Abstract

Groundwater Recharge in the Lake Chad Basin

Sara Vassolo

Bundesanstalt für Geowissenschaften und Rohstoffe, email: s.vassolo@bgr.de

Key words: Lake Chad basin, groundwater recharge, quaternary aquifer

Introduction

The Lake Chad Basin is located in the central part of Northern Africa and occupies an area of about 2 300 000 km^2 (Fig. 1). It is an extended plain mostly covered by medium to fine-grained sands with heights that vary between 180 and 380 metres above mean sea level (mamsl) in the centre that rise up to 3,000 mamsl to the borders.

The basin is characterized by two different landscapes subdivided by the 14°N parallel: sand dunes and the absence of surface water sources are typical for the northern part (Kanem region), while the south is richly watered by two main rivers that discharge in the lake. They are the Chari-Logone that supplies about 95 percent of the annual volume of water that reaches the lake and the Komadugu-Yobe that provides about 3 percent of the annual inflow into the lake (Fig. 1). The precipitation over the lake surface completes the remaining 2 percent.

Within the basin there are very important and well-known swamp regions: the Yaérés in the extreme north of Cameroon, the Lake Chad itself, Lake Fitri, the Massénya and the Salamat to the south and southeast of the Lake Chad respectively, and the Komadugu-Yobe to the north-east of Nigeria (shaded areas in Fig. 1).

Climatically the basin is characterized by three different zones: hyper-arid to arid in the north, semi-arid in the centre and subtropical in the south. Mean annual rainfall varies from less than 50 mm in the north to above 1000 mm in the south. High temperatures throughout the year, very low humidity except during the rainy season from June to August, intense solar radiation and strong winds lead to a high annual potential evapotranspiration of around 2,200 mm (Carmouze 1976).

Geology and Hydrogeology of the Basin

Most of the Lake Chad Basin is covered by Quaternary sands (Fig. 2) of different depositional origins. The dunes in the northern part of the basin are the effect of an aeolic deposition. Fluviatile, lacustrine and deltaic depositions in the south result in alternating sequences of thin layers of sand and clay that produce mainly clayey soils. These Quaternary sands act as an unconfined to semi confined aquifer with low hydraulic conductivity, especially vertical, due to the sequences of sand and clay in the south. Furthermore, due to its flatness and low gradient (in average 0.0005), the horizontal flow is very slow.

At a depth of some 75 metres appears a thick clay layer of some 280 m of Upper Pliocene age. This layer is almost impermeable and separates the Quaternary sands above from the Lower Pliocene sand and sandstone aquifer below and gives place to a widespread artesianism, especially in the central part of the basin. The Lower Pliocene has a thickness of 30 m and is underlain by the sandstones of the Continental Terminal (CT) of Tertiary age with a thickness of some 150 m. According to Eberschweiler (1993), both the Lower Pliocene and the CT have similar good hydrogeological properties and comparable water chemistry, therefore they can be considered as one aquifer. The deepest aquifer is expected to be the sandstone of the Continental Hamadien (Cretaceous), but it has not been studied yet and its extension and hydrogeological properties are almost unknown. Granitic rocks of the basement build the basis of the basin.

www.borntraeger-cramer.de
2012/0171/$ 1.50

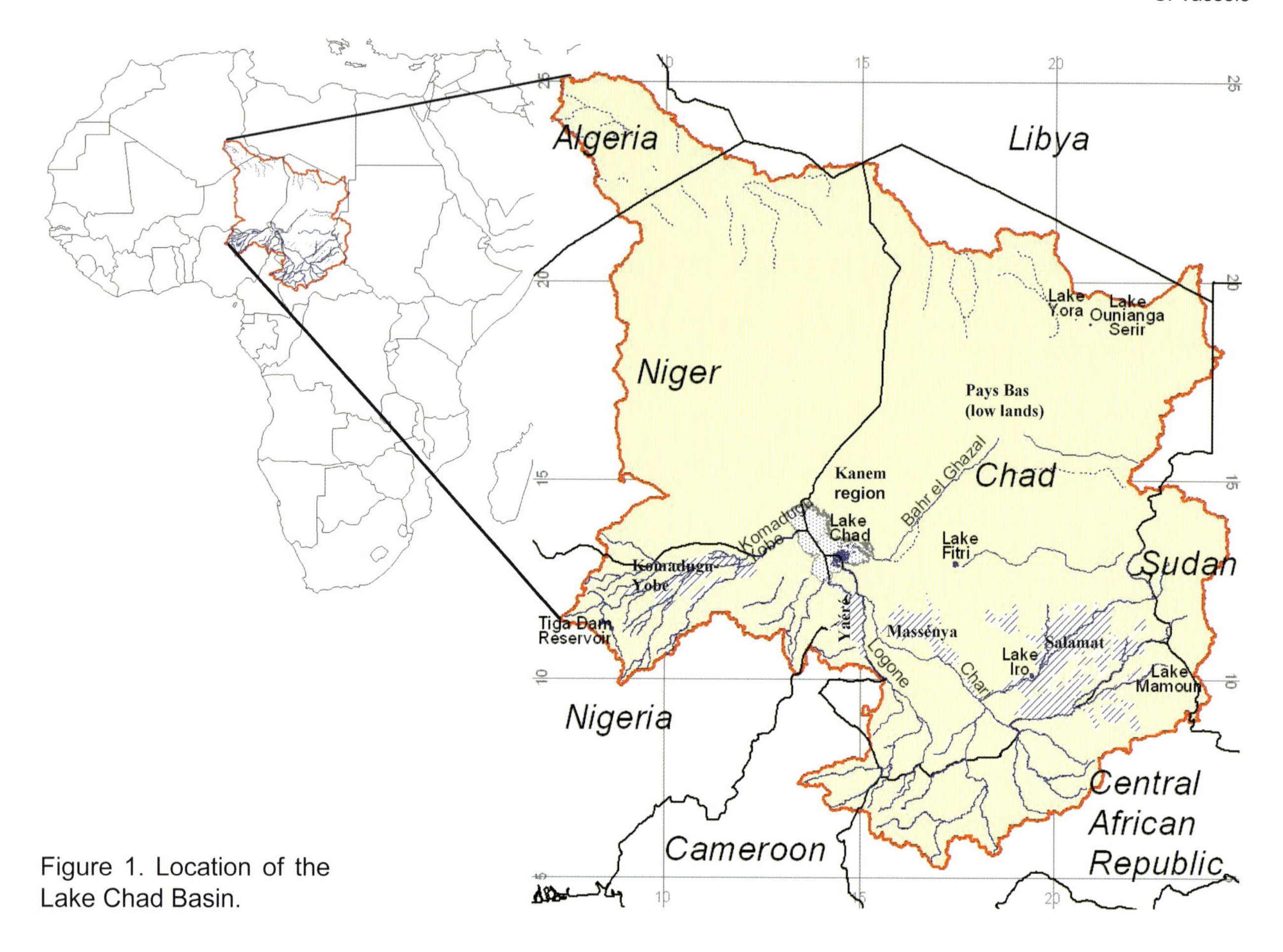

Figure 1. Location of the Lake Chad Basin.

Groundwater Recharge Investigations

Although the annual renewal is very important to the water balance, it is almost unknown both in quantity and distribution. Based on calibration results of a numerical model, Boronina et al 2007 estimate mean annual recharge figures in the ranges of 1 to 4 mm for the dunes of the Kanem region and 0.3 to 1 mm for the riverbeds and swamp areas in the south.

To investigate more about groundwater recharge distribution, measurements were performed in the frame of a regional BMZ/LCBC project, especially concerning the Quaternary aquifer in the Chad Republic. Measurements include groundwater level and water quality in 443 water points, oxygen-18 and deuterium concentrations in 383 boreholes and tritium in 54 wells.

Results and Discussion

Groundwater contours

The water levels were used to elaborate a groundwater contour map (Fig. 3) that allows determining flow directions. The map indicates groundwater flow from the Massénya swamp toward north, from the Logone River and the Yaéré swamp towards the Lake Chad, and from the Lake Chad itself towards the east-south-east. Further, groundwater flows from the Kanem region towards the south-east.

Considering that the higher groundwater contours are the result of surface water or precipitation percolation into the aquifer, recharge is caused by the Massénya swamp, the Logone River, and the Yaéré swamp, but the Chari River appears to be disconnected with the aquifer. Further, the Lake Chad seems to leak into the aquifer towards the east-south-east. Some sort of recharge should also take place in the Kanem region, due to direct percolation of precipitation through the aeolic sands.

SO_4 Concentration

The sulfate concentration in groundwater is presented in Figure 4. The map shows sulfate concentrations of less than 5 mg/l in the southern part of the study area. It is assumed that these low sulfate concentrations are the result of percolation of surface water, especially considering that surface water in the region contains less than 1 mg/l of sulfate. In other words, the Massénya and Yaéré swamps as well as the Logone River appear to recharge the aquifer, confirming the results of the contour map.

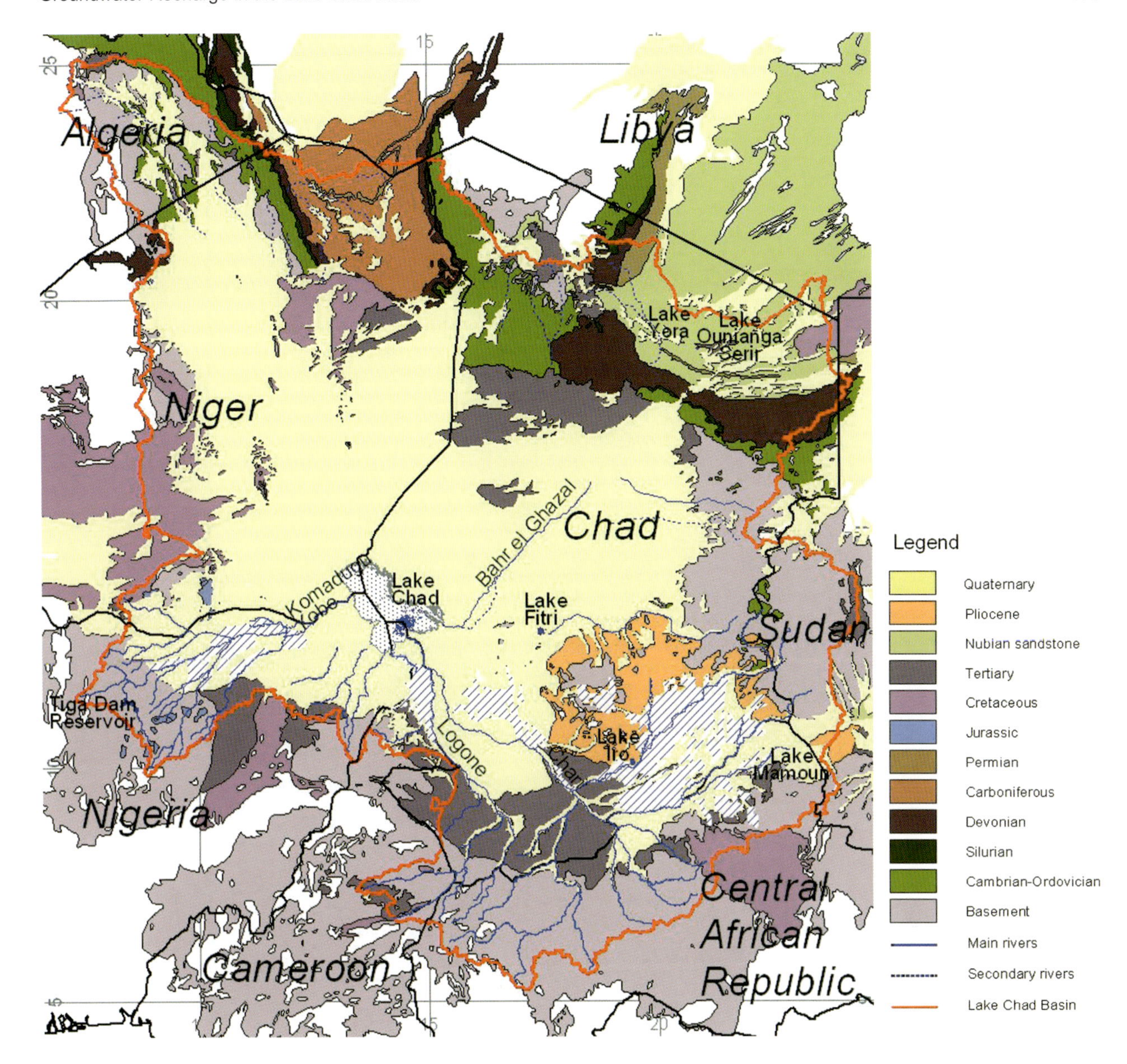

Figure 2. Geology of the Lake Chad Basin.

However and contradicting the results of the contour map, here the Chari River seems to contribute to the renewal of the Quaternary aquifer.

The Kanem region to the north of the Lake shows higher sulfate concentrations that could be the result of renewal from precipitation, but from an older period. Therefore, groundwater has had time to absorb sulfate from the underground.

The elevated sulfate concentration of groundwater to the SE of the Lake Chad indicates that, if leakage from the lake water at less than 1 mg/l of sulfate takes place, groundwater flow velocity must be very slow. The residence time is than very large giving groundwater time to absorb sulfate from the underground.

Isotopes

Due to the prevailing elevate temperatures, the surface water (lakes, rivers and swamps) experience high evaporation rates that concern mainly the light oxygen isotopes leaving behind the heavy isotopes. Therefore, if groundwater is recharged by surface water, it will be characterized by heavy oxygen isotopes.

The mapping of ^{18}O (Fig. 5) shows "heavy" groundwater as the result of renewal along the Logone River and the Yaéré swamp, and under the Massénya swamp, confirming the findings from above. The Chari Rivers seems to be disconnected from the Quaternary aquifer. It also shows heavy ^{18}O isotopes under the Lake Chad that could be

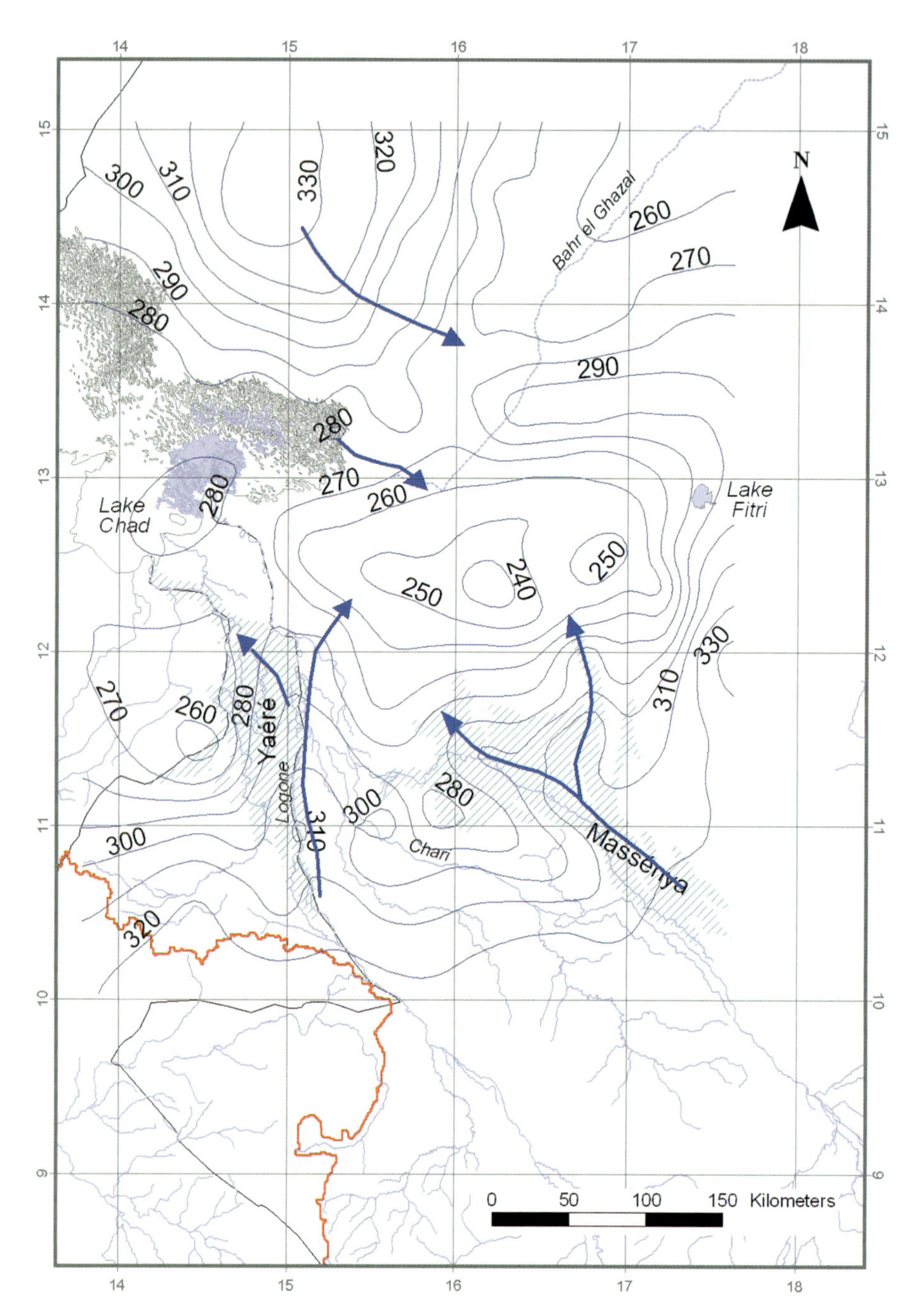

Figure 3. Map for groundwater surface. The arrows indicate water flow direction.

explained by direct infiltration of lake water. Further, heavy groundwater appears also to the east-south-east of the lake confirming the findings from above.

If the Kanem region gets recharge, it is due to direct percolation of precipitation during downpours. Groundwater in the area shows values of ^{18}O close to –4 that are typical for heavy precipitation rates in the area. However, long-term precipitation data does not show large amounts of precipitation in the area since the beginning of the 70ies. Thus, recharge must be rather old, what explains the relatively high sulfate concentration found in the area.

Conclusions and Recommendations

The general conclusions of the study can be summarized as follows:

- The Quaternary aquifer receives recharge from different sources, at least in the Chadian part.

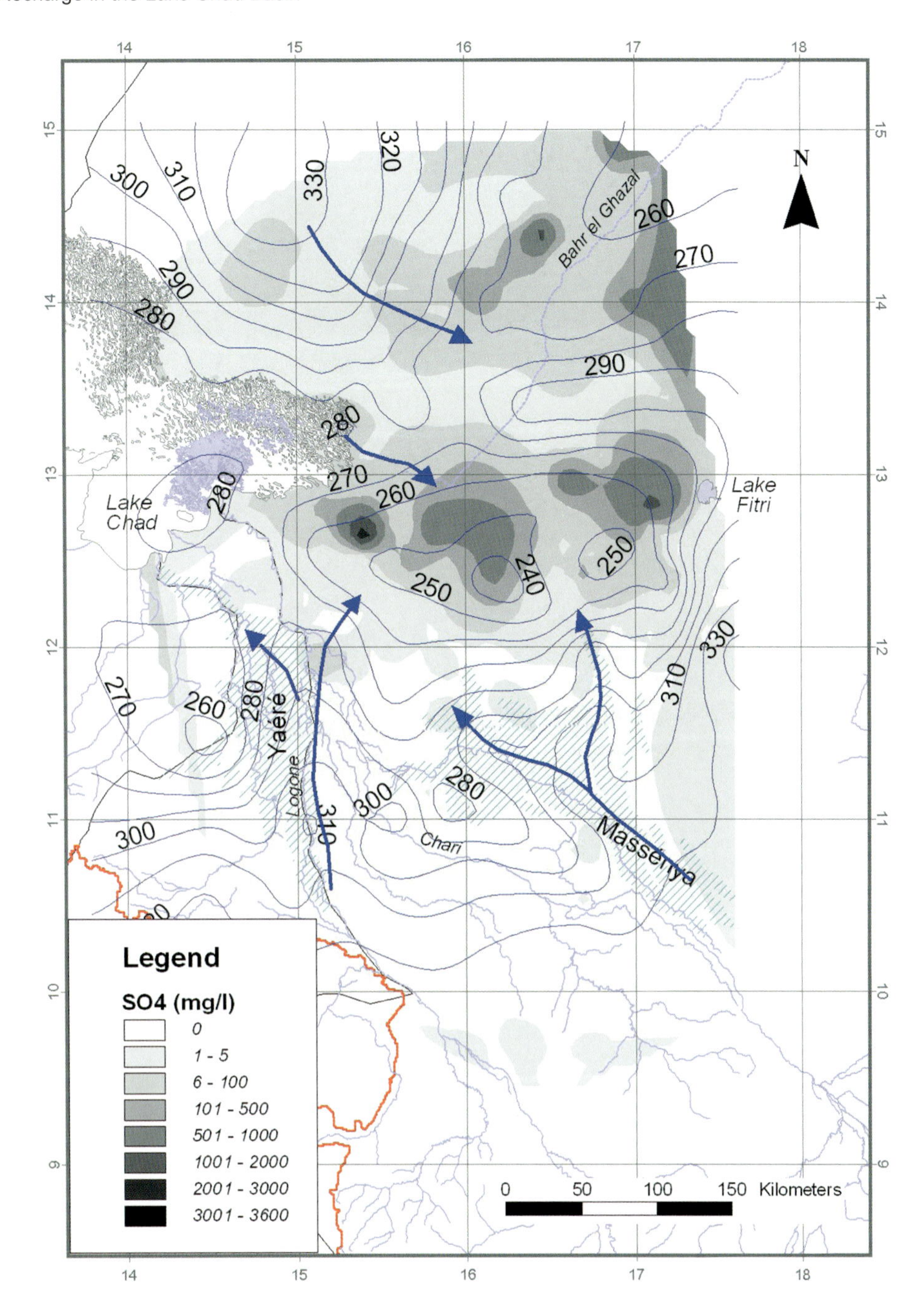

Figure 4. Contours of sulfate concentrations in groundwater for the Quaternary aquifer in the Chad Republic. The arrows indicate flow directions.

Sources of recharge are the Yaéré and Massénya swamps as well as the Logone River, but the Chari River seems to be disconnected from the aquifer.

- The Lake Chad leaks directly into the aquifer and water is "lost" towards the east-south-east at a very low flow velocity.
- Recharge in the Kanem region is relatively old. It was caused by direct percolation of downpours that took place before the 70ies.

As a result of the study it is highly recommended that the swamps in the area as well as the Logone River be protected concerning retention capacity and water quality, as they are the main source of recharge for the Quaternary aquifer.

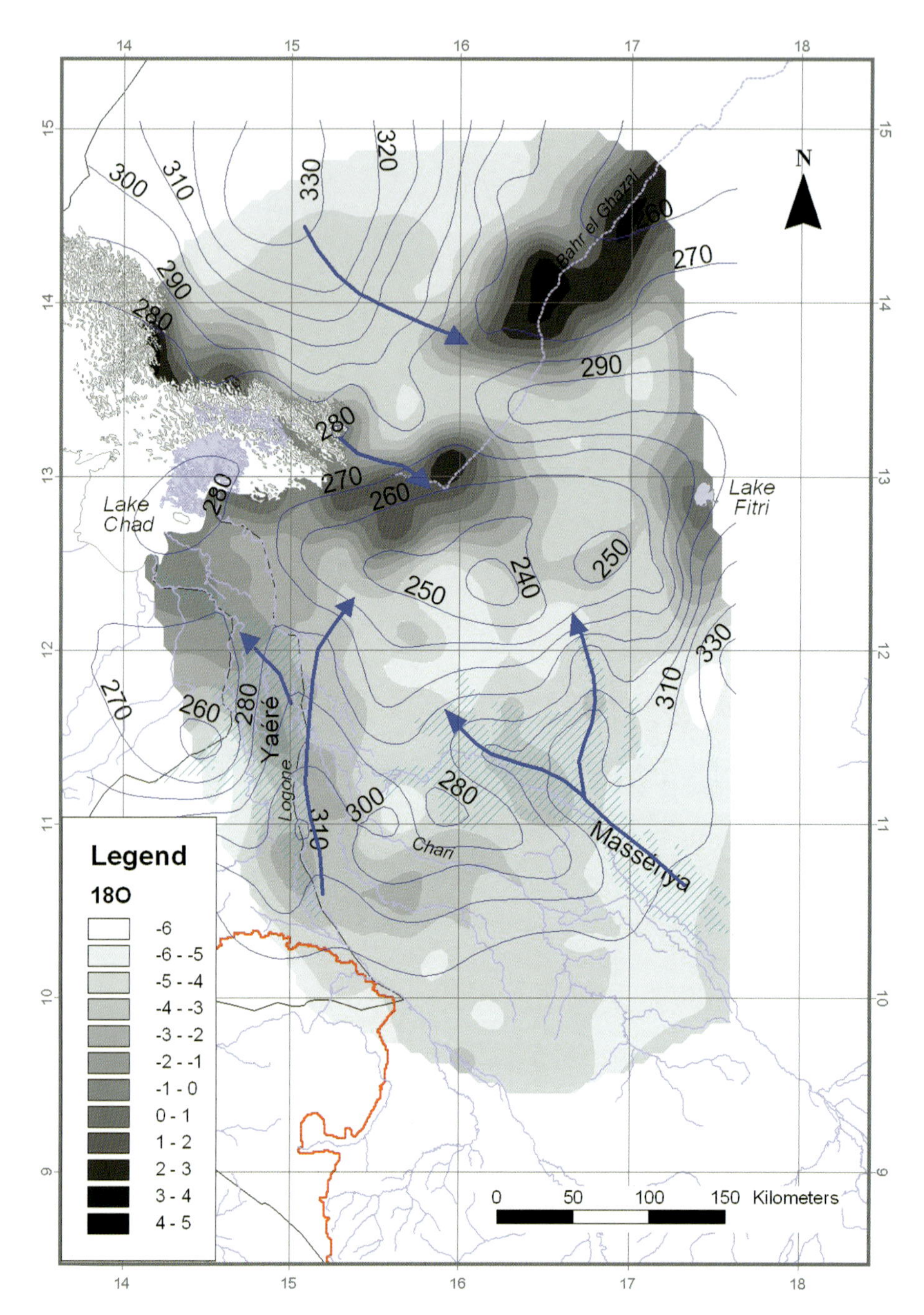

Figure 5. Map of oxygen-18. The arrows indicate flow directions.

References

Boronina, A., Favreau, G., Coudrain, A., Leduc, Ch. & Dieulin, C., 2007: Hydrogeology of the phreatic aquifer in the Lake Chad basin: groundwater modeling in conditions of data scarcity. HydroSciences, IRD-Université de Montpellier II, Case courrier MSE 300, avenue Emile Jeanbrau, 34095 Montpellier Cedex 5, France.

Carmouze, J.-P., 1976: Les grands traits de l'hydrologie et de l'hydrochimie du Lac Tchad. – Cahier O.R.S .T.O.M., sér. Hydrobiologie, vol. X, N° 1, 1976, pp. 33–56.

Eberschweiler, Ch., 1993: Monitoring and management of groundwater resources in the Lake Chad Basin. Final Report. BRGM-LCBC. Funding: Fonds d'Aide et de Coopération de la République Française Convention N° 98/C88/ITE.

Schneider, J. L. & Wolff, J. P., 1992: Carte géologique et cartes hydrogéologiques à 1/1,500,000 de la Républic du Tchad : Mémoire explicatif. Document du BRGM N° 209, Vol. 2. Éditions du BRGM.

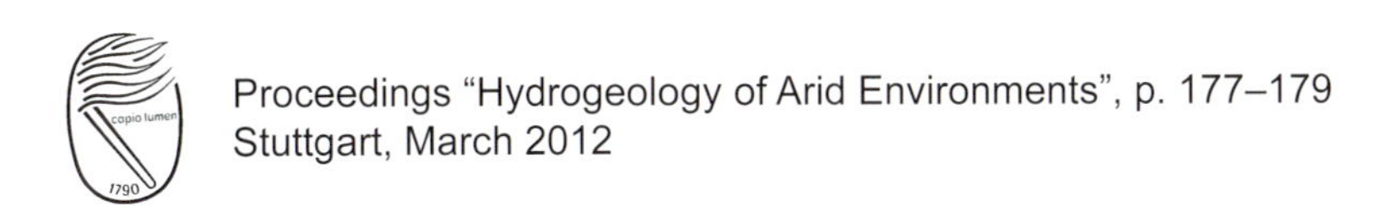

Proceedings "Hydrogeology of Arid Environments", p. 177–179
Stuttgart, March 2012

Extended Abstract

Monitoring of Water Resources in Arid Regions

Kai Vogel[1], Johannes Döhler[2]

[1] SEBA Hydrometrie GmbH & Co. KG, Gewerbestrasse 61a, 87600 Kaufbeuren, Germany, email: vogel@seba.de
[2] Dornier Consulting GmbH, P.O. Box 2730, Riyadh 11461, Kingdom of Saudi-Arabia,
email: johannes.doehler@dornier-consulting.com

Key words: monitoring networks, groundwater, meteorology, telemetry, SEBA, time & cost efficiency

Monitoring and managing of limited and generally non-renewable groundwater resources in arid regions is a vital and essential task. Its negligence can cause a severe threat to the economic development and social peace of the region. A central obstacle for the sustainable management of scarce water resources is the lack of reliable and long-term data series and information for planning future allocations driven by progressively limited water quantities.

In the last decade, several large scale groundwater resources projects were implemented in the Middle East, which included also the review, rehabilitation, technical upgrade, and extension of existing monitoring networks. For the design of a state-of-the-art and custom-tailored monitoring network, it is inevitable not only to consider hydrological and hydrogeological aspects but also economic and infrastructural facets. These facets are essential in respect of a long-term and sustainable network operation and management (O&M). Particularly in arid regions the technical implementation of monitoring networks is challenging due to the prevailing harsh environmental conditions. These are reflected by high ambient air-temperatures and -fluctuations (day/night), high air-humidity (coastal plains) / -aridity, extreme weather events (e.g. flash-floods in wadis) as well as specific hydro-chemical groundwater conditions. Technical modifications of monitoring system

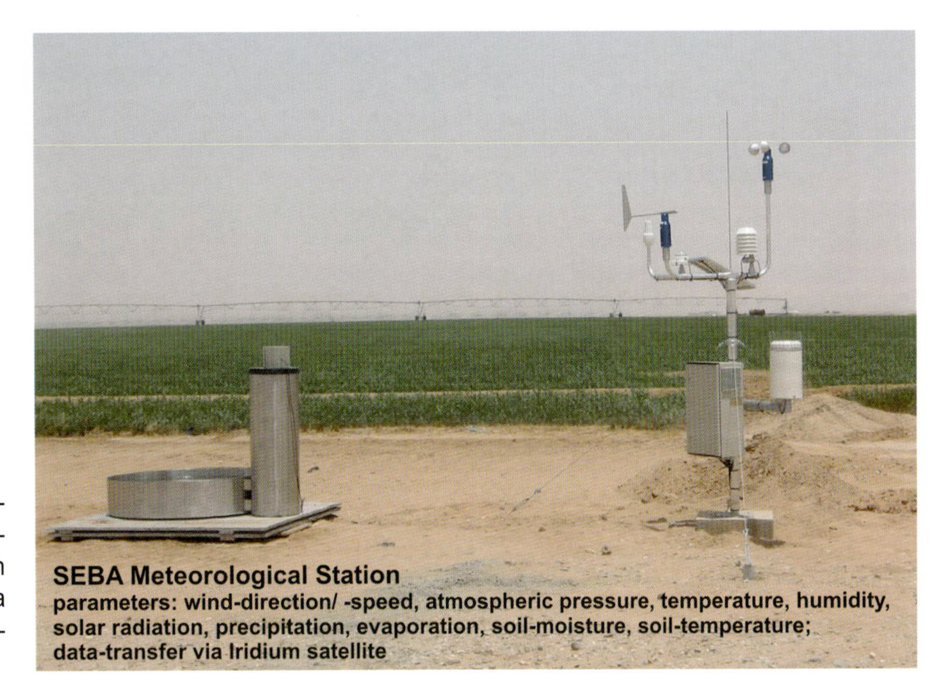

Figure 1: Kingdom of Saudi Arabia – SEBA Telemetric meteorological station incl. evaporation pan (data transfer via Iridium satellite).

www.borntraeger-cramer.de
2012/0177/$ 0.75

modules with e.g. extended temperature ranges and/or improved protection against electrostatic charges (caused by hot, dry winds) are therefore indispensable. In arid regions a high technical reliability of a monitoring system is essential since they are often required in remote areas in order to e.g. determine the spatial, temporal and seasonal distribution of precipitation, potential evapotranspiration, surface runoff and natural groundwater recharge.

All these parameters are indispensable for any hydrogeological model, which is an important tool for groundwater management.

An additional important aspect is the general protection of monitoring systems against vandalism (= worldwide phenomena) which is a crucial factor in terms of a long term perspective of a fully operational network. Depending on the surrounding conditions, there are several possibilities to increase the safety of a monitoring system by e.g. the re-use of an existing station infrastructure (e.g. Jordan) the setup of well-houses (e.g. Saudi Arabia) or the general minimization of the deployment of solar panels in unprotected locations (e.g. Jordan).

Associated with a state-of-the-art network are the telemetric data transfer and communication between the SEBA monitoring station and the central data base (SEBA DEMASdb). This telemetric set-up has various advantages, especially for arid regions. Besides an up-to-date information on the measured parameters, the telemetric data transfer serves also as an early warning system on critical parameters like strong precipitation events which can result in heavy flash floods with a disastrous impact on the affected local population (flood and disaster management).

In addition, the telemetry provides permanent up-to-date information on the current technical status of the network. In regard to large spatial networks (e.g. Saudi Arabia, UAE, Qatar) with remote station locations the telemetry especially enables a time and cost efficient operation and maintenance of the entire monitoring network. Malfunctions of stations can immediately be identified and eliminated and the downtime of the system minimized. The set up of telemetric features are realized by SEBA monitoring stations which combine intelligent energy management with state-of-the-art data transfer functionalities (e.g. ftp-push technology) as well as integrated alarm management. A direct access to the measured values via the internet or intranet for defined users can be realized by means of the Hydrocenter (SEBA).

An essential part of any monitoring network rehabilitation/extension project is a custom-tailored knowledge transfer and capacity building concept.

Figure 2: Jordan – SEBA Telemetric groundwater monitoring station (ftp-push). SEBA-equipment integrated into existing protection housing (SEBA SlimCom2 with multi-parameter sensor)

By means of regular workshops throughout the project it is important to enable the end-user (e.g. ministry experts) to handle the hard/software and incoming data as well as conduct maintenance services which can, if available, be supported by the local partner.

Figure 3: Kingdom of Saudi Arabia – SEBA Telemetric "combined" groundwater and rainfall monitoring station in new well-house (SEBA UnilogCom with multi-parameter sensor and rain gauge RG50).

Figure 4: Sultanate of Oman – SEBA Telemetric rainfall station (ftp-push) incl. solar panel (SEBA UnilogCom with rain gauge RG100).

Proceedings “Hydrogeology of Arid Environments”, p. 180–184
Stuttgart, March 2012

Extended Abstract

Modeling Ecological Water Releases to the Lower Tarim River

Haijing Wang[1,3], Pengnian Yang[2] and Wolfgang Kinzelbach[3]

[1] hydrosolutions, Technoparkstrasse 1, CH-8005, Zurich, Switzerland. Wang@hydrosolutions.ch
[2] College of Water Conservancy and Civil Engineering Xinjang Agricultural University, Urumqi, 830052, Xinjiang, China. Ypn10@163.com
[3] Institute of Environmental Engineering, IfU, ETH Zurich, Wolfgang-Pauli-Strasse 15, 8097, Zurich, Switzerland. wang or kinzelbach@ifu.baug.ethz.ch

Key words: ecological releases, Tarim river, groundwater model

Introduction

The Tarim River is the longest inland river in China. Its lower reach used to feed the so-called Green Corridor, a riverine forest belt of *Populus euphratica*, covering a total length of 428 km upstream of Taitema Lake, and separating the Kuluk Desert from the biggest desert in China, the Taklimakan Desert. Since the 1950s increasing agricultural activities in the upper and middle reaches of Tarim river left less and less water for the downstream. Especially after the construction of Daxihaizi reservoir in 1972, the 375 km long lower river reach between Daxihaizi and Taitema Lake had carried no flow for almost 30 years (Deng 2009). The groundwater table in the region dropped continuously, which caused severe degradation of the ecosystem along the lower Tarim River. In many places the Green Corridor has been narrowed down to thin and fragile pieces and the two deserts have started to merge. In order to improve the ecological condition of the green corridor in the lower Tarim river basin, between 2000 and 2006 eight ecological water releases from Daxihaizi reservoir to Taitema Lake have been made.

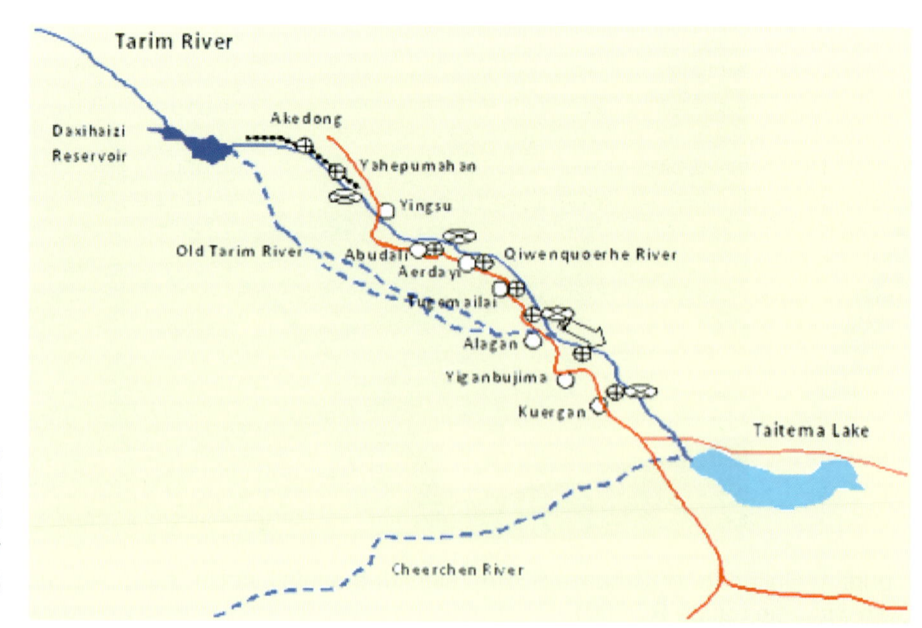

Figure 1: The lower Tarim River between Daxihaizi Reservoir and Taitema Lake and the groundwater monitoring stations along Qiwenquoer River.

www.borntraeger-cramer.de
2012/0180/$ 1.25

Table 1: Total amount of water (in 10^6 m^3) discharged through each monitoring cross-section during the eight ecological releases (summarized from Table 18.1 in Deng 2009).

Release	Daxihaizi	Yinsu	Alagan	Yiganbujima	Taitema Lake	Old Tarim
1	99.23	15.07				
2	226.55	96.36	5.92			
3.1	184.34	100.17	20.62	8.33		
3.2	197.91	144.71	73.43	33.60	7.48	
4	331.29	253.76	103.93	47.73	14.18	
5.1	340.28	98.53	127.03	60.59	23.15	220.01
5.2	279.98	114.02	70.63	24.83	6.62	133.11
6	105.28	88.95	31.30	12.47		
7.1	52.37	27.98	5.29			
7.2	229.98	77.98	54.22	30.72	4.97	98.82
8	234.00	91.36	16.26	3.37		

Table 2: Surface topography of Yiganbujima cross-section.

Distance (m)	100	300	500	750	1050
Surface Elevation (m asl)	809.81	810.94	810.52	810.32	809.76

The response of the groundwater table along the lower Tarim River reflects the interplay of infiltration, groundwater storage, and evapotranspiration, and thus reveals the effectiveness of the water release in supporting the vegetation.

In this paper we concentrate on the last section of the lower Tarim River, which covers the 153 km long narrow green corridor between Alagan and Taitema Lake. In this section the vegetation had been under the most severe water stress and ecological conditions were the most fragile. Before the releases, there were only randomly distributed *Populus euphratica*, and *Tamarix chinensis* on fixed or moving sand dunes in the region, but hardly any other living ground vegetation (Song et al. 2000, Zhang et al. 2004). The eight ecological releases discharged in total 2,281 million tons of water (Deng 2009) from Daxihaizi Reservoir to the lower Tarim River. Among those, releases 3.2, 4, 5.1, 5.2 and 7.2 reached Taitema Lake. The total amounts of water passing through each river monitoring cross-section are listed in Table 1. In addition to the river monitoring stations, there are in total eleven groundwater monitoring stations. But in the last section of lower Tarim River, only at Yiganbujima both groundwater measurements and river water level measurements are sufficient for calibration of a model. So we choose the river cross-section at Yiganbujima to develop a model and analyze the changes in groundwater table during the ecological releases.

HydroGeoSphere model setup

HydroGeoSphere (HGS) (Therrien et al. 2010) is an integrated modeling tool for watershed management. It allows modeling surface and variably saturated subsurface flow while at the same time taking into account the evapotranspiration of plants. We used HydroGeoSphere to model the effect of ecological releases in cross-sections of the lower Tarim River. In this paper we create a 2D vertical model at Yiganbujima where data from three old groundwater observation wells (HL1-HL3) for the releases 3 to 5.2 and from six new groundwater observation wells (H1–H6) for the releases from 5.2 to 8 are available. The two dimensional vertical model has a thickness of one unit length in Y direction (river axis). The X direction is horizontally perpendicular to the river axis, while Z denotes the vertical direction. The shortest distance from a well to the river characterizes the X-coordinate of the well.

Due to data restrictions our model for Yiganbujima starts from the fourth release. The monitoring period in Yiganbujima between the fourth and eighth releases covers a period of about 1600 days. The surface topography of Yiganbujima is interpolated from the 5 points given in table 2. The positions of the observation wells are shown on a DigitalGlobe image from Google Earth (Fig. 2).

The initial condition of the groundwater table at Yiganbujima cross-section was defined based on

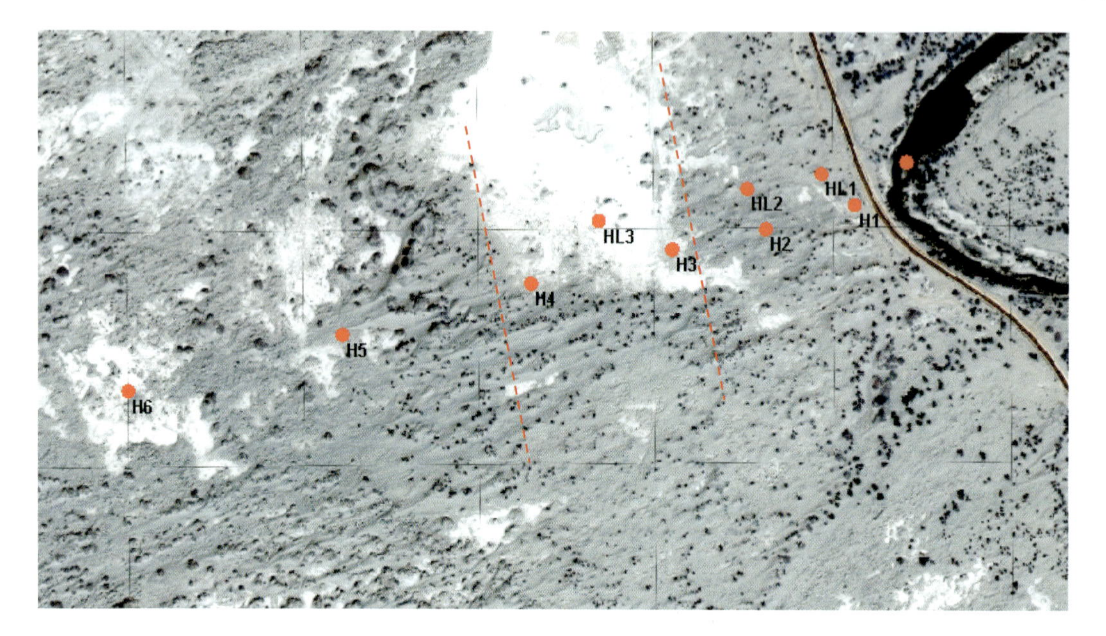

Figure 2: Locations of groundwater observation wells in Yiganbujima on a Digital Globe image.

the measurements at the available observation wells before the fourth ecological release. The groundwater table measurements in Yiganbujima at three old wells and six new wells are plotted together with the river water level in Figure 3. The groundwater heads decrease with distance from the river.

At the Yiganbujima cross-section some synchronous flow rate and water level measurements are available, which we used to establish a relation between river flow rate Q and river water level H, which is: $H = -0.0111Q^2 + 0.2937Q + 806.8$ (with Q in m^3/s and H in m asl). Thus the river flow rate on each record day can be used to estimate a model input for the river water level in meter above sea level.

The average potential evapotranspiration rate at Yiganbujima is 0.0077m/day or 2.81 m/year. Yiganbujima has almost no ground vegetation and very few trees in the region where groundwater observation wells are located, as shown in Fig. 2. We therefore chose very small leave area index (LAI) values (around 0.01). The actual distribution of vegetation in Yiganbujima shows roughly three different kinds of land cover features: close to the river bank (within 250m), there are some trees with almost no ground vegetation; further away (250–500m) there is a salt pan with very few trees by the edge of the pan; even further (500–1500m) away there are more shrubs with still fewer trees. These vegetation features are incorporated into the model via the parameters of the Kristensen-Jensen model for transpiration embedded in HGS. The values for the three vegetation zones are given in table 3. *Populus euphratica*

Table 3: List of input parameters of the three zones for evapotranspiration calculation in Yiganbujima.

Zone Nr.	C1	C2	LAI	C2+C1*LAI	C3
1	10	0.05	0.02	0.25	0.5
2	10	0.05	0.01	0.15	0.5
3	10	0.10	0.01	0.20	0.5

in Yiganbujima cross-section are mainly distributed along the river bank. Further away the vegetation consists mostly of *Tamarix chinensis*. Therefore we defined the root depths according to the average root depths of Populus and Tamarix for the three zones as 5, 7 and 9 m, respectively.

The aquifer in Yiganbujima region is composed of very fine and silty sand. We used a porosity of 0.25 and a hydraulic conductivity of 3.56 m/day based on previous work in Yingsu (Schilling 2011).

Model results and discussions

The flood waves caused by the releases influence the groundwater table two to three hundred meters away from the river considerably. Further away the influence is weaker and only reflected in a slow and steady increase of groundwater levels. At a distance of one thousand meters there is almost no effect left. The total of eight releases causes a few centimeters increase in groundwater table at a distance of 1450 m from the river in the cross-section. The choice of using 1500 m width for modeling is adequate. The

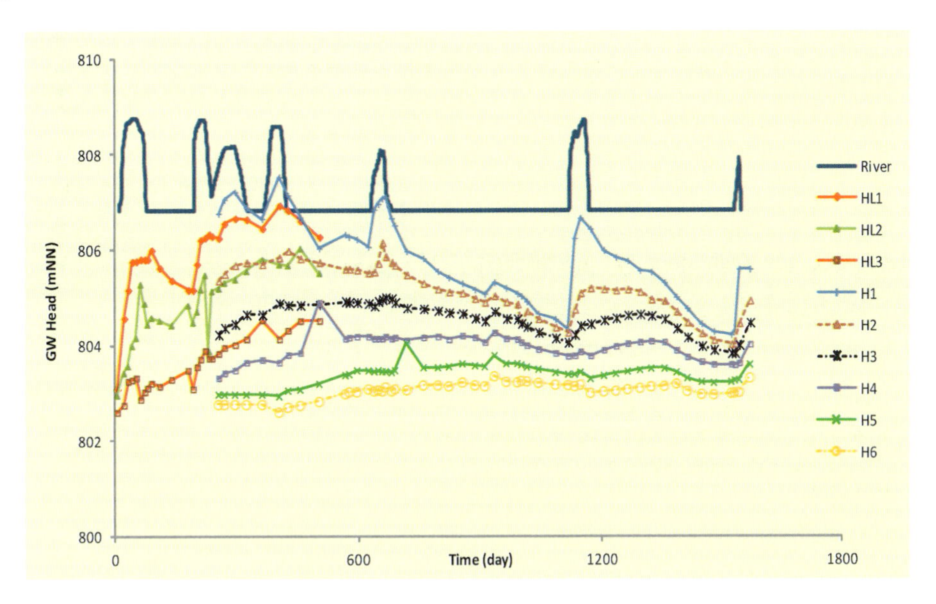

Figure 3: Comparison of water head measurements at groundwater observation wells with input river water levels.

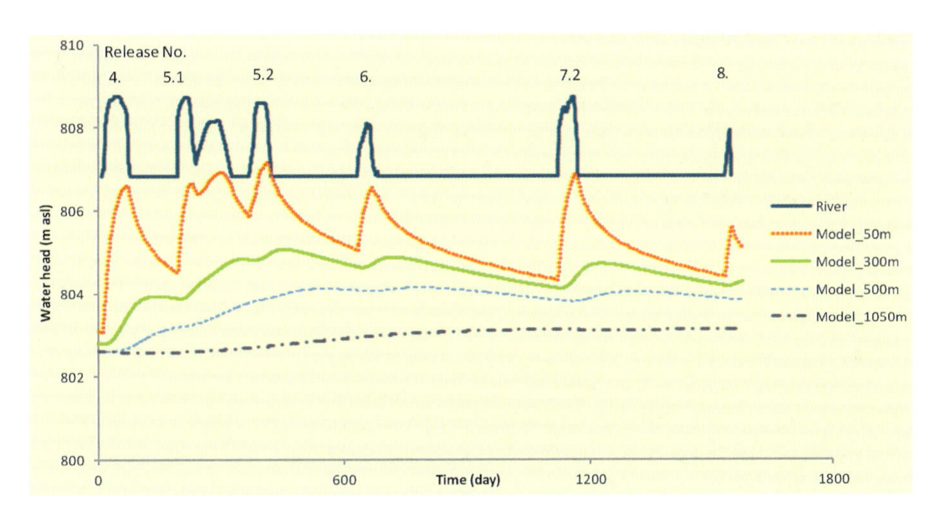

Figure 4: Modeled groundwater levels for different distances from the river in comparison with river water levels for all the modeled releases (releases 4 to 8).

modeled groundwater heads at different distances in comparison with river water levels during and after releases are shown in Fig. 4. They agree very well with the measured heads.

It is remarkable that with only very few fit parameters the time series of 1600 days could be reproduced very well. Still, the parameters are neither independent nor unique. So some uncertainty in the interpretation exists. What can, however, be deduced with much less uncertainty is the ratio of infiltration and evapotranspiration in the cross-section.

The total cumulative transpiration in the cross-section at the end of the releases is about 15% of the total infiltration. Most of the infiltration contributes to groundwater storage in a 1000 m strip of aquifer on both sides of the river. The experiment, which is probably the largest artificial groundwater recharge experiments worldwide shows that artificial recharge on a large scale using an existing river bed is well feasible. The efficiency of providing water to plants is limited, depending on the amount of living vegetation still available. The Green Corridor at the last section of the lower Tarim River can be only be revived if the local groundwater table would be lifted up to the Populus root depth level with more frequent floods.

References

Chen, Y.-N., Chen, Y.-P., Xu, C., Ye, Z., Li, Z., Zhu, C. & Ma, X., 2010: Effects of ecological water conveyance on groundwater dynamics and riparian vegetation in the lower reaches of Tarim River, China. – Hydrological Processes **24**: 170–177.

Deng, M., 2009: Tarim River in China: Theories and Practices of Water Resources Management. – Science Publisher, Beijing (in Chinese).

Schilling, O., 2011: Quantifying the dynamics of water flow between surface water, groundwater and vegetation at the lower Tarim River, China. – Master Thesis, ETH Zurich.

Therrien, R., McLaren, R. & Sudicky, E., 2010: HydroGeoSphere: a three-dimensional numerical model describing fully-integrated subsurface and surface flow and solute transport. – Groundwater Simulations Group, University of Waterloo, Waterloo, Ontario, Canada.

Xu, H., Ye, M., Song, Y. & Chen, Y., 2007: The natural vegetation responses to the groundwater change resulting from ecological water conveyances to the lower Tarim River. – Environmental Monitoring and Assessment **131**(1-3): 37–48.

Yang, P.-N., Zhang, S. & Dong, X., 2007: Study on the characteristics of water conversion after conveying stream water to the lower reaches of the Tarim River for regenerating the ecology. – Arid Zone Research (in Chinese).

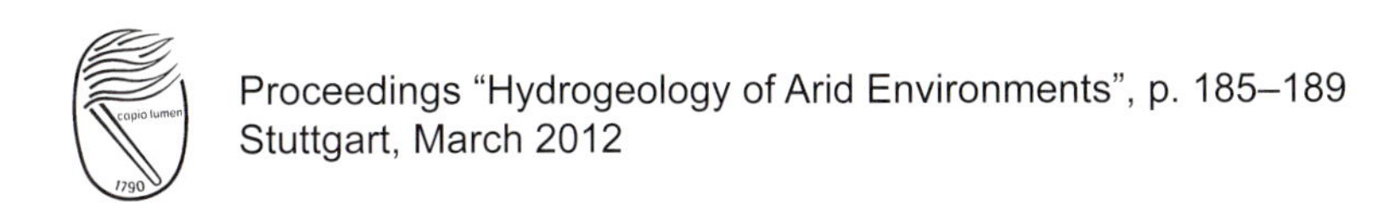
Proceedings "Hydrogeology of Arid Environments", p. 185–189
Stuttgart, March 2012

Extended Abstract

Tademait Plateau: A Regional Groundwater Recharge Area in the Centre of the Algerian Sahara

K. Udo Weyer and James C. Ellis

WDA Consultants Inc., 4827 Vienna Drive NW, Calgary, AB, T3A 0W7; email: weyer@wda-consultants.com

Key words: groundwater flow systems, Hubbert's force potential, Tademait Plateau, CO_2 sequestration, In Salah, groundwater recharge

Introduction

The injection of CO_2 at 'In Salah' (Krechba gas field) in the Algerian Sahara tests the behaviour of the sequestered CO_2 in the subsurface. Figure 1 shows the occurrence of the Krechba gas reservoir in approximately 1850 m depth in 20 m of Carboniferous sandstone. It is overlain by 900 m of Carboniferous mudstone, 700 m of Lower Cretaceous sandstone, an aquifer, and 200 m of Middle and Upper Cretaceous mudstone. In a newly drilled observation well the water level from the aquifer rose to about the middle of the overlying Cretaceous mudstone. When drilling through the Carboniferous mudstone, loss of circulation was frequently encountered in the upper 400 m and lower 200 m of the mudstone, a caprock in oil field terminology, an aquitard in hydrogeological terminology. It is under debate whether the circulation losses were caused by pre-existing fractures or by hydraulic fracturing (Iding & Ringrose 2009).

The Cretaceous aquifers of the Tademait Plateau belong to the 'Aquifère du Continental Intercalaire' system (Castany 1982). Traditionally much of

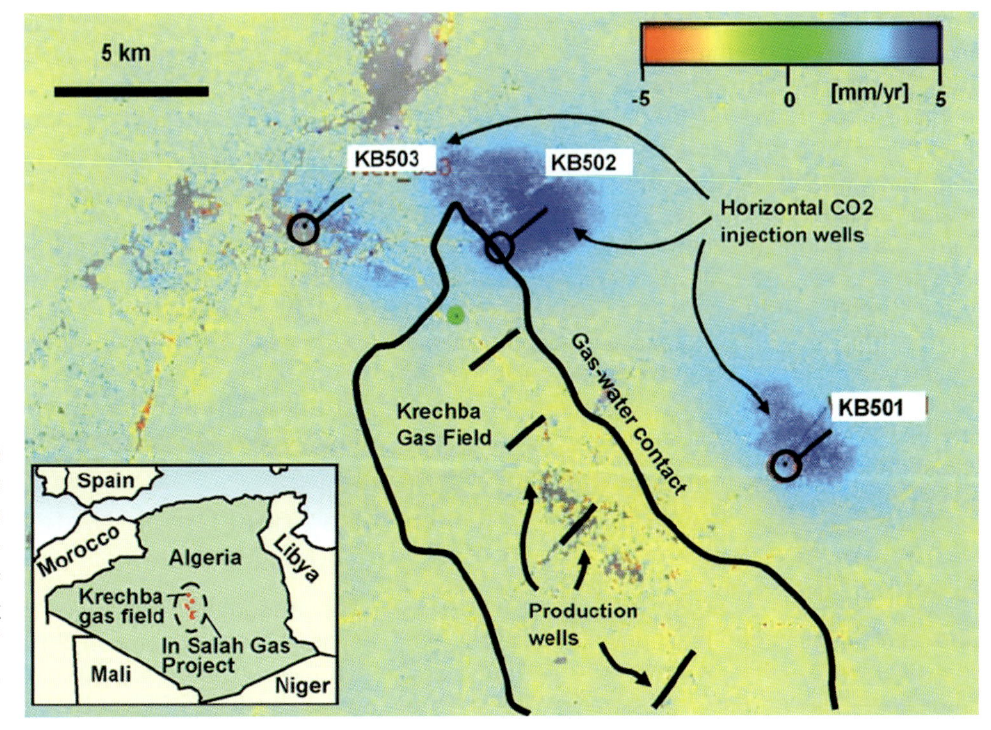

Fig. 1. InSAR data of average distance change (close to vertical displacement) evaluated by TRE from August 2004 to March 2007 (from Rutqvist et al. 2010, Fig. 2).

www.borntraeger-cramer.de
2012/0185/$ 1.25

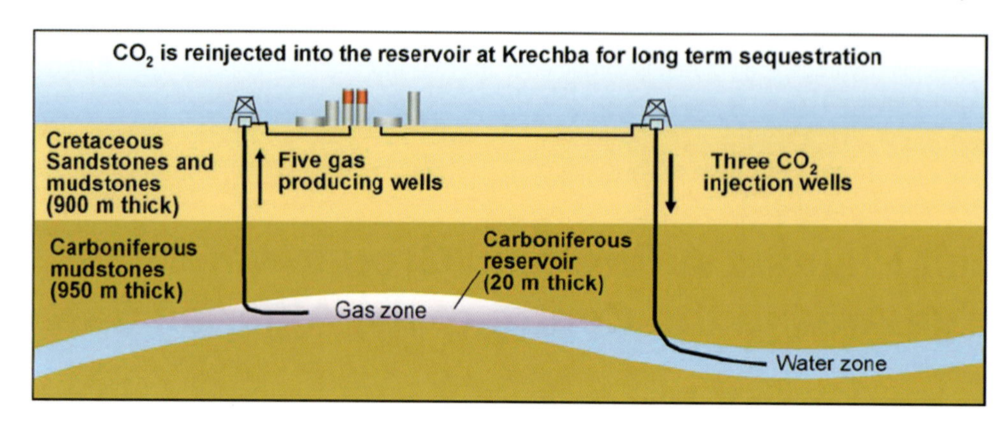

Fig. 2. General geology and technical installations at the Krechba gas reservoir (from Rutqvist et al. 2010, Fig. 1).

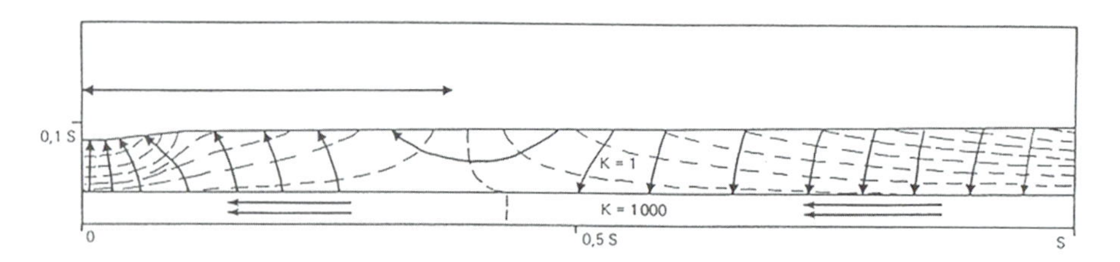

Fig. 3. Groundwater flow through an aquitard to the underlying aquifer and back to the discharge area indicated by the horizontal arrows. (after Freeze & Witherspoon 1967). The permeability contrast is 1000.

the groundwater flow in the Sahara Basin was seen as originating in the Atlas Mountains and shown to underflow the Tademait Plateau partially from NE to south and partially from NE to SW (Ben Dhia 1991, his Figs. 4 and 5). The guiding concept was the conceptual model that groundwater flow would be limited to aquifers themselves and, in this case, to an aquifer system with an outcrop and thereby recharge area in the Atlas Mountains. In Groundwater Flow Systems theory aquitards at the surface (Fig. 3) were shown to be natural recharge areas for deeper aquifers by Freeze & Witherspoon (1967). Tóth (1962) had introduced the concept of Groundwater Flow Systems with recharge and discharge areas whereby the penetration depth can exceed 5 km (Tóth 2009). In a recharge area the flux of groundwater crosses the groundwater table into the saturated domain; in a discharge area the flux of groundwater is directed from the groundwater body into surface waters or to the surface for evaporation.

Behaviour of sequestered CO_2 in the Krechba field

Unexpectedly the geo-mechanical behaviour and flow direction of the injected CO_2 did not follow predictions. Firstly, rises of surface elevations of several centimeters have so far been measured by satellites (Fig. 1 and 2). Secondly, the areal extent of these uprising areas showed that in about 2000 m depth the CO_2 migrates down dip and, in a northwestern direction away from the pressure sink of the gas production area which is located up dip of the CO_2 injection sites (Fig. 1). Both, gas reservoir and CO_2 injection site are located within the same Carboniferous sandstone of approximately 20 m thickness (Fig. 2).

The encountered flow direction cannot be explained by buoyant flow behaviour as had been expected from the supercritical CO_2 fluid with a density of about 0.7 g/cm^3 in a salty host fluid of a density probably exceeding 1.1 g/cm^3. The hydrodynamic behaviour of the CO_2 can be explained, however, by applying Hubbert's (1940, 1953) force potential and groundwater flow systems theory (Tóth 1962, Freeze & Witherspoon 1967).

Regional Groundwater Flow in the Tademait Plateau

The Tademait Plateau is a distinctive mountain system in the centre of the Algerian Sahara. It is wedged between the Atlas Mountains to the northwest and the Tefedest Mountains to the southeast

(Fig. 4). Elevation differences between the main part of the Tademait Plateau and the surrounding lowlands to the SW reach up to 550 metres, with the length of the flow systems exceeding 200 km.

The southwestern edge of the Plateau is highlighted by the occurrence of 81 oases (black dots in Fig. 4). According to surface topography and thereby the approximate topography of the groundwater table, 53 of these oases are located on the slopes of the Tademait Plateau with an additional 26 arranged in the down slope area of the Atlas Mountain system. Oases occur in groundwater discharge areas; hence a rim of discharge areas occurs to the west, southwest, and east of the Tademait Plateau. The geometry of these occurrences and hydrodynamic reasons (Weyer 2010) identify both recharge and discharge areas in this area. The Tademait Plateau is an active and efficient recharge area for regional groundwater flow systems discharging at the oases and in the deeper low lands to the west, southwest, and east of the plateau. At the Krechba site the Cretaceous aquifer system contains fresh water while the Carboniferous reservoir sandstone contains salt water with a TDS of more than 100 g/l (A. S. Mathieson, oral communication, October 2011).

All oases along the southern rim of the Tademait Plateau appear to be located on lower Cretaceous layers, some of them possibly on Jurassic and Triassic layers (Ben Dhia 1991, his Fig. 2 and 4A). At the Krechba site the Cretaceous layers carry fresh water. The five 'In Salah' oases, in the past, probably used to be fed by fresh, good water from foggaras (qanats) as the name 'good well' implies in Arabic. Since the installation of boreholes for water supply the water is known for 'its rather unpleasant, salty taste' (Wikipedia contributors, 2011). It is therefore probable that these boreholes draw saline water from Carboniferous layers and possibly Triassic salt layers existing in the general area. This would imply that the regional groundwater flow systems also penetrate the Carboniferous layers as they should due to hydraulic reasons. More detailed investigations about flow of recharged groundwater into the Carboniferous layers and the reservoir are under debate.

In any case, the subsurface hydraulic force fields are determined by the groundwater table in fresh water systems and the migration of all fluids is governed by these fresh water force fields and the pressure potential force of the fluid under consideration; hydrous fluids usually pass freely through caprocks (Hubbert 1953). In all likelihood the migration behaviour of the sequestered CO_2 is caused by fresh water force fields which also exist within brines, regardless of the presence of fresh water. The northwesterly flow direction of the injected CO_2 coincides with the general flow direction of fresh groundwater recharged in the Tademait Plateau and does not coincide with the general southeasterly flow directions shown by Ben Dhia (1991, his Fig. 4A) who assumed groundwater recharge in the Atlas Mountains.

Infiltration and Groundwater Recharge in Arid Environments

The conceptual model of sustained groundwater recharge in the middle of the Sahara seems to contradict traditional knowledge. For decades the assumption prevailed that, in a desert environment, most of the precipitation would evaporate. In the presence of plants the suction of the root system creates very negative pressures at a depth from 1 to 5 metres (Phillips et al. 2004). This strongly unsaturated zone, permanently maintained by evapotranspiration, prevents, where it exists, substantial recharge to the groundwater system, even if larger amounts of precipitation infiltrate the upper soil layers.

Research at the Yucca Mountain (in the Death Valley area of the Southwestern US) identified soil infiltration rates for soil, plant and exposure conditions through field studies and mathematical modeling. Depending upon the thickness of soil, plant density, and exposure conditions at the sites, infiltration rates reached from 5–10 mm/year up to >250 mm/year (Flint et al. 2001). Higher recharge rates occur where soil over fractured bedrock is less than 0.5 m thick and in topographic depressions such as ephemeral streams (Phillips et al. 2004). Wilson & Guan (2004) confirm that significant recharge can occur where soils are thin or absent over fractured bedrock.

Characteristically most of the area of Tademait Plateau is without continuous plant cover and much of it seems to have thin soil cover over fractured bedrock leading to the conclusion that much of the precipitation may infiltrate the soil and recharge the groundwater body. In addition, at the Krechba site the water table appears to be less than 100 m below surface while it is up to 500 m below surface at the Yucca Mountains implying active infiltration into the soil and recharge to the groundwater body. The actual infiltration rates in the past maintained the water supply of the oases at the rim of the Tademait Plateau system.

Conclusions

The Tademait Plateau has been shown to be an extended and active recharge area for groundwater flow towards a belt of 53 oases to the west, south-

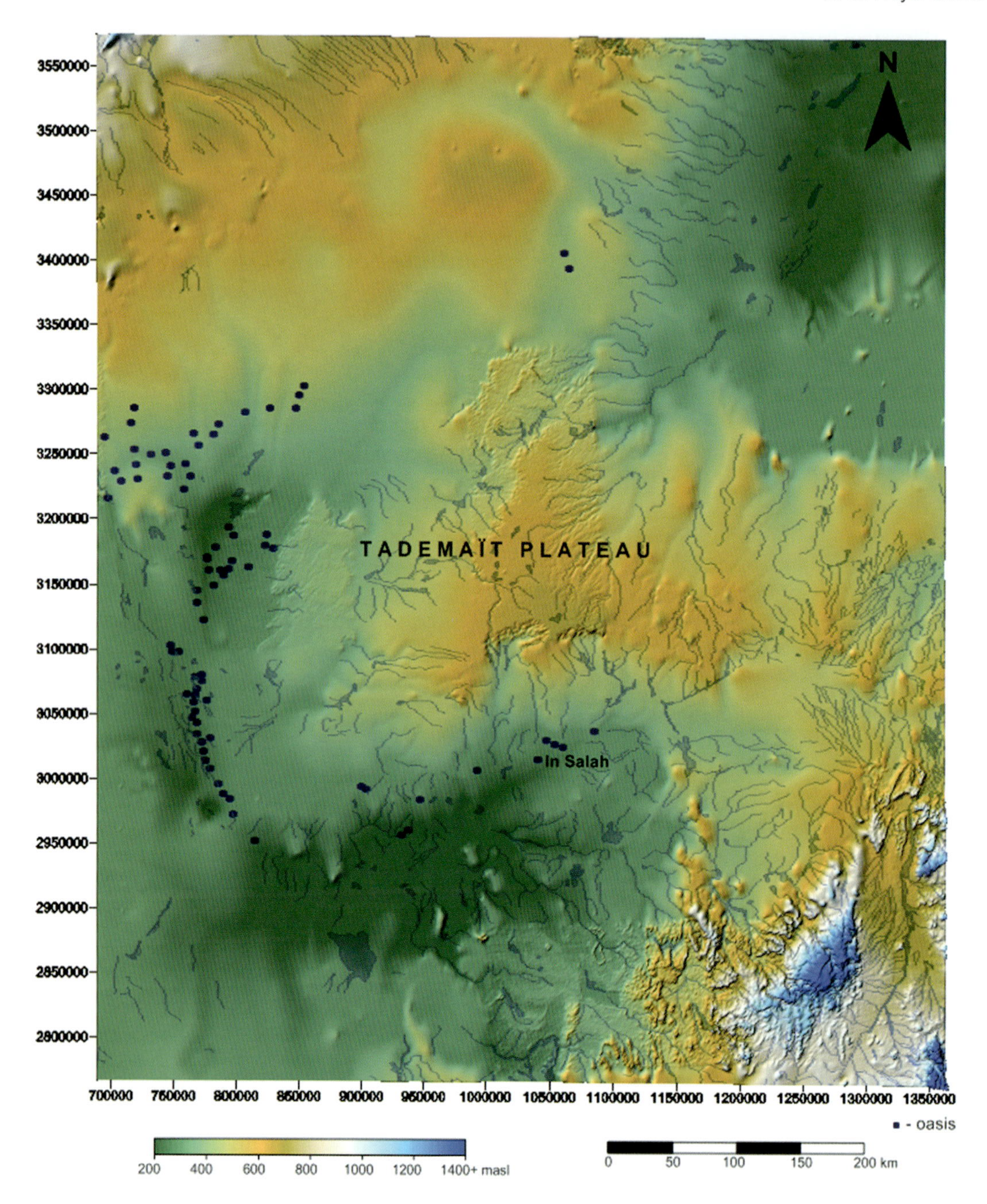

Fig. 4 Topography of the Tademait Plateau (centre) and parts of the Atlas Mountains to the northwest and the Tefedest Mountains to the southeast. A belt of 53 oases are located on the slope of the Tademait Plateau; 26 oases are located on the slope of the Atlas Mountain system. This DEM was based on the USGS' GTOPO30 DEM, transferred into UTM using AutoCAD, and then re-gridded (1 km grid spacing) using SURFER . The locations of the oases were determined from a map with a scale of 1:1,700,000 (World Mapping Project, Algeria).

west, and south. Recharge occurs through a surface aquitard. The depth penetration of the regional groundwater flow system may be several kilometres but has not yet been determined. In any case, the force field of the fresh groundwater ultimately determines the flow directions of other fluids present in the Carboniferous, including that of the sequestered CO_2. The northern line of 26 oases is supplied with groundwater originating in the Atlas Mountains.

Acknowledgements

We thank Allan S. Mathieson for making, at the occasion of a 2011 SPE Forum on CO_2 sequestration, the primary author aware of the unusual migration pattern of sequestered CO_2 at the 'In Salah' injection site, for discussing some of the particulars, and for encouraging us to evaluate available data. More detailed evaluation of additional data is under debate.

References

Ben Dhia, H., 1991: Thermal regime and hydrodynamics in Tunisia and Algeria. – Geophysics **56**(**7**): 1093–1102.

Castany, G., 1982. Bassin sédimentaire du Sahara septentrional (Algérie – Tunisie). Aquifère du Continental Intercalaire et du Complex Terminal. –Bull. B.R.G.M. Sec 3. **2**: 127–147.

Flint, A. L., Flint, L. E., Kwicklis, E. M., Bodvarsson, G. S. & Fabryka-Martin, J. T., 2001: Hydrology of Yucca Mountain, Nevada. – Reviews of Geophysics **39**: 447–470.

Freeze, R. A. & Witherspoon, P. A., 1967: Theoretical analysis of regional groundwater flow: 2. Effect of water table configuration and subsurface permeability variation. – Water Resources Research **4** (**3**): 581–590.

Phillips, F. M., Hogan, J. F. & Scanlon, B. R., 2004: Introduction and overview. – In: Hogan, J. F., Phillips, F. M. & Scanlon, B. R. (eds.): Groundwater Recharge in a Desert Environment: The Southwestern United States. American Geophysical Union, Washington, D.C., 294 p.

Hubbert, M. K., 1940: The theory of groundwater motion. – J. Geol., **48** (8): 785–944.

Hubbert, M. K., 1953: Entrapment of petroleum under hydrodynamic conditions. – The Bulletin of the American Association of Petroleum Geologists **37** (8): 1954–2026.

Iding, M. & Ringrose, P., 2009: Evaluating the impact of fractures on the performance of the In Salah CO2 storage site. – International Journal of Greenhouse Gas Control, March 2010, **4** (2): 242–248.

Rutqvist, J., Vasco, D. W. & Myer, L., 2010: Coupled reservoir-geomechanical analysis of CO2 injection and ground deformations at In Salah, Algeria. – Int. Journal of Greenhouse Gas Control **4** (2010): 225–230.

Tóth, J., 1962: A theory of groundwater motion in small drainage basins in Central Alberta, Canada. – J. Geophys. Res. **67** (1): 4375–4387.

Tóth, J., 2009: Gravitational systems of groundwater flow; Theory, Evaluation, Utilization. – Cambridge University Press, 297 pp.

Weyer, K. U., 2010: Differing physical processes in offshore and on-shore CO2 storage. – Poster presented at GHGT-10, Amsterdam, The Netherlands, September 2010. Available from http://www-wda-consultants.com.

Wikipedia contributors: In Salah [Internet] - Wikipedia, The Free Encyclopedia; 2011 Nov 16, 04:13 UTC [cited 2011 Dec 16]. http://en.wikipedia.org/wiki/In_Salah.

Wilson, J. L. & Guan, H., 2004: Mountain-Block Hydrology and Mountain Front Recharge. – In: Hogan, J. F., Phillips, F. M. & Scanlon B. R. (eds.): Groundwater Recharge in a Desert Environment: The Southwestern United States. American Geophysical Union, Washington, D.C., 294 p.

Proceedings "Hydrogeology of Arid Environments", p. 190
Stuttgart, March 2012

Poster Abstract

Surface Karst Features Mapping for the Groundwater Catchment Area of Jeita Spring

Jean Abi Rizk[1], Armin Margane[2]

[1, 2] Federal Institute for Geosciences and Natural Resources BGR,
email: abirizk.jean@gmail.com, armin.margane@bgr.de

Key words: karst, groundwater protection, Lebanon, Quaternary glaciation

Jeita spring emerges from a thick and highly karstified limestone aquifer in the Mount Lebanon mountain range near Beirut, Lebanon, and provides 70% of the drinking water for the Greater Beirut area. The absence of wastewater collection and treatment systems in its groundwater catchment causes a high pollution load at the spring.

In the framework of the bilateral Technical Cooperation project Protection of Jeita Spring groundwater protection zones were delineated using groundwater vulnerability mapping.

For vulnerability mapping two methods specifically developed for karst areas were used: EPIK (Saefl 2000) and COP (Vias et al. 2002). Surface karst features are the most important component determining the level of vulnerability and had to be mapped in detail for both methods. Karst features are divided into two groups; large-scale and small-scale features. For every type of features its genesis and role in infiltration was described. The study differentiates between infiltrating and non-infiltrating karst features. The surface karst features mapping gives an opportunity to better understand the behavior of the karst hydrogeological system and shows where high-risk areas for pollution are located. This will help to more efficiently protect the water resources in this highly karstic area.

Karstification is most extensive where limestone had been exposed over a long period of time at elevations between 1,200 and 2,600 m and where actual rainfall is between 1,300 and 2,000 mm/a. There are two geological units in the Jeita catchment which are highly karstified. The uppermost part of the Jurassic geological unit J4 (Kesrouane Formation; total thickness 1,000–1,500 m) shows large karren fields, dolines and sinkholes at elevations between 1,000 and 1,400 m. The Upper Cretaceous Sannine Formation (C4) is exposed in the high plateau of the Mount Lebanon mountain range (>1,800 m). There are practically no surface water drainage features developed on this plateau because rainfall and snow almost completely infiltrates into extended fields of dolines. The snow cover between December and May plays an important role for development of this extreme karstification of the Sannine Formation.

Esker-like structures were found at elevations between 800 and 1,200 m. It is believed that large parts of the Mount Lebanon mountain range were covered by glaciers during the Quaternary glacial periods. Similar extents of glaciations were reported from Turkey (Sarikaya et al. 2011) but not yet from Lebanon. This Quaternary glaciation has probably significantly promoted karst formation in the upper Mount Lebanon mountain range.

References

Saefl, 2000: Practical Guide Groundwater Vulnerability Mapping in Karst Regions (EPIK). – Report, 57 p., Bern/CH.

Sarikaya, M. A., Ciner, A. & Zreda, M., 2011: Quaternary Glaciations of Turkey. – In: Ehlers, J., Gibbard, P. L. & Huges, P. D. (Eds.): Quaternary Glaciations – Extent and Chronology. – Developments in Quaternary Science, 15, pp. 393–403, London (Elsevier).

Vias, J. M., Andreo, B., Perles, M. J., Carrasco, F., Vadillo, I. & Jimenez, P., 2002: Preliminary proposal of a method for contamination vulnerability mapping in carbonate aquifers. – In: Carasco, F., Duran, J. J. & Andreo, B. (Eds.): Karst and Environment, pp 75-83.

www.borntraeger-cramer.de
2012/0190/$ 0.25

Proceedings "Hydrogeology of Arid Environments", p. 191
Stuttgart, March 2012

Poster Abstract

Potential of Managed Aquifer Recharge of Treated Wastewater in the Governate of Muscat, the Sultanate of Oman

S. Al Jabri[1], M. Ahmed[2], A. Al Maktoumi[3], S. A. Prathapar[4]

[1] Sultan Qaboos University, email: salemj@squ.edu.om
[2] Sultan Qaboos University, email: ahmedm@squ.edu.om
[3] Sultan Qaboos University, email: ali4530@squ.edu.om
[4] International Water Management Institute, email: s.prathapar@cgiar.org

Key words: Managed Aquifer Recharge (MAR), Treated Water (TE), Muscat, Oman

The Sultanate of Oman suffers from water scarcity at all levels. Annual available water is estimated to be 1,267 million cubic meters (MCM), while the demand is about 1645 MCM. This makes an annual water deficit of about 378 MCM (MRMWR 2005). Alternative sources for water must be explored and considered. Treated wastewater (TE) is considered a promising option to meet water demand of many practices, such as irrigating agricultural and urban lands, injecting it into groundwater aquifers (to enhance storage capacities or to mitigate seawater intrusion), mixing it with municipal water, or simply disposing it to the ocean. The Oman wastewater Company (Haya) is the one that treats wastewater and manages facilities for TE in the Governate of Muscat (Oman). TE in the Governate of Muscat is expected to increase from about 8 MCM in 2003 to 80 MCM in 2035 (Haya 2004). This concept paper discusses the potential uses of excess TE, after irrigating landscape amenities, in Muscat area. This may include injecting aquifers on coastal areas or utilizing it to irrigate field and/or greenhouse crops. We used HYDRUS (Šimůnek et al. 2011) to simulate infiltration form surface ponds on two soils; sandy loam and loamy sand, that are located on nearby facilities of Haya. HYDRUS simulations show that areas with sandy loam soils are suited for infiltration ponds. However, sandy loams are not suitable for surface ponding and, therefore, deep injection may be necessary. Steady infiltration rates were from HYDRUS simulations were used as inputs into Visual MODFLOW (Waterloo Hydrologic 2004) to simulate groundwater recharge into coastal aquifers nearby the Haya facilities in Al Khod area of Muscat. MODFLOW simulations showed that upper and lower catchments are unsuitable for recharge. This is because of limited aquifer thickness in the upper catchment and closer proximity to the sea for the lower catchment. The middle zone gives a reasonable compromise where the aquifer is relatively thicker and is distant from the coastal zone. However, it is shown that sites with a hydraulic conductivity (K) of 10 m/day will yield 40-m watertable mounds. Locating more permeable sites with higher K values (about 30 m/d) will reduce the mound. In addition, more than one recharge location can be used in the middle zone to control the watertable mound. This can also enhanced by a controlled injection rates. Another option is to utilize the excess TE for irrigation of commercial field and greenhouse crops. This is likely to be financially attractive and socially acceptable. However, a socio-economic analysis must be performed with this option.

References

Haya (Oman Wastewater Services Company), 2004: personal communication.

Ministry of Regional Muncaipalites and Water Resources, 2005: Water resources in Oman. – MRMWAR, Muscat, 127 pp.

Šimůnek, J., Th. Van Genuchten, M. & Šejna, M., 2011: The HYDRUS software package for simulating two- and three dimensional movement of water, heat, and multiple solutes in variably-saturated media. Version 2.0, PC Progress, Prague, Czech Republic.

Waterloo Hydeogeologic Inc., 2004: Visual MODFLOW Pro. Waterloo, Ontario, Canada.

www.borntraeger-cramer.de
2012/0191/$ 0.25

Proceedings "Hydrogeology of Arid Environments", p. 192
Stuttgart, March 2012

Poster Abstract

Application of Groundwater Modeling for Water Resources Management in Arid Environments

Jihad Al Mahamid[1]

[1]Head of Integrated Water Resources Management Program, The Arab Center for the Studies of Arid Zones and Dry Lands (ACSAD). Jmahameed_1968@yahoo.com

Key words: groundwater, model, MODFLOW, sustainable

Groundwater constitutes an important source of water for agricultural, industrial, domestic and environmental supply. With rapid economical development and population growth, demands for more water with higher quality are continuously increasing. Over-exploitation of groundwater has caused serious problems of aquifer depletion, quality deterioration and side effects on the environment. Management of groundwater resources has become highly important issue than other resources.

A model is any device that represents an approximation of a field situation. It is a simplified version of the real (here groundwater) system that approximately simulated the excitation – response relations of the latter. The simplification is introduced in the form of a set of assumption that expresses our understanding of the nature system and its behavior. The main benefit of the model is to increase our understanding of the interaction of simultaneous processes and influences, formulate the present problems to minimize and give alternative solutions of the current problems.

Groundwater models play an important role. Groundwater flow and transport models can be used as: (1) supporting tools for planning field investigation; (2) predictive tools for predicting future conditions or the impacts of waste disposal actions; (3) screening tools for evaluating the remediation alternatives; (4) interpretive tools for studying groundwater system dynamics and understanding the physical processes; and (5) management tools for identifying optimal strategies for groundwater resources development and protection.

The popular groundwater model MODFLOW, PMPATH, MT3D and a graphic user interface GMS, VM and PMWIN are used in the application of groundwater models for aid and semi-arid areas from the Arab countries. The model results have been validated and are using to help the decision makers to manage water resources in proper and sustainable way. In addition, models prediction were using for water resources planning and aquifer storage prediction.

www.borntraeger-cramer.de
2012/0192/$ 0.25

Proceedings "Hydrogeology of Arid Environments", p. 193
Stuttgart, March 2012

Poster Abstract

The Hydrogeology of Al Hassa Springs in the Kingdom of Saudi Arabia – A Case Study for the Depletion of an Aquifer

Ali Saad Al Tokhais[1], **Randolf Rausch**[2], **Heiko Dirks**[3]

[1] Member of Majlis Ash Shura, Riyadh, Saudi Arabia, email: tkhais@hotmail.com
[2] GIZ International Services, Riyadh, Saudi Arabia, email: randolf.rausch@gizdco.com
[3] Dornier Consulting, Riyadh, Saudi Arabia, email: heiko.dirks@gizdco.com

Key words: Al Hassa oasis, Saudi Arabia, Umm Er Radhuma aquifer

Al Hassa Oasis with its capital Al Hofuf is the largest oasis in the Kingdom of Saudi Arabia and one of the largest spring-fed oases in the world. In Al Hassa Oasis the groundwater came up from about 280 springs. In the middle of the last century the total spring discharge was about 315 MCM/a (10 m^3/s). Today the springs are no longer flowing. Due to overexploitation of the groundwater resources the groundwater levels have fallen dramatically and the springs are dried up. At present time, the total groundwater abstraction in the region of Al Hassa is about 712 MCM/a (22.6 m^3/s).

Spring discharge in Al Hassa Oasis originates from the Umm Er Radhuma aquifer system. The aquifer system consists of four partly interconnected aquifers, from top to bottom: Neogene aquifer complex, Dammam aquifer complex, Umm Er Radhuma aquifer, and Aruma aquifer. The aquifers are partly interconnected. Because of intensive fracturing along the Ghawar anticline, preferential flow paths are developed. They connect the different aquifers and forced the groundwater to discharge in Al Hassa Oasis. From hydrochemical data, isotope hydrogeological data, temperature data, and hydraulic information it can be concluded that the water discharging from the karst springs at Al Hassa from the Neogene formation originates from the deeper Umm Er Radhuma aquifer.

As a consequence of the overexploitation of the resources, a decline in groundwater levels is observed. The springs in the Al Hassa Oasis were running dry. The observed maximum drawdowns during the last 30 years are about 150 m. The vertical groundwater flow direction has changed. Now, groundwater from the upper parts of the aquifer system flows down to the deeper parts of the aquifer system. The diameter of the cone of depression reaches 100 km. If current groundwater abstraction continues at the present rate, more wells in the Neogene and the Dammam aquifer complex will fall dry and the cone of depression will extend. To satisfy the water demand, deeper wells tapping the underlying aquifers must be drilled. Furthermore, a deterioration of groundwater quality will occur, caused by up coning of saline groundwater.

The groundwater of the Umm Er Radhuma aquifer system forms a large and important resource on the Arabian Peninsula; it is definitely a non-renewable resource. The groundwater resource was replenished over a long period of time in the distant past. The exploitation of these reserves can last only several decades. Therefore, a wise and smart groundwater management of the remaining resources in Al Hassa area and in the Kingdom of Saudi Arabia is a must.

www.borntraeger-cramer.de
2012/0193/$ 0.25

Proceedings "Hydrogeology of Arid Environments", p. 194
Stuttgart, March 2012

Poster Abstract

Hydrogeochemistry of Groundwaters of the Sana'a Basin Aquifer System, Yemen

Ahmed Al-ameri[1], Michael Schneider[1], Silvio Janetz[2]

[1] Freie Universität Berlin, Institute for Geological Sciences, Hydrogeology Group, Malteserstraße 74-100, 12249 Berlin, Germany, email: ahmed.alameri@fu-berlin.de
[2] Brandenburg University of Technology Cottbus, Department of Environmental Geology, Erich-Weinert-Strasse 1, 03046 Cottbus, Germany, email: janetz@tu-cottbus.de

Key words: Sana'a Basin, hydrogeochemistry, water classification

In the present study, physico-chemical parameters were applied to characterize and classify ground- and spring water samples collected from the Sana'a Basin (Yemen). A total of 24 groundwater samples from deep wells and 13 spring water samples were collected from the Sana'a basin between September and October 2009. Major anions (Cl^-, HCO_3^-, NO_3^-, SO_4^{2-} and Br^-) and major cations (Ca^{2+}, Mg^{2+}, Na^+ and K^+) were measured. The physical parameters, which include water temperature, electrical conductivity and pH-value, and determination of hydrogen-carbonate, were measured on site.

The ground- and spring water samples collected from the Sana'a basin were classified in groups according to their major ions (anions and cations) content. The classical use of the groundwater in hydrology is to produce information concerning the water quality.

The classification was based on several hydrochemical methods, such as Ca^{2+} and Mg^{2+} hardness, Sodium Absorption Ration (SAR), Magnesium hazard (MH), saturation indices (SI) and Piper diagram. To ensure the suitability of ground- and spring water in the Sana'a basin for drinking purposes, the hydrochemical parameters were compared with the guidelines recommended by the World Health Organisation (WHO) and the National Water Resources Authority – Yemen (NWRA) standards. In order to check the suitability of ground- and spring water for irrigation purposes the samples were classified based on MH and calculated SAR.

The data were plotted on the United State Salinity Laboratory (U.S.S.L) diagram.

www.borntraeger-cramer.de
2012/0194/$ 0.25

Proceedings "Hydrogeology of Arid Environments", p. 195
Stuttgart, March 2012

Poster Abstract

Water Security and Water Quality Impacts in Yemen, with Special Reference to Fluorosis

Abdulmohsen Saleh Al-Amry

Department of Engineering Geology, Faculty of Oil and Minerals, Aden University, Yemen,
email: alamry1972@yahoo.com

Key words: water quality, water security, fluorosis, Yemen

In recent times, Yemen has fallen into a water crisis caused by a rapid mining of groundwater, resulting in extreme water supply shortages in the major cities, and limited access of the population to safe drinking water. The over abstraction of groundwater, affected both quantity and quality of the water source, and made a question mark for future water security of the country. Water scarcity is leading people to use water regardless of its quality, and this situation leads to water related diseases.

Recent research studies on the water quality indicates, markedly increasing in fluoride content in groundwater in districts of some Yemeni governorates such as Sana'a, Ibb, Dhamar, Taiz, Al-Dhalei and Raimah. Endemic fluorosis has been nearly recognized as a major public health problem in six governorates. Since the groundwater forms a major source of drinking water in rural areas, rural populations are facing a major health problem in these governorates. More and more areas are being discovered regularly that are affected by fluorosis in different parts of the country.

www.borntraeger-cramer.de
2012/0195/$ 0.25

Proceedings "Hydrogeology of Arid Environments", p. 196
Stuttgart, March 2012

Poster Abstract

Interdependent Hydrogeology of Vadose Zone and Shallow Groundwater in Agricultural Ecosystems for Estimation of Irrigation Requirement

Hosein Alizadeh[1], S. Jamshid Mousavi

[1] PHD candidate, Department of Civil and Environmental Engineering, Amirkabir University of Technology, Tehran, Iran, email: h.alizadeh@aut.ac.ir, hos.alizadeh@gmail.com
[2] Associate Professor, Department of Civil and Environmental Engineering, Amirkabir University of Technology, Tehran, Iran. email: jmosavi@aut.ac.ir

Key words: vadose zone, shallow groundwater, stochastic physically based model, irrigation requirement

Saturated and unsaturated zones of soil in agricultural ecosystems are inextricably linked. While shallow groundwater is fed by deep percolation according to both effective rainfall and irrigation, capillary rise causes upfluxes to enter the unsaturated layer. In arid and semi-arid climates, irrigation plays a key role in feeding shallow water table such that the more is the applied irrigation water volume during the growing season, the shallower will be the water table and as a result the more upfluxes enter the root zone. Nevertheless, besides the important participation of water table in meeting crop water requirement, due to high salinity of the groundwater, capillary rise could have harmful effects on root zone from soil salinization point of view.

This study presents a modified physically-based stochastic soil water model at root zone to assess interdependent hydrogeology of root zone and water table in agricultural systems. The modified model is based on the famous analytical model of Rodriguez-Iturbe & Porporato (2004) which was modified for effects of irrigation dependent shallow groundwater in this study. All effective phenomenon including rainfall, runoff, evapotranspiration, leakage, capillary upflux, and irrigation are incorporated in a comprehensive modeling framework. Rainfall is taken into account as a source of uncertainty while considering close relationship between irrigation and water table. A generalized demand-based irrigation scheme is employed which can distinguish between performances of different irrigation technologies including surface irrigation, sprinkler and micro-irrigation. Also effects of taking different irrigation policies from intensive irrigation to rainfed agriculture including extensive irrigation are analyzed. Due to complex interaction of unsaturated zone and shallow water table a semi-analytical approach has been taken which consists of two steps; at first, assuming fixed water table condition, analytical expressions were derived for probability density function of soil water at root zone and long-term average of each water balance component; then the numerical bisection method was utilized to modify value of water table level.

The presented model was applied in a real case with semi-arid climate in Iran, viz Dasht-e-Abbas Irrigation District which encounters presence of shallow water table due to adverse soil situation. Results show that the shallow water table could effectively contribute in satisfaction of crop water requirement in the area. Also sensitivity analysis on effects of different parameters of climate, soil, crop and irrigation policy has been done.

References

Rodriguez-Iturbe, I. & Porporato, A., 2004: Ecohydrology of Water controlled Ecosystems: Soil Moisture and Plant Dynamics. – Cambridge Univ. Press, Cambridge, U. K.

www.borntraeger-cramer.de
2012/0196/$ 0.25

Proceedings “Hydrogeology of Arid Environments”, p. 197
Stuttgart, March 2012

Poster Abstract

Hydrological Modelling in Arid and Semi-Arid Regions Using a Physics Based Model as a Tool for Water Budget Assessment

William Alkhoury, Martin Sauter

Department of Applied Geology, Georg-August-University of Göttingen, Goldschmidtstr. 3, 37077, Germany, email: wkhoury@gwdg.de, martin.sauter@geo.uni-goettingen.de

Key words: hydrological modelling, physics based models, semi-arid regions.

Due to the critical state of Jordan's water resources, numerous groundwater management studies have been conducted with the aim of improving freshwater yield while reducing the adverse impact of overabstraction on Jordanian aquifers. Although only available episodically, surface water runoff might also play a role in the water budget considerations. Surface water contributes locally in wadis to short term and intensive groundwater recharge or is collected by large scale embankments at the wadi outflow to the Jordan valley.

Regarding the prediction of the future development of available water from runoff, the water sector in Jordan has witnessed large steps forward in monitoring and gauging several wadis, mainly in the western Wadi catchments, which drain to the Jordan Valley as they account for most of the potential surface water resources. Still, gaps in hydrographic records exist and the prediction of surface water runoff and a quantitative approach as to the availability of this type of resource has received little attention.

The focus of this study is on the investigation of runoff generation mechanisms and transmission losses along the wadis in order to assess the total quantity of available water and groundwater recharge in arid and semi-arid regions based on high resolution hydrological data and mathematical modelling. Due to previous data scarcity and the diversity of natural conditions in the study area, a consistent reliable and up-to-date database was prepared as well as a systematic reasoning as to the selection of the appropriate modelling tool. After successful calibration and validation of the model, it was demonstrated in how far the calibrated model can be employed to investigate the effect of climatic and land use changes on surface water availability.

The results of this research showed a distinctive overestimation of the evaporated amounts in the semi-arid region of Wadi Kafrein and a large underestimation of the diverted rainfall volumes to runoff and recharge in earlier studies.

www.borntraeger-cramer.de
2012/0197/$ 0.25

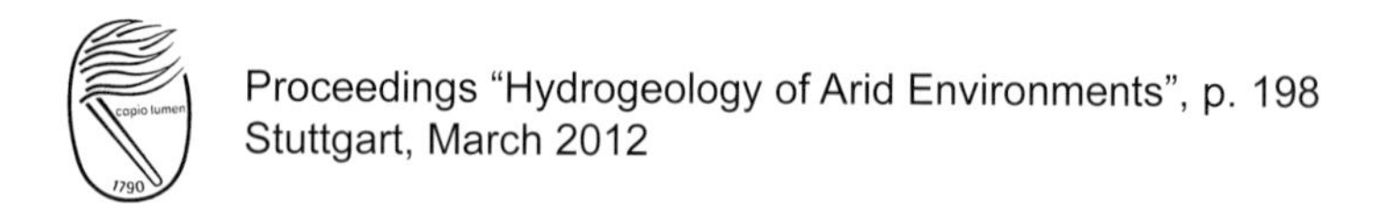
Proceedings "Hydrogeology of Arid Environments", p. 198
Stuttgart, March 2012

Poster Abstract

Groundwater Resources in Nagorno-Karabakh

Tatevik Avagimyan[1*], Boris Petrosyan[1], Grigori Sargsyan[1]

[1*] Eastern Survey SNEO of NSSP Agency Ministry of Emergency Situations of the Republic of Armenia, Araqelyan str., Stepanakert, Nagorno-Karabakh Republic
* Corresponding Author: Ms. Tatevik Avagimyan, Eastern Survey SNEO of NSSP Agency Ministry of Emergency Situations of the Republic of Armenia, Araqelyan str., Stepanakert, Nagorno-Karabakh Republic, email: bioinfo65@gmail.com

Key words: groundwater resources, zone of tectonic infringement, spectral-seismic profiling method

The water resources are national wealth of any country and one of the major bases of its economic development. They provide all spheres of a life and economic activities of people, define possibilities of development of the industry and agriculture, the organization of rest and improvement of people. Access to safe water is the basic factor not only for good health, but also for satisfactory means of subsistence, human advantage and prospects of economic growth. Nowadays, in Nagorno-Karabakh accesses to qualitative water resources are not available and the most expedient scheme of water supplies is the groundwater sources. In the present work there have been investigated the groundwater resources in the territory of Nagorno-Karabakh. Three sites (№1 filtration station of Stepanakert, Dracus and Qarahundj villages) have been investigated, in each of them have been aimed to detect the place for drilling of chinks and receive deep waters. The investigations of the sites gave affirmative results and the water which has been found there is in good quality and suitable for drinking. The supplies of fresh waters in Nagorno-Karabakh are limited, but thanks to Glikman's spectral-seismic profiling method (SSP) has been increased the level of water supplies of people, and it will rather positively affect conditions of residing at the most remote from the centre of a civilization points. SSP has shown high efficiency on set of objects at a solution of different problems. Today it is possible to tell with confidence two things. The first, on the Earth there are no areas without water, which are named zone of tectonic infringement (ZTI). And the second SSP method is currently a unique hardware, which help to carry out the searching of water supplies. So, the SSP method is alternative to traditional seismic methods and is based on use of some new, before unknown physical effects.

www.borntraeger-cramer.de
2012/0198/$ 0.25

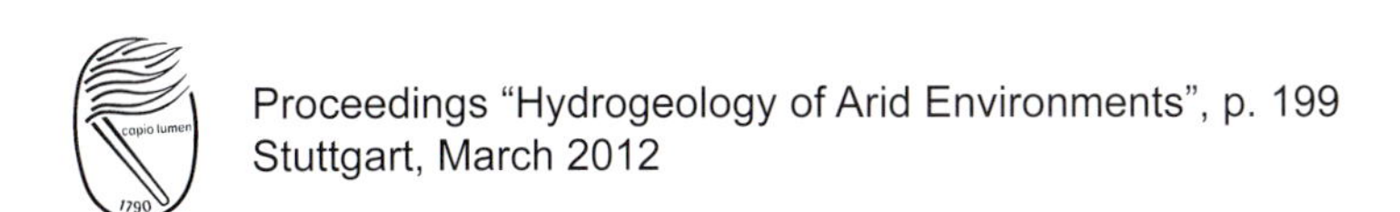

Proceedings "Hydrogeology of Arid Environments", p. 199
Stuttgart, March 2012

Poster Abstract

Water Resources Protection Plan for Zara – Ma'een – Mujib Project Watershed

Refaat Bani-Khalaf[1], Abbas Kalbouneh[2]

[1] Water Authority of Jordan, email: refaat_waj@hotmail.com
[2] Water Authority of Jordan, email: Kallbouneh_Abbas@yahoo.com

Key words: watershed, delineated, Protection, Zone

Jordan has one of the lowest levels of water resource availability per capita in the world. Water scarcity will become an even greater problem over the next two decades as the population doubles and climate change potentially makes precipitation more uncertain and variable. Management of water resources is therefore a key issue facing national government authorities. This also includes the supply of clean drinking water, which can be achieved by protecting available resources from pollution (Tamer & Fayez 2009).

Within the Project 'Mujib-Zara-Main Watershed Protection ,conducted in Cooperation between the Water Authority of Jordan (WAJ) and USAID, the Drinking Water Resources Protection Guidelines (2006) of Jordan were applied on 5 sub-watersheds in the study area. The aim of the Project was to enhance the water quality by protecting the drinking water resources of the watersheds, as it is used for drinking water supply to Amman and Madaba Governorates (USAID-CDM 2009).

The Zara – Maieen – Mujib (ZMM) Watershed is located in the central part of Jordan, including the majority of Madaba Governorate and limited southern parts of Amman Governorate. The area of this watershed covers about 1019 km^2 merging between the Jordan valley (central zone) and the eastern escarpment.

Water protection Zone-1 and Zone-2 for the dams, wadis and springs were designed according to the National Water Resources Protection Guidelines, issued in 2006.

For Mujib and Wala Dam, the border of Protection Zone 1 is formed by the shoreline of the lakes. For the wadis, the main course of the riverbed represents the border of Zone 1. Protection Zone 2 of the major wadis in the watershed was delineated the following way.

Special attention was drawn to Wadi Mujib and Wadi Walla/Hiddan sub-watersheds, as they provide the major part of the water supply from the project area. Protection zone 2 for those wadis was divided into part A and B. Zone-2A protection is delineated with a line of 350 m from the wadi edges and along the Wadi course, while Zone-2B is delineated to cover the direct topographical sub catchments for those wadis, which is formed by the area below the contour line of 700 m.a.s.l. Accordingly, the following is recommended:

Mining activities should be prohibited below the contour line of 700 m.a.s.l.

In the area between the Mujib dam and the boundaries of the Mujib natural reserve the usage of chemical pesticides and fertilizers should be reduced and replaced with the concept of organic farming.

For Wadi Abu Khusheibah and Zarqa Maieen springs: Zone-2 water protection was divided into two parts, A and B. Zone-2A protection is delineated with a line of 350 m from the wadi edges and along the Wadi course. While Zone-2B is delineated to cover direct topographical sub catchment of each wadi. This detailed delineation was implemented to protect and conserve springs located along the wadi course, in addition to the aquifer outcrops that is the source for the spring's water.

References

JICA/WAJ, 1987: Hydrogeological and water use study of the Mujib watershed. – Final Report, Appendix (I), Amman-Jordan.

Tamer, A. A. & Fayez, A. A, 2009: Artificial groundwater recharge to a semi-arid basin: case study of Mujib aquifer, Jordan. – Environmental Earth Sciences **60**, 4: 845–859.

USAID-CDM, 2009: Zara – Main – Mujib Watershed Management Plan, PP76.

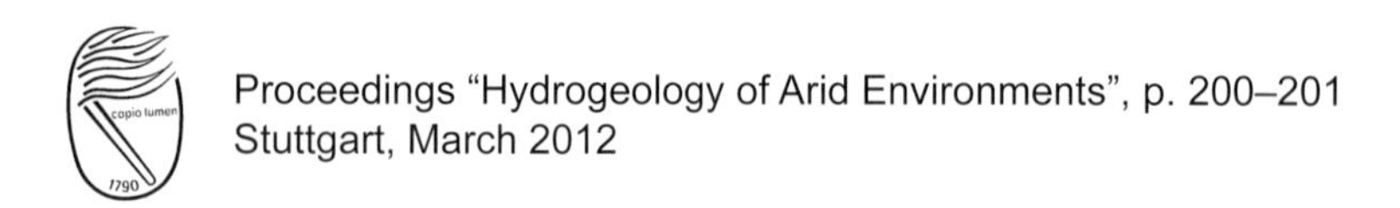

Proceedings "Hydrogeology of Arid Environments", p. 200–201
Stuttgart, March 2012

Poster Abstract

Radioactive Anomalies in the Paleozoic-Mesozoic Aquifers of the Arabian Platform

A. Bassis[1], M. Hinderer[1], R. Rausch[2], M. Keller[2], C. Schüth[1], H. Al-Ajmi[3], N. Michelsen[1]

[1] Technische Universität Darmstadt, Institut für Geowissenschaften, Schnittspahnstr. 9, 64287 Darmstadt, Germany, alex.bassis@hotmail.com , hinderer@geo.tu-darmstadt.de, schueth@geo.tu-darmstadt.de, n.michelsen@gmx.de
[2] giz International Services / Dornier Consulting P.O. Box 2730, Riyadh 11461, Saudi Arabia, Randolf.Rausch@gizdco.com, martin.keller@gizdco.com
[3] Ministry of Water & Electricity, Riyadh, Saudi Arabia, hussain.alajmi@yahoo.com

Key words: radioactivity, groundwater, aquifer, Saudi Arabia

The Saudi Arabian sandstone aquifer systems of the Upper and Lower Wajid Sandstone, Wasia-Biyadh Sandstone and Saq Sandstone show increased natural groundwater radioactivity, with some samples exceeding the WHO limit of 100 mBq/l for ^{228}Radium (WHO 2008). ^{228}Ra is a major contributor to natural groundwater radioactivity and contamination rates vary strongly: Maximum activities are 220 mBq/l in the Biyadh and 465 mBq/l in the Wasia aquifers (oral communication N. Michelsen), 1,575 mBq/l in the Upper and 3,550 mBq/l in the Lower Wajid aquifers (MoWE 2011) and 4,750 mBq/l in the Saq aquifer (BRGM/ATC 2008). Water samples, borehole and outcrop spectral-γ-ray logs show erratic peaks which are yet hard to predict. ^{228}Ra is a decay product of ^{232}Thorium, with a half-life of 5.75 a. ^{232}Th is contained as trace element in heavy minerals, e.g. Monazite, Titanite, Zircon, Thorite and Thorianite (Wickleder et al. 2006), for which the Arabian Shield is a possible source. Enrichments of these Th-bearing heavy minerals are thought to be responsible for local radioactivity peaks. Regarding the facts that both ^{228}Ra and ^{232}Th are poorly soluble in water, ^{228}Ra having a rather short half-life and low flow rates within the aquifers (Wagner 2011), it can be assumed that ^{232}Th-sources are at or near the sites of ^{228}Ra activity peaks. Based on boreholes and outcrop-analogue studies, we aim to link groundwater radioactivity, heavy mineral enrichments and specific sedimentary facies. We also intend to prove a suspected connection between geographic/stratigraphic distance to the Arabian Shield and contamination rates. This will be achieved by identifying individual 'hot spots' of increased radioactivity in a field scale using data from MoWE projects, various Diploma/Master theses and own samples and logs. ^{232}Th-bearing heavy minerals will be identified with geochemical/petrographical methods. α-recoil is the process believed responsible for the transfer of the poorly soluble ^{228}Ra into the groundwater. Investigation of this process will include microprobe and thermochronological techniques. Spectral-γ-ray- and lithofacies data will be used to localize ^{232}Th-rich heavy mineral associations in the three sandstone successions. An ongoing drilling campaign of MoWE offers a unique opportunity to compare existing groundwater data with natural-γ-ray logs and drilling cuttings. The results of these studies will enhance the predictability of radioactive anomalies in these sandstone aquifers and thus contribute to an efficient groundwater management in the Kingdom of Saudi Arabia. We expect to draw conclusions also for other arid regions that rely on fossil groundwater from sandstone aquifers.

References

BRGM/ATC – Bureau de Recherches Geologiques et Minieres/Abunayyan Trading Corporation, 2008: Investigations for Updating the Groundwater Mathematical Model(s) for the Saq and Overlying Aquifers, Vol. 8, Groundwater Quality. – Riyadh (Ministry of Water and Electricity), unpublished.

www.borntraeger-cramer.de
2012/0200/$ 0.50

MoWE – Ministry of Water & Electricity, 2011: Detailed Water Resources Studies of Wajid and Overlying Aquifers, Vol. 11, Groundwater Quality - Hydrochemistry. – Riyadh (Ministry of Water and Electricity), unpublished.

Wagner, W., 2011: Groundwater in the Arab Middle East. – Springer, Berlin Heidelberg, 443 pp.

WHO – World Health Organization, 2008: Guidelines for Drinking-Water Quality, Vol. 1, 3rd edition incorporating 1st and 2nd addenda. – Geneva, 515 pp.

Wickleder, M. S., Fourest, B. & Dorhout, P. K., 2006: Thorium. – In: Morss, L. R., Edelstein, N. M. & Fuger, J. (eds.): The Chemistry of the Actinide and Transactinide Elements. Springer, Dordrecht, pp. 52–160.

Proceedings “Hydrogeology of Arid Environments”, p. 202
Stuttgart, March 2012

Poster Abstract

Sedimentological and Petrophysical Outcrop Analogue Studies of the Cretaceous Wasia-Biyadh and Aruma Aquifers in Saudi Arabia

D. Bohnsack[1], M. Keller[2], M. Hinderer[3], J. Hornung[3], P. Witte[1], H. Al-Ajmi[4], C. Schüth[3], R. Rausch[2]

[1] GZN, GeoZentrum Nord-Bayern, Friedrich-Alexander Universität Erlangen-Nürnberg, Germany
[2] GIZ, Gesellschaft für Internationale Zusammenarbeit, Office Riyadh, Kingdom of Saudi Arabia
[3] Technische Universität Darmstadt, Institut für Geowissenschaften, Schnittspahnstr. 9, 64287 Darmstadt, Germany, email: hinderer@geo.tu-darmstadt.de
[4] Ministry of Water and Electricity, Kingdom of Saudi Arabia

Key words: Reservoir quality, sedimentary facies, permeability

In recent years, outcrop analogue studies have become a powerful tool in sedimentology for the assessment of reservoirs, both in hydrocarbon and aquifer studies. Data from exploratory drilling campaigns can significantly be augmented by observations on the outcrop of the corresponding stratigraphical interval with the objective to validate the borehole information through direct observation. In addition, through the physical separation of the outcrop area and the subsurface, the increased spatial coverage of a reservoir and its equivalents provide additional information about facies and their changes and thus on reservoir properties.

We carried out analogue outcrop studies on the major cretaceous sedimentary aquifers in Saudi Arabia (Wasia-Biyadh, Aruma) in order to better assess the storage volume of fossil groundwater which is of fundamental importance for the hyper-arid kingdom. Special attention is paid to the overall siliciclastic Wasia-Biyadh aquifer. A typical workflow in these strata starts with detailed lithologic logging of a section and the subsequent mapping of the facies across the outcrop. Due to the variability of marginal marine and fluvial-lacustrine deposits many different sections are logged in order to control lateral changes of facies. Together with bedding and bed forms, these are the basic elements of a 3-dimensional architectural framework of sedimentary formations on a regional scale. In addition, we logged spectral gamma ray emissions at the same interval (usually 30 cm) as in the boreholes. In a subsequent step, samples have been taken for measurements of porosity and permeability, and for detailed lithologic description of thin sections under the microscope. Selected samples were investigated with a scanning electron microscope (SEM) for further statements about the depositional style, cementation and diagenesis. Finally, the structural inventory along the outcrop is documented.

In this contribution we focus on measurements of porosity and permeability of ca. 150 samples and interpret reservoir quality in terms of sedimentary facies and its diagenetic imprint. Both, porosity and permeability are varying but in general are high (Biyadh: 1 to 36% and $2 \cdot 10^{-6}$ to 6.5 Darcy; Wasia: 3 to 42% and $2 \cdot 10^{-6}$ to 5 Darcy; Aruma: 1 – 38% and 10^{-6} to 0.15 Darcy). This let us to conclude that the storage volume and hydraulics of these regional aquifers are not only controlled by their fracturing but also by their matrix porosity. Permeability varies about an order of magnitude among samples or between vertical and horizontal permeability within some samples. This variation can be well explained by heterogeneity due to sedimentary facies, e.g. cross bedding and bioturbation. In some areas the kind of cementation and its intensity have a large effect on the permeability. The data obtained enhance the quality of the hydraulic interpretations of this aquifer system.

www.borntraeger-cramer.de
2012/0202/$ 0.25

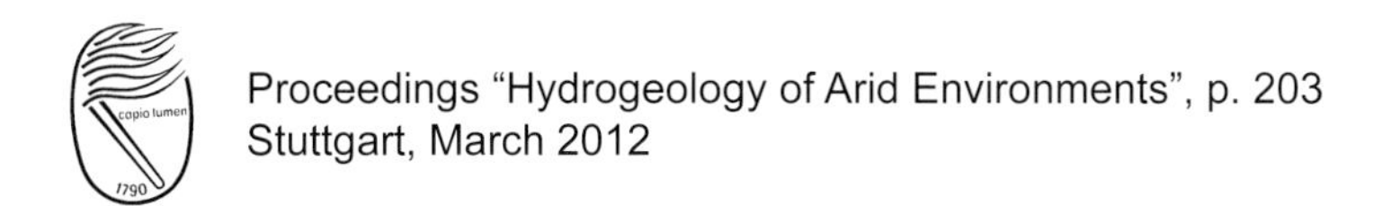
Proceedings "Hydrogeology of Arid Environments", p. 203
Stuttgart, March 2012

Poster Abstract

Evaluation of Water Resources in the Dakhla Oasis (Egypt)

B. Burges[1], W. Gossel[2], P. Wycisk[3]

[1, 2, 3] Martin-Luther-University Halle-Wittenberg, Institute of Geosciences and Geography, Hydrogeology and Environmental Geology, Von-Seckendorff-Platz 3, 06120 Halle (Saale): email: benjamin.burges@student.uni-halle.de, wolfgang.gossel@geo.uni-halle.de, peter.wycisk@geo.uni-halle.de

Key words: oasis, Dakhla, water management, WEAP

Prior work has been done to create numerical groundwater flow models for the Nubian Sandstone Aquifer System (NSAS) to quantify the drawdown in different scenarios (Sefelnasr 2007). The aim of this work is to implement management aspects which have not been considered as of yet. Agriculture is estimated to be the major consumer of water, approx. 87% (Näther 2008) of all extracted water, especially with regard to the low efficiency of flood irrigation and open channel transport. A differentiated WEAP model for the Dakhla oasis was created to emphasize a recommendation for the region. The Oasis is located in the Western Desert (Egypt) and situated above the NSAS with an arid climate and a high fluctuation in temperature (–4 to 50 °C). With almost no rain (<1mm/a) nor a river, the oasis and its approx. 12000 ha of cultivated land are completely dependent on the NSAS to satisfy their water demands. The confining conditions determine the supply of water in the Dakhla oasis. More and more artesian wells are drying up due to drawdown of the groundwater level. Although the NSAS is one of the biggest aquifers in the world the inflow to the wells is limited by the hydraulic conductivity. Future extraction rates will cause a further drop of the groundwater head to below economic limit (100 m bgl).

WEAP is able to work even on sparse data and let the user decide the degree of detail in analysis based on the availability of input data. The model area was divided into 17 subareas according to administrative units that were modeled in a reference scenario. Possible future scenarios were created particularly related to crop cycle, irrigation technology, different developments and improvements. Expansion of agricultural land, population growth and increasing efficiency in water use were integrated. Evaporation in this arid region leads to salinization processes, which were modeled to provide changes in soil quality from insufficient drainage. Salinization is a considerable danger for agricultural productivity by degrading fertile soils.

In particular wet rice, a common crop in Dakhla, provides seepage water that hinders salinization. The results showed that even traditional economic plants (dates) seem not to be very advantageously related to their water demand. Accessibility of data referring to Dakhla has been difficult; accordingly a lot of data had to be derived from indirect sources; other data had to be imported as estimates from similar areas. Prospective results can be improved greatly by increasing the percentage of direct input data to overcome uncertainties arising from imported estimates. The validity of the model is sufficient to illustrate how water management affects the supply requirements based on scenarios in the Dakhla region.

References

Sefelnasr, A., 2007: Development of Groundwater Flow Model for water resources management in the development areas of the Western Desert, Egypt. – Dissertation, Martin-Luther-Universität Halle-Wittenberg.

Näther, B., 2008: Ägypten: Entwicklung durch nachhaltige Wasserpolitik. – In: Janosch, M. & Schomaker, R. (eds.): Wasser im Nahen Osten und Nordafrika, Waxmann, Münster, p. 213.

www.borntraeger-cramer.de
2012/0203/$ 0.25

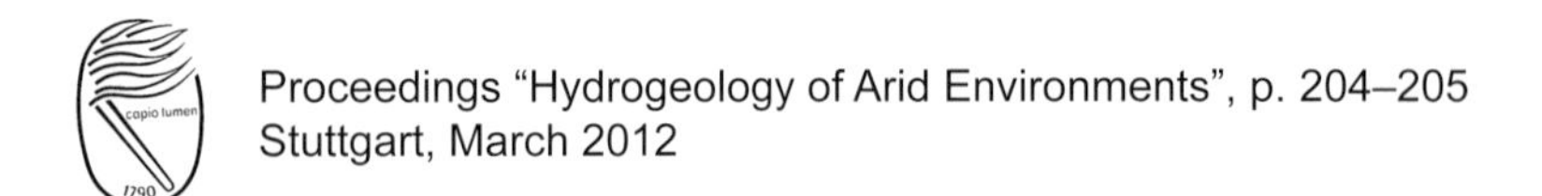

Proceedings "Hydrogeology of Arid Environments", p. 204–205
Stuttgart, March 2012

Poster Abstract

Combined Uses of WTF, CMB and Environmental Isotopes to Investigate Groundwater Recharge in the Thiaroye Sandy Aquifer (Dakar, Senegal)

S. Cissé Faye[1], O. C. Diouf[1], M. Kaba[1], M. Diedhiou[1], S. Faye[1], S. Wohnlich[2], C. B. Gaye[1], A. Faye[1]

[1] University Cheikh Anta Diop Faculty of Science & Technique Dakar Senegal, nabouthies@yahoo.com, ouscolydiouf@yahoo.fr, kaba_mariama@yahoo.com,diedhioumathias@yahoo.fr, serigne_faye@yahoo.com, cheikh.gaye@gmail.com, abfaye@sentoo.sn
[2] Institute of Geology, Mineralogy and Geophysics Faculty of Geosciences Ruhr University Bochum, Stefan.Wohnlich@ruhr-uni-bochum.de

Key words: water table fluctuations, groundwater recharge, Quaternary aquifer

The sahalien climate in the Dakar region is characterized by low mean annual rainfall (450–500 mm) occurring exclusively during the rainy season which lasts 3–4 months (July to October), high temperature (between 21 and 29 °C) and high evapotranspiration (561 mm). About 1/8th of the total precipitation is estimated to percolate through the sandy soils; the remainder is lost through evaporation and run off to local depression zones. Climate conditions and increasing water demand for the Dakar city require accurate estimation of the groundwater recharge for proper and sustainable management. In this respect, methods focusing on a daily Water Table Fluctuations at different observations wells (P3.1 and PSQ1) in response to precipitation during time period (2010–2011) are investigated to evaluate the recharge rate. The groundwater level monitored data from February 2010 to Mars 2011 are collected using "Thalimède" recorders and the specific yield for the sandy matrix is between 15 and 32%. The computed recharge rate during June, August and September 2010 using this approach varies between 197 to 205 mm/year with 331 mm annual rainfall recorded in the Thiaroye basin. Analysis of the results can, however, suggest an overestimated recharge rate which range from 0 to 81 mm/year by other methods. The unsewered urban zone is likely to induce groundwater quality deterioration (Cissé Faye 2004) and induced recharge which can lead to water table increase. Despite that, uncertainties method may be related to the accuracy of the specific yield and the validity of the assumptions considered. For this reason Chloride Mass Balance (CMB) and environmental isotopes (^{2}H, ^{18}O, ^{3}H) are used to validate results. The CMB approach widely used for estimating low recharge rates in arid regions (Cook et al. 1994) was considered in this context to assess the recharge by rainwater. Results give range of recharge values between 9 to 73 mm/year. The $\delta^{18}O$ and $\delta^{2}H$ in precipitation range from –7.6‰ to –4‰ and from –51‰ to –25‰, with mean values of –5.7‰ and –36‰. They are linearly similar to the world meteoric water line (WMWL) (Craig 1961), with an equation of $\delta^{2}H = 7.40\delta^{18}O + 5.82$ ($r^2 = 0.98$); which appears to be similar to the local water meteoric line (LWML) ($\delta^{2}H = 7.93\delta^{18}O + 10.09$) defined by Travi (1987). Groundwater samples have $\delta^{18}O$ values in the range of –5.1‰ to –1.5‰ with a mean of –3.8‰ and $\delta^{2}H$-values range from –38.1‰ to –19‰ with a mean of –31.1‰. It shows a significant depletion in $\delta^{2}H$ in relation to the WMWL and deviation is observed with best fit curve of $\delta^{2}H = 4.02\delta^{18}O - 15.9$ ($r^2 = 0.80$). As reported by Fontes et al. (1991), this trend is typically characteristics of the sahalien aquifers. Relationship of $\delta^{18}O$ and $\delta^{2}H$ together with low ^{3}H values (0.8–5.3 TU) shows clearly that the Thiaroye shallow groundwater is related to modern rainfall recharge.

www.borntraeger-cramer.de
2012/0204/$ 0.50

References

Cissé Faye, S., Faye, S., Wohnlich, S. & Gaye, 2004: An assessment of the risk associated with urban development in the Thiaroye area. – Journal of Env. Geology 45: 312–325.

Cook, P. G., Jolly, I. D., Leany F. W. & Walker, G. R., 1994: Unsaturated zone tritium and chlorine 36 profiles from southern Australian: their use as tracers of soil water movement. – Water Resource Research 30: 1709–1719.

Craig, H., 1961: Isotopic variations in meteoric water. – Science **133**: 1702–1703.

Fontes, J. C., Andrews, J. N., Edmunds, W. M., Guerre, A. & Travi, Y., 1991: Palaeorecharge by the Niger River (Mali) deduced from groundwater geochemistry. – Water Resource Research **27**: 199–214.

Travi, Y., 1987: Hydrogéologie et Hydrochimie des aquifères du Sénégal. – Thèse es Sciences, Pub. Univ. Louis Pasteur & NCRS, 150 pp.

Proceedings "Hydrogeology of Arid Environments", p. 206
Stuttgart, March 2012

Poster Abstract

Hydrogeology of the Wasia-Biyadh Aquifer, Saudi Arabia

Andreas Deckelmann[1], Ahmet Al Khalifa[2], Randolf Rausch[3]

[1] Dornier Consulting, Riyadh, Saudi Arabia, email: andreas.deckelmann@gizdco.com
[2] Ministry of Water & Electricity, Riyadh, Saudi Arabia, email: krm_1403@yahoo.com
[3] Gesellschaft für Internationale Zusammenarbeit IS, Riyadh, Saudi Arabia, email: randolf.rausch@gizdco.com

Key words: Cretaceous, Wasia-Biyadh aquifer, aquifer parameters and dynamics

The Cretaceous Wasia-Biyadh sandstones form a major aquifer on the Arabian Peninsula and were investigated within a detailed water resources study. Basic themes of the study include the analysis of the aquifer parameters (transmissivity, hydraulic conductivity, specific yield and storage coefficient), the groundwater quality, and the groundwater dynamics. The processing of the obtained data within a hydrogeological model and water management considerations completed the study.

Within the study, 13 wells (8 exploration and 5 observation wells) with a total length of approximately 15 km were drilled and hydraulically tested. In addition to the drilling, geophysical borehole logging and pumping tests on existing wells were conducted. The study area covers approximately 700,000 km^2 of the central and eastern part of Saudi Arabia. The outcrops of the Middle Cretaceous Wasia and the Early Cretaceous Biyadh are located in a 60 km wide stripe along the western border of the study area between latitudes 20 °N and 25 °N. Only close to the outcrop the Wasia and the Biyadh sandstones are in direct contact and form one hydrostratigraphic unit. Further to the east, the Shu'aiba aquitard, which mainly consists of limestone and dolomite, separates the Wasia and the Biyadh into two aquifers. The Wasia sandstones are overlain by shales and limestone, which form a prominent aquiclude towards the east. Over large areas the aquifer consists of an alternation of medium to coarse grained, poor cemented sandstones with minor shaly intercalations. The primary porosity through the rock matrix is supposed to dominate over the secondary porosity through fractures. Consistently high hydraulic conductivities with an average of $K = 3 \cdot 10^{-4}$ m/s and a specific yield between S_y = 10 and 15 % were obtained for the aquifer.

The general groundwater flow direction is towards the east. A difference of the potentiometric surface of less than 100 m from the outcrop area close to Riyadh to the Arabian Gulf coast results in a low average hydraulic gradient of $3 \cdot 10^{-4}$. Despite the overlying aquiclude, a discharge from the Wasia-Biyadh aquifer into Late Cretaceous and Tertiary limestone aquifers takes place, especially at anticline structures along the flow path and at the Arabian Gulf.

The water quality of the aquifer worsens considerably towards the east. Groundwater salinity increases from 1,500 ppm near the outcrop areas to more than 10,000 ppm in the eastern part of the study area. Highest salinities of more than 150,000 ppm are associated with oil field structures. Compared to other principal aquifers on the Arabian Platform, anthropogenic groundwater abstraction from the Wasia-Biyadh aquifer was limited in the past, however it was concentrated on a few main abstraction centers. Total abstraction for agricultural, domestic, and industrial consumption amounts to 632 MCM per year.

www.borntraeger-cramer.de
2012/0206/$ 0.25

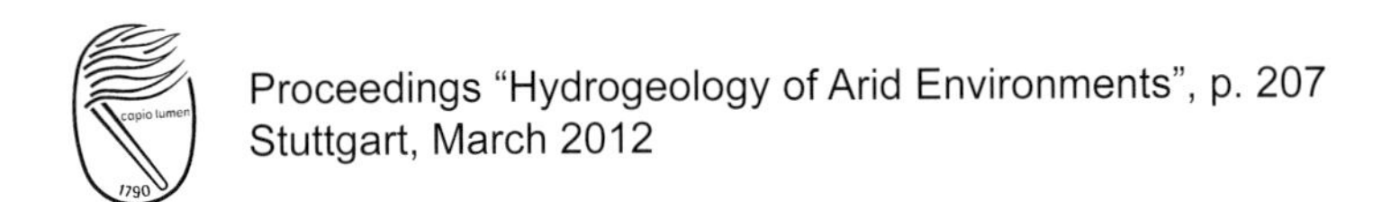
Proceedings “Hydrogeology of Arid Environments”, p. 207
Stuttgart, March 2012

Poster Abstract

Geophysical Exploration of Groundwater in a Sedimentary Terrain: A Case Study from Uli over Benin Formation in Niger Delta Basin, Nigeria

Emmanuel Dioha[1], Laura Scherer[1] and Mohammed Shiru[1]

[1] University of Göttingen, Faculty of Geoscience and Geography, Goldschmidtstr. 3, 37077 Göttingen, Germany, Corresponding email: e.dioha@stud.uni-goettingen.de

Key words: Benin formation, resistivity sounding, groundwater table

Resistivity sounding using Schlumberger array was performed in Uli town within geologic terrain often referred to as Benin Formation in the Niger Delta Basin, southern Nigeria. The interpretation of three resistivity curves indicates that the area has a high groundwater potential.

The Benin Formation is Miocene to recent sediments of alluvium and sandstone. The delineated geoelectric sections include lateritic top soil, clay, sandstone and gravel/pebbles which is consistent with the known lithologies in the area. The sandstone and gravel/pebble unit which is the aquiferous zone has a mean thickness of at least 102 m for the three main boreholes of the study area. However, the aquifer is expected to extend further below the measurement depth as been confirmed by literatures. The depth to the water table is at least 35 m as at the time of the survey, which was performed during the dry season. A fair correlation with lithologic logs in the study area was also carried out.

This study has provided information on the depth to the groundwater table and an estimate of the aquifer thickness at some locations in the study area. This information is going to be very relevant to the development of an effective water scheme for the area and possibly beyond other areas underlain by the Benin formation.

www.borntraeger-cramer.de
2012/0207/$ 0.25

Proceedings "Hydrogeology of Arid Environments", p. 208
Stuttgart, March 2012

Poster Abstract

Assessment of Transport Parameters in a Karst System under Various Flow Periods through Extensive Analysis of Artifical Tracer Tests

Joanna Doummar[1], Armin Margane[2], Martin Sauter[1], Tobias Geyer[1]

[1] Georg-August University, Goldschmidtstraße 3, 37077, Göttingen, Germany, email: jdoumma@gwdg.de
[2] Bundesanstalt für Geowissenschaften und Rohstoffe (BGR), Hannover, Germany

Key words: tracer tests, karst, transport, 2NREM

It is primordial to understand the sensibility of a catchment or a spring against contamination to secure a sustainable water resource management in karst aquifers. Artificial tracer tests have proven to be excellent tools for the simulation of contaminant transport within an aquifer before its arrival at a karst spring as they provide information about transit times, dispersivities and therefore insights into the vulnerability of a water body against contamination (Geyer et al. 2007). For this purpose, extensive analysis of artificial tracer tests was undertaken in the following work, in order to acquire conservative transport parameters along fast and slow pathways in a mature karst system under various flow conditions. In the framework of the project "Protection of Jeita Spring" (BGR), about 30 tracer tests were conducted on the catchment area of the Jeita spring in Lebanon (Q= 1 to 20 m3/s) under various flow conditions and with different injection points (dolines, sinkholes, subsurface, and underground channel). Tracer breakthrough curves (TBC) observed at karst springs and in the conduit system were analyzed using the two-region non-equilibrium approach (2NREM) (Toride & van Genuchten 1999). The approach accounts for the skewness in the TBCs long tailings, which cannot be described with one dimensional advective-dispersive transport models (Geyer et al. 2007). Relationships between the modeling parameters estimated from the TBC were established under various flow periods. Rating curves for velocity and discharge show that the flow velocity increases with spring discharge. The calibrated portion of the immobile region in the conduit system is relatively low. Estimated longitudinal dispersivities in the conduit system range between 7 and 10 m in high flow periods and decreases linearly with increasing flow. In low flow periods, this relationship doesn't hold true as longitudinal dispersivities range randomly between 4 and 7 m. The longitudinal dispersivity decreases with increasing flow rates because of the increase of advection control over dispersion and increasing dilution. Therefore variance of the TBC is controlled on the hand by dispersivity during high flow periods and on the other hand by increasing mobile phase in low flow periods due to an increase of the portion of immobile zones (pools and ripples) as water level decreases. For tracer tests with injection points at the surface, longitudinal dispersivities are found to be of higher ranges (8–27 m) and highly reflective of the compartments in which the tracer is flowing (unsaturated rock matrix, conduits or channel). The comparison of tracer tests with different injection points shows clearly that the tailing observed in some of the breakthrough curves is mainly generated in the unsaturated zone before the tracer arrives to the main channel draining the system and decreases gradually within the channel.

References

Geyer, T., Birk, S., Licha, T., Liedl, R. & Sauter, M., 2007: Multitracer Approach to Characterize Reactive Transport in Karst Aquifers. – Groundwater **35**,1: 35–45.

Toride, N., Leij, F. J. & van Genuchten, M. T., 1999: The CXTFIT code (version 2.1) for estimating transport parameters from laboratory or field tracer experiments. – U.S. Salinity Laboratory Agricultural Research Service, U.S. Department of Agriculture Riverside, California. Research Report **137**.

www.borntraeger-cramer.de
2012/0208/$ 0.25

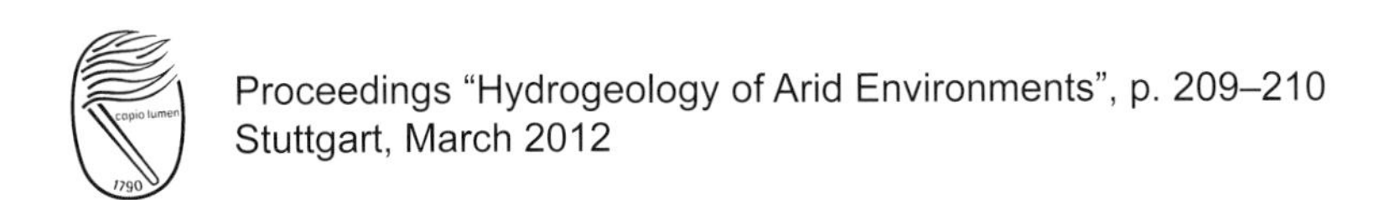
Proceedings "Hydrogeology of Arid Environments", p. 209–210
Stuttgart, March 2012

Poster Abstract

Determining the Water Budget of the Gunt (Semi-Arid Tajik Pamir) Using Stable Water Isotopes, Hydrochemical- and Remote Sensing Data

Christiane Ebert[1], Stefan Geyer[1], Tino Rödiger[1], Wolfgang Busch[1], Malte Knoche[1], Richard Gloaguen[2], Eric Pohl[2], Karsten Osenbrück[3], Jamila Baidulloeva[4], Stephan M. Weise[1]

[1] Helmholtz-Centre for Environmental Research GmbH – UFZ, Theodor-Lieser-Str. 4, D-06120-Halle, email: christiane.ebert@ufz.de
[2] TU Bergakademie Freiberg, Institute for Geology, Remote Sensing Group, Bernhard-v.-Cotta Str. 2, D-09596-Freiberg, email: gloaguen@geo.tu-freiberg.de
[3] Universität Tübingen, Institute for Geosciences, Water & Earth System Sciences Research Center, Keplerstr. 17, D-72074-Tübingen, email: karsten.osenbrueck@uni-tuebingen.de
[4] State administration for hydrometeorology, 47 Shevchenko Str., TJ-734025-Dushanbe, email: hydrometcenter@gmail.com

Key words: Isotopes; Hydrochemistry; semi-arid region; Pamir

Renewable water resources (groundwater, rivers) in Western Tibet and the arid Central Asian lowlands are vulnerable due to their unsustainable exploitation and susceptibility to climatic variations, resulting in water scarcity issues and deteriorating water quality.

Large rivers such as the Panj and Amu-Darya, whose water is excessively exploited for irrigation purposes, are mainly fed from snow- and glacial melt occurring in the Pamir and Tien Shan mountains (elevations over 5000 m) covering parts of Tajikistan and western China (Barlow & Tippett 2008). Unlike other Central Asian mountainous regions such as the Himalaya or Hindu Kush, which are influenced by the summer monsoon, the Pamir Mountains receive their precipitation as snow in winter and spring due to westerly winds originating in the Atlantic (Barlow & Tippett 2008).

The main research objectives are to understand the current and future key hydrological processes, such as streamflow generation and groundwater recharge in an exemplary drainage system in the Tajik Pamir and to use remote sensing data to enable a regionalisation of the results over the whole of western Tibet. A combination of remote sensing techniques, isotope-hydrological methods and enhanced hydrological simulation models will be used to understand the hydrological system of the area of replenishment on catchment scale.

This includes detailed groundwater studies, investigation of seasonal dynamics of runoff components and how this is connected to climate variability.

Remote sensing data is essential for monitoring the ice-cover and changes in glacial extent of the last decades and allows mapping the geometry of stream beds to outline areas of potential water infiltration, accumulation of sediments and regions of high hydraulic energy. Investigations focus on the Gunt catchment (ca. 14,000 km^2) in the semi-arid Tajik Pamir, which is representative for the entire region.

As a first step towards estimation of the origin, interaction and dynamics of stream and subsurface water components, samples for hydrochemical and isotopical analyses are and will be taken from river water and groundwater. Groundwater recharge and discharge, streamflow components as well as water residence times will be characterised and quantified using hydrochemistry (major and trace elements) and stable and radioactive environmental isotopes (e.g. ^{2}H, ^{3}H, ^{7}Li, ^{18}O, ^{13}C, ^{14}C, ^{87}Sr).

In the two field campaigns in summer and autumn 2011 we sampled both water of the stream,

www.borntraeger-cramer.de
2012/0209/$ 0.50

of selected tributaries and groundwater. Among others the stable water isotopes (^{2}H and ^{18}O) of the samples were analysed and as result the isotopes highlight variations between the different end members. The collected hydrochemical data show also differences between the various samples.

References

Barlow, M. & Tippett, M. K., 2008: Variability and predictability of Central Asia river flows: Antecedent winter precipitation and large-scale teleconnections. – J. of Hydrometeorology **9**: 1334–1349.

Proceedings "Hydrogeology of Arid Environments", p. 211–212
Stuttgart, March 2012

Poster Abstract

Deep Exploratory Drilling for Wajid Sandstone Reservoirs

Mohammed El Shazly Mahmoud, Ahmed Saeed Abu Degen, Saleh Hassan Abu Degen

Abu Degen Company for Drilling Wells, P.O. Box: 5350, Code: 11422, Riyadh, Kingdom of Saudi Arabia, e-mail: shazm46@yahoo.com, ghamshok@hotmail.com, asa5350@hotmail.com

Key words: Wajid sandstone, exploration wells, well logging

Most of Kingdom of Saudi Arabia is arid. The average rainfall ranges from 25 mm to 150 mm compared to the average annual evaporation that ranges from 2,500 mm to about 4,500 mm. The socio-economic development of the country has been supported in large measure by its intensive use of groundwater including non-renewable groundwater resources.

The Wajid Sandstone represents a principal reservoir. The Wajid Sandstone is a succession of Paleozoic siliciclastic deposits in the south western part of Kingdom of Saudi Arabia.

During 2007–2009 extensive exploratory drilling campaign conducted by Abu Degen Company for Drilling Wells (as a contractor for GIZ/DCo) within the scope of Wajid Water Resources Studies (Ministry of Water and Electricity – GIZ/DCo: Consultant). The overall objective was to obtain basic information on the **lithostratigraphy** and hydrogeology of the Wajid Sandstone Sequence.

The drilling program comprised total of 24 drilling sites with accumulated drilling length of 13,000 meters. The depth range of the wells was between 500–1500 m including observation and exploration wells.

In some cases twin wells were drilled tapping the same aquifer as (7a, 7b) to provide data on the hydraulic parameters of the Wajid Sandstones (transmissivity, storage coefficient). Detailed geophysical logging was conducted for each well for several runs during the course of the drilling. Pumping tests were carried out in step drawdown tests (4 stages) and long duration tests for 72 hrs to obtain hydraulic data.

The data clarified that the Wajid reservoir is hydraulically separated into an upper and lower sequence by the shale of Qusaiba member of the Qalibah formation. This has been proven for all the exploratory wells drilled. The noticeable total gamma ray logs supported by the resistivity as well the density logs patterns of the Qalibah formation can be seen as a marker key bed for the **lithostratigraphic** relation between wells.

The separation of the Wajid Sandstone into two individual reservoirs has encouraged the Ministry of Water and Electricity to locate two new well fields tapping the Upper Wajid Sandstone (depth 650 m), to supply two cities with groundwater namely Najran and Yadmah.

The Upper Wajid reservoir has been selected for immediate groundwater abstraction due to its more economic and favorable hydrogeologic parameters in terms of groundwater production and water quality. The lower Wajid reservoir might be a future target for water supply.

References

Abu Degen Company, 2009: Final Technical Reports and Well Logging Interpretation for Wajid Exploratory-Observation Wells [Unpublished].

Evans, D. S., Lathon, R. B., Senalp, M. & Connally, T. C., 1991: Stratigraphy of the Wajid Sandstone of South-Western Saudi Arabia. – Society of Petroleum Engineers, SPE Middle East Oil Show, Bahrain, 16–19 November 1991, Paper SPE 21449: 947–960.

Filomena, C. M., 2007: Sedimentary Evolution of a Paleozoic Sandstone Aquifer: The Lower Wajid Group in Wadi Ad Dawasir, South Western Saudi Arabia: Diploma Thesis, University of Tübingen, 126 pp. – [Unpublished].

www.borntraeger-cramer.de
2012/0211/$ 0.50

ITALCONSULT (1969): Water and Agricultural Development Survey for Areas II and III. – Ministry of Agriculture [Unpublished].

Kellogg, K. S., Janjou, D., Minoux, L. & Fourniguet, J.,1986: Explanatory Notes to the Geologic Map of the Wadi Tathlith Quadrangle, Sheet 20 G, Kingdom of Saudi Arabia. – Ministry of Petroleum and Mineral Resources, Deputy Ministry for Mineral Resources, 27 pp.

Stump, T. E. & Van Der Eem, J. G., 1995: The stratigraphy, Depositional Environments and Periods of Deformation of he Wajid outcrop belt, South-Western Saudi Arabia. – Journal of African Earth Sciences **21**: 421–441.

Stump, T. E. & Van Der Eem, J. G., 1996: Overview of the Stratigraphy, Depositional Environments and Periods of Deformation of the Wajid outcrop belt, south-western Saudi Arabia. – In: Al-Husseini, M. I. (ed.): Geo '94, the Middle East Petroleum Geosciences, selected Middle East Papers from the Middle East geoscience conference: 867–876.

Proceedings "Hydrogeology of Arid Environments", p. 213
Stuttgart, March 2012

Poster Abstract

Proposing the Best Groundwater Utilization for Haddat Al Sham Arid Region Aquifer, Western Saudi Arabia

A. S. El-Hames, A. Al Thobaiti

Dept. of Hydrogeology, King Abdulaziz University, Jeddah, Saudi Arabia, email: a_hames@hotmail.com

Key words: groundwater aquifer arid Region MODFLOW

Haddat Al Sham area is located in the Western region of Saudi Arabia which is categorized as arid and thus suffers from shortage of water. It relies on local groundwater resources to supply water to the local inhabitants and the surrounding villages as well as some commercial water companies. Nevertheless excessive exploitation of the available groundwater has resulted in degradation in its quality and quantity. This study lays guidelines for the best management practice for utilizing this aquifer by exploring areas with high groundwater potential and the safe and optimum pumping rate which can be operated with minimal influence on the groundwater aquifer at the medium term. In order to accomplish these objectives, advanced hydrogeological modeling practice was utilized with different pumping scenarios by utilizing MODFLOW.

In order to supply MODFLOW with the required data, intensive fieldwork has been carried out through which well inventory in addition to pumping tests have been performed. Results of the scenario simulations show that the aquifer can be divided into three areas each of which shows identical behavior to the groundwater pumping. These areas are located in the upper, middle, and lower parts of the aquifer. It is observed that the aquifer in the upstream area is very shallow and it is inclined towards the downstream, while the downstream area acts as a pond that receives the incoming water from the upstream of the aquifer. It can also be concluded that the most promising area in terms of water quantity is that located in the downstream of the aquifer and the worst is that located in the upstream area. The study has shown that it is best to utilize the downstream area for pumping purposes with maximum pumping rate not exceeding 400 m^3/day for the wells located in this area.

www.borntraeger-cramer.de
2012/0213/$ 0.25

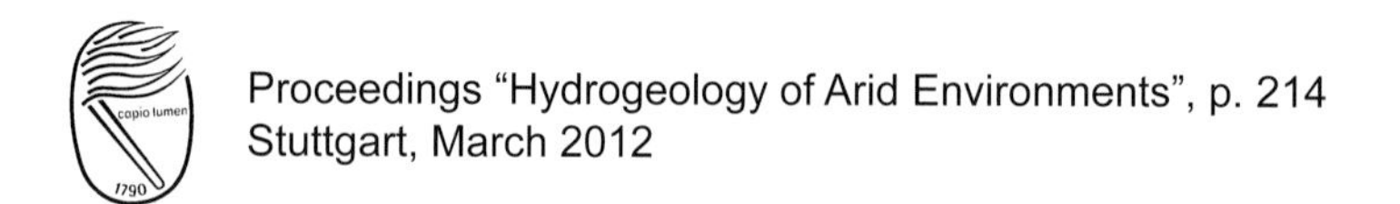

Proceedings "Hydrogeology of Arid Environments", p. 214
Stuttgart, March 2012

Poster Abstract

Hydrochemical Characterization and Groundwater Quality in Delta Tokar Alluvial Plain, Red Sea Coast – Sudan

Adil Balla Elkrail[1], Bashir A. Obied[2]

[1,2] Department of Hydrogeology, Faculty of Petroleum & Minerals, Al Neelain University, Khartoum, Sudan.
P. O. Box 12702, email: adilmagboul321@yahoo.com

Key words: quifer, salinization, chemical facies

The leaching processes along the flow path and over abstraction of the alluvial aquifer, the principal aquifer in delta Tokar, by the agricultural and domestic sectors and natural factors, have led to its salinization which may be due to interaction between geological formations and adjacent brackish and saline water bodies as well as seawater transgression. The main objectives of this study are to assess the hydrochemical characteristics of the groundwater and to delineate the locations and the sources of aquifer salinization. Water samples in the project area were chemically analyzed for major cations and anions at the laboratory by the standard analytical procedures. Chemical data were manipulated using GIS techniques for hydro chemical maps and Piper diagram for chemical facies and SPSS software for statistical analyses such as basic statistics (mean and standard deviation) and Spearman's correlation matrix. A hydrochemical study identified the locations and the sources of aquifer salinization and delineated their areas of influence. The investigation indicates that the aquifer water quality is significantly modified as groundwater flows from the southwestern parts of the study area, where the aquifer receives its water by lateral underflow from Khor Baraka flood plain, to the central and northeastern parts, with few exceptions of scattered anomalous pockets in the deltaic plain. Significant correlation between TDS and/or EC with the major components of Na^+, Cl^- and SO_4^{-2} ions is an indication of seawater influence on the groundwater salinity. Moreover, Cl^-, $SO^{2-}{}_4$ and Na^+ are predominant ions followed by Ca^{2+} and $HCO^-{}_3$. Hence, four types of groundwater can be chemically distinguished: Na-Ca-SO_4-Cl-facies; Na-Cl-SO_4-HCO_3-facies, Na-Ca-Mg-SO_4-Cl-HCO_3-facies and Na-Ca-Mg-Cl-SO_4-facies. The processes that govern changes in groundwater composition as revealed by chemical and statistical analyses, are mainly associated with over-abstraction, biodegradation, marine intrusions and carbonate saturation.

www.borntraeger-cramer.de
2012/0214/$ 0.25

Proceedings “Hydrogeology of Arid Environments”, p. 215
Stuttgart, March 2012

Poster Abstract

Hydrogeological Characteristics of Djbel Es Senn Turonian Limestone Aquifer of Tebessa Area (North East of Algeria)

Chelih Fatha[1], Fehdi Chemseddine[2]

[1] Université de Tébessa, Département de Géologie, Tébessa 12002, Algeria. email: fatha_geo@yahoo.fr
[2] Université de Tébessa, Département de Géologie, Tébessa 12002, Algeria. email: fehdi@yahoo.fr

Key words: aquifer, hydrogeology, hydrochemistry, Algeria

The study area (Massif de Djebel Es Senn, Troubia), located southwest of the region of Hammamet, is one of the high plains areas of eastern Algeria on the border between Algeria and Tunisia. The study area is constituted in the major part by cretaceous formations, forming a succession of anticlines and synclines. The stratigraphic sequence is presented in the form of alternation of carbonated formations of limestones, marly-limestones and argillaceous marls.

The plio-quaternary and quaternary terrains occupy the central part; they are consisted by actual and recent alluvial deposits, conglomerates, gravels, sandstones, etc.

The summary analysis of the stratigraphic column of the study area shows the presence of three aquiferous formations among them the formation of Plio- quaternary one .This aquifer of great extension occupies the major part of the tectonic basin, limited at the West and at the East by two great faults of NW-SE orientation.

Groundwater Aquifer of Hammamet have important changes in mineralization is increasing from south to north and for West to East in the direction of the flow. The conductivity is generally high and varies between 900 µS/cm and 2100 µS/cm. The study areas are in direct contact with the Triassic saline rocks. This is mainly controlled by salinity, chlorides and sodium. The use of major and minor chemical elements allowed us to understand the process of mineralization of the water. Thus, the mineralization come from the dissolution-precipitation of the rock aquifer, evaporates and by cationic exchange reactions.

www.borntraeger-cramer.de
2012/0215/$ 0.25

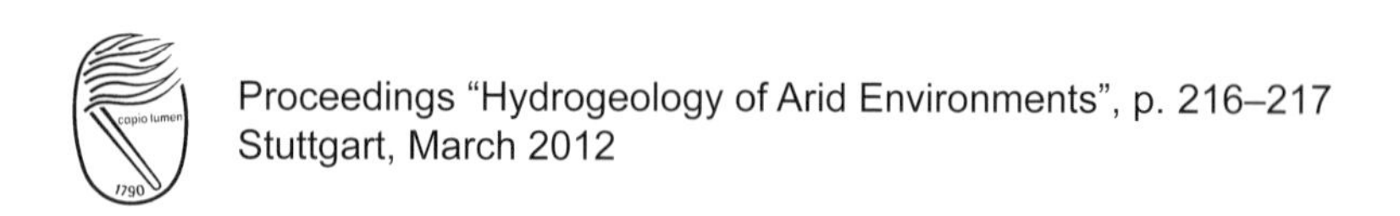

Proceedings "Hydrogeology of Arid Environments", p. 216–217
Stuttgart, March 2012

Poster Abstract

The Upper Permian, Triassic and Jurassic Formations of the Eastern Arabian Peninsula – Drawing Up a Geological 3D-Model for the Khuff-Jilh-Minjur-Dhruma Aquifer System

Gero Friedrich[1], Klaus Reicherter[1], Randolf Rausch[2], Heiko Dirks[3], Hussain Al-Ajmi[4], Olaf Kolditz[5], Karsten Rink[5]

[1] RWTH Aachen University, Institute for Neotectonics and Natural Hazards, Lochnerstr. 4-20, 52056 Aachen, gero.friedrich@rwth-aachen.de
[2] GIZ IS, P.O. Box 2730, Riyadh 11461, Kingdom of Saudi Arabia
[3] Dornier Consulting GmbH, P.O. Box 2730, Riyadh 11461, Kingdom of Saudi Arabia
[4] Ministry of Water & Electricity, Saud Mall Center, Riyadh 11233, Kingdom of Saudi Arabia
[5] UFZ, Environmental Informatics, Permoserstr. 15, 04318 Leipzig

Key words: 3D geological model, Arabian Peninsula, Triassic, Jurassic

Introduction

The Arabian Peninsula is located on the Arabian Plate, which is surrounded to the southwest by a divergent boundary, to the northwest and southeast by a convergent boundary, and a transform boundary to the northeast (Sharland et al. 2001). Geologically, the Arabian Plate can be divided into two main structural units: the western Arabian Shield, which is mainly built of igneous and metamorphic rocks; and the eastern Arabian Platform consisting of a sedimentary sequence that reaches up to 12 km thickness (Powers et al. 1966).

Based on data from groundwater resources studies carried out by the Ministry of Water & Electricity, a geological 3D model was created for the formations of Upper Permian, Triassic and Jurassic formations. Additionally, different kinds of lithologies have been characterized hydrogeologically and have been divided into seven hydrofacies zones. The hydrofacies zones represent zones of relatively homogeneous aquifer properties (hydraulic conductivity, storage coefficient) and are the prerequisite for the development of a regional groundwater model.

Methodology

The geological model is based on literature- and borehole data as well as on a digital elevation model (SRTM) of the Arabian Peninsula. Relevant references have been investigated with respect to extension, thickness and facies of affected formations. Cartographical information about distribution and facies (Ziegler 2001) of investigated formations as well as the general tectonical features of the Arabian Plate have been digitized, georeferenced and stored as vector data using ArcGis 9.2. Providing that sufficient coordinates are given, thickness information of formations has been integrated into a borehole database as point data. The borehole database has been actualized, filtered and reformatted as a result of software requirements. On the basis of literature and borehole data, structure contours were drawn for formations of Khuff, Jilh, Minjur, and Dhruma. Taking the borehole data into account, all other formations have been integrated into the developed base model using thickness ratio provided by Le Nindre (2003).

www.borntraeger-cramer.de
2012/0216/$ 0.50

Results

The results show a structure contour map and an isopach map as well as a lithofacies map and a hydrofacies map for all formations. The geological model was 3D-visualized and visually validated by using OpenGeoSys, an application developed by the UFZ Leipzig. Structural elements like the Rub'Al-Khali basin or the Ghawar anticline system are well displayed. Also, strata-comprehensive hydrofacies bodies are well observed in the visualization, which allows an enhanced understanding of the hydraulic system.

Following the existing geological model, a complete model for the Phanerozoic sediments of the Arabian Peninsula may be developed in subsequent studies. Additionally, this model will be the basis for a detailed groundwater flow model.

References

Le Nindre, Y., Vaslet, D., Le Métour, J., Bertrand, J. & Halawani, M.,2003: Subsidence modelling of the Arabian Platform from Permian to Paleogene outcrops. – Sedimentary Geology **156**: 263–285.

Powers, R. W., Ramirez, L. F., Redmond, C. D. & Elberg Jr., E. L., 1966: Geology of the Arabian Peninsula: sedimentary geology of Saudi Arabia. U.S. Geological Survey, Prof. Pap. 560 (D), 147 pp; Washington.

Sharland, P. R., Archer, R., Casey, D. M., Davies, R. B., Hall, S. H., Heward, A. P., 2001: Arabian Plate Sequence Stratigraphy. – Gulf PetroLink – GeoArabia Special Publication **2**: 1–369.

Ziegler, M. A., 2001: Late Permian to Holocene Paleofacies Evolution of the Arabian Plate and its Hydrocarbon Occurrences. – Gulf PetroLink – GeoArabia **6** (3): 445–504.

Proceedings "Hydrogeology of Arid Environments", p. 218
Stuttgart, March 2012

Poster Abstract

Hydrogeology of the Late Permian, Triassic & Jurassic Aquifers of Saudi Arabia

Tobias Fuest[1], Ibrahim Al-Shabibi[2], Heiko Dirks[1], Randolf Rausch[3]

[1] Dornier Consulting GmbH, P.O. Box 2730, Riyadh 11461, Saudi Arabia. email: tobias.fuest@gizdco.com
[2] Ministry of Water & Electricity of the Kingdom of Saudi Arabia, P.O. Box 57616, Riyadh 11584, Saudi Arabia.
[3] GIZ International Services, P.O. Box 2730, Riyadh 11461, Saudi Arabia.

Key words: Saudi Arabia, groundwater flow, aquifer properties

Introduction

Within the framework of a large groundwater assessment study in Saudi Arabia, aquifers of Late Permian, Triassic and Jurassic age are investigated. The study area comprises approximately 230,000 km^2 from Latitude 28°N to Latitude 18°N along a north-south trending band in central Saudi Arabia. Five major aquifers can be distinguished, which are named as follows (from bottom to top): Khuff (Late Permian), Jilh, Minjur (Triassic), Dhruma, and Arab aquifer (Jurassic).

To its top, the aquifer system is separated from the overlying formations by the Upper Jurassic Hith aquiclude. To the bottom, it is bound by impermeable basement rocks in the central part of Saudi Arabia, while to the north and south of the study area the aquifer system is underlain by Paleozoic sandstones. From them, vertical groundwater inflow occurs into the aquifer system. In addition, minor inflow takes place from groundwater recharge in the outcrop areas. However, most of the groundwater stored in the aquifer system can be considered as fossil and has been recharged before 6,000 years B.P., when arid conditions started to prevail on the Arabian Peninsula.

Aquifer properties & Groundwater flow

The general groundwater flow direction follows the dip of the geological formations from west to east. In areas of heavy groundwater abstraction, e.g. around Riyadh (Minjur aquifer), Kharj (Arab aquifer), and As Sulayyil (Dhruma and Minjur aquifers), flow direction is locally altered through large cones of depression.

The first four (Khuff, Jilh, Minjur, Dhruma) of the above mentioned aquifers can be characterized as fractured sandstone aquifers, while the Arab aquifer is characterized as a fractured limestone aquifer. Major shifts in lithofacies within all geological formations from the north to the south of the study area have a strong influence on the aquifer properties. The highest average hydraulic conductivity of $K = 1.34 \cdot 10^{-4}$ m/s was found for the Dhruma aquifer, the lowest for the Jilh aquifer ($K = 1.8 \cdot 10^{-6}$ m/s).

In general, the hydraulic separation by aquitards is imperfect within the aquifer system, and the aquifers are partly connected. This is indicated by the piezometric head differences between the aquifers along the flowpath. Head differences of 100 m in the outcrop areas decline to less than 50 m in the east of the study area. In some areas aquifers are completely connected and form a single aquifer, such as the Minjur and the Dhruma in the southern part of the study area, where the confining layers of the Marrat shale are missing in between.

Outlook

The groundwater assessment study of the central Saudi Arabian aquifers is still underway. Additional field data from drilling, pumping tests, and geophysical borehole logging will improve the knowledge of the aquifers, especially in areas, where information from literature is scarce. In a final step the entire aquifer system will be described by a comprehensive regional groundwater flow model.

www.borntraeger-cramer.de
2012/0218/$ 0.25

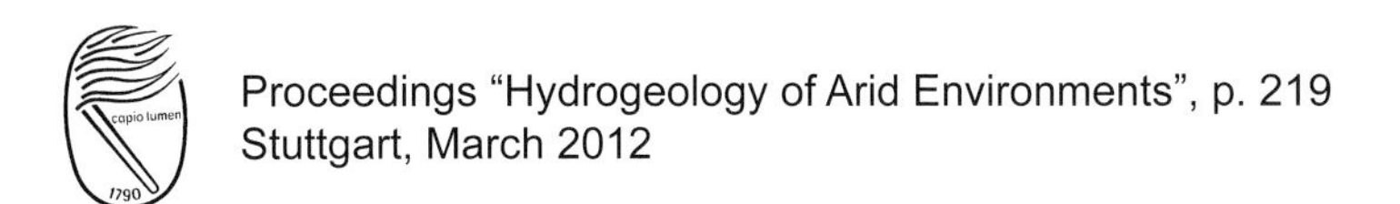
Proceedings "Hydrogeology of Arid Environments", p. 219
Stuttgart, March 2012

Poster Abstract

Characterising Recharge through an Arid Zone River Using an Environmental Tracer Approach

Simon Fulton[1], Daniel Wohling[2], Andrew Love[3]

[1] Northern Territory Department of Natural Resources, Environment, The Arts and Sport, PO Box 496 Palmerston NT Australia 0830, simon.fulton@nt.gov.au
[2] South Australian Department for Water, Level 11 25 Grenfell St Adelaide SA Australia, 5000, Daniel.Wohling@sa.gov.au
[3] Flinders University, Sturt Rd Bedford Park SA Australia 5042, andy.love@flinders.edu.au

Key words: preferential recharge, Finke River, environmental isotopes

The Great Artesian Basin (GAB) is Australia's largest water resource extending over 1.7 million square kilometers or 22% of the continent. The southwest of the basin is situated in arid central Australia and represents the only significant and reliable source of water in the region. Assessing recharge to the GAB, both spatially and temporally, is essential in determining the water requirements of environmental assets, such as the iconic mound springs, and balancing these needs with increasing demand for consumptive groundwater use. Along the western margin of the GAB average annual rainfall is between 150–200 mm/year and is minimal in comparison to potential evaporation, which can exceed 3,000 mm/year. This deficit limits the potential for direct recharge to the aquifer and under current climatic conditions diffuse recharge is considered negligible. Rainfall is dominated by summer monsoonal systems and intense rainfall events can trigger large flood events in ephemeral desert rivers. Where these drainage lines cross the margin of the GAB there is potential for the direct infiltration of flood water into the aquifer. This paper investigates this process using environmental tracers to characterise the spatial and temporal variation in recharge to the GAB aquifer through the Finke River.

The Finke River is a large ephemeral drainage system, which crosses the outcropping GAB aquifer around Finke Community 200 km southeast of Alice Springs. Estimating recharge in the system is particularly difficult due to limited groundwater and surface water monitoring infrastructure, a lack of time series water level data and the irregularity of flood events. The study concentrated on the use of isotopic and geochemical sampling from existing groundwater bores to describe the recharge conditions and estimate recharge rates. Groundwater close to the Finke River has a depleted stable isotope signature ($\delta^{18}O$ = –9.37 to –10.35 ‰) implying a direct recharge process consistent with the ephemeral flood model. In contrast, groundwater distant to the Finke River has an enriched stable isotope signature ($\delta^{18}O$ = –7.83 to –6.09 ‰) indicating longer residence in the soil zone and a different recharge mechanism. ^{14}C concentrations were elevated in bores around the Finke River (78–101 PMC) and show a clear decline in bores more distant to the river (3–13 pMC). Results strongly support the operation of the Finke River as a recharge sink. Elevated ^{14}C concentrations (>85 pMC) suggest a thermonuclear component and indicate groundwater recharged in the last 40 years has reached the watertable. Recharge rates were calculated from ^{14}C derived groundwater velocities using an approach based on Vogel (1967). Rates ranged from 450 to 1000 mm/year and compared favourably with annualised rates calculated using hydraulic data from a recent flood event in October 2010 (500 mm/year).

References

Vogel, J. C., 1967: Investigation of groundwater flow with radiocarbon. – In: Isotopes in Hydrology, International Atomic Energy Agency, Vienna pp. 355–369.

www.borntraeger-cramer.de
2012/0219/$ 0.25

Proceedings "Hydrogeology of Arid Environments", p. 220–221
Stuttgart, March 2012

Poster Abstract

Groundwater Resource Protection in Jordan, a Case Study from AWSA and Heedan Well Fields

Niklas Gassen[1], Ibraheem Hamdan[1], Ali Subah[2], Ayman Jaber[2], Tobias El-Fahem[1]

[1] Federal Institute of Geoscience and Natural Ressources, P.O. Box 926238, 11190 Amman, Jordan, email: niklas.gassen@bgr.de, [2] Ministry of Water and Irrigation

Key words: Groundwater Protection, Jordan, Heedan Well Field, AWSA Well Field

Water resources are extremely scarce in Jordan. Rapid increase in groundwater development due to high population growth and high agricultural activities has led to groundwater level declines over the past 2 decades (World Bank 2009). In 2009/2010 annual consumption in the country exceeded the safe yield by 182 MCM (MWI 2010). Concerning this situation, regularly occurring contaminations of water resources pose a serious risk to the sustainable supply of drinking water to Jordan.

In order to protect available water resources from pollution, the Jordanian Ministry of Water and Irrigation (MWI) and the German Federal Institute for Geoscience (BGR) work together on the establishment of water resource protection zones in selected areas. A national Guideline for Drinking Water Resource Protection was introduced in 2006 and updated in 2011 to strengthen the legal basis for the delineation of water protection zones.

Until now protection zones for 7 well fields, 7 springs and 2 dams have been delineated (Margane et al. 2010). Within the current cooperation project 'Water Aspects in Land-Use Planning' (Project Phase: 2009–2014), 2 well fields (Heedan and AWSA well fields) are being observed in respect to groundwater protection.

Heedan Well field is located 45 km southwest of Amman. It produces 12 MCM/a from the A7/B2 aquifer, supplying drinking water to Amman, Madaba and the surrounding region (Margane et al. 2008). Regular problems of high turbidity and microbiological contamination occur in the well field. Water Quality problems mainly occurring during the rainy season (November – April (Ta'any& Al Atrash 2010)) and are therefore most likely related to flood events. Contamination sources during flood events are seeping surface water upstream of the well field and poorly sealed wellheads. In order to prevent the contamination events leading to a disruption of the water supply, a detailed concept for the delineation of protection zones within the well field is developed in the course of this study.

AWSA Well field is located in the Aria of the Azraq Oasis (Zarqa Governorate), 85 km southeast of Amman. It produces 23–25 MCM/a from the B4/B5 Aquifer. The B4/B5 Aquifer is hydraulically connected with the overlying Basalt aquifer, which is highly fractured and therefore vulnerable to pollution, making the delineation of protection zones in this area necessary (Ta'any 1996).

References

Ministry of Water and Irrigation (MWI), 2010: Water Budget 2010 – unpublished arabic internal report.

Margane, A., Borgstedt, A., Hamdan, I., Subah, A. & Hajali, Z., 2008: Delineation of Surface Water Protection Zones for the Wala Dam. – Technical Cooperation Project 'Environmental Geology for Regional Planning', Technical Report No. 12, prepared by BGR & MWI, BGR archive no. 012xxxx, 126 p.; Amman.

Margane, A., Subah, A., Hamdan, I., Almomani, T., Hajali, Z., Ma'moun, I., Al-Hassani, I. & Smadi, H., 2010: Delineation of Groundwater Protection Zones for the Lajjun, Qatrana, Sultani and GhweirWellfields. – Technical Cooperation Project 'Groundwater Resources Management', Technical Report No. 9, prepared by BGR & MWI, 292 p.; Amman.

Ta'any, R. A. & Al Atrash, M., 2010: Effects of Wala Dam on the Groundwater of Wala Catchment Area. – Unpublished Report prepared at Al Balqa Apllied University and Ministry of Water and Irrigation

www.borntraeger-cramer.de
2012/0220/$ 0.50

Ta'any, R. A.,1996: Hydrological and Hydrogeological Study of the Azraq Basin. – A Thesis submitted to the College Of Science University Of Baghdad, in partial fulfillment of the requirements for the degree of Doctor Of Philosophy in Geology (Hydrology).

World Bank (2009): Hashemite Kingdom of Jordan – Country Environmental Analysis. Report No. 47829-JO.

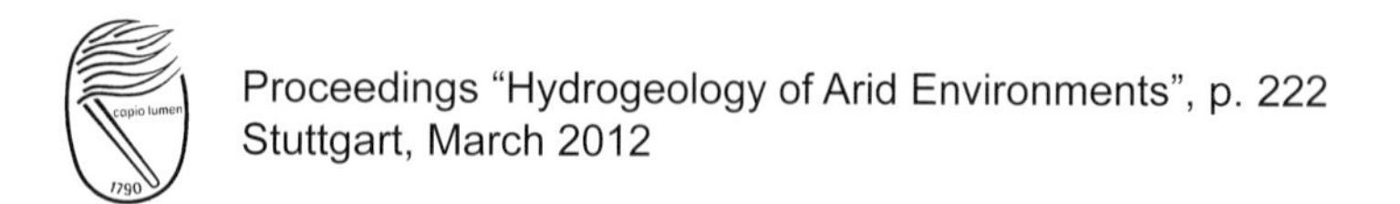

Proceedings "Hydrogeology of Arid Environments", p. 222
Stuttgart, March 2012

Poster Abstract

Environmental Isotopes to Understand the Groundwater Flow Dynamics in Raya Valley, Northern Ethiopia

Merhawi Gebreegziabher[1], Tenalem Ayenew[1], Seifu Kebede[1], Sileshi Mamo[2]

[1] Addis Aaba University, Department of Earth Science, P.O. Box 1176, Ethiopia. email: meregebre@gmail.com (M. Gebreegziabher), tenalema@yahoo.com (T. Ayenew), seifukebede@yahoo.com (S. Kebede).
[2] Geological Survey of Ethiopia. email: sileshi.mamo@yahoo.com (S. Mamo)

Key words: Raya valley basin, stable isotopes, semiarid, alluvial aquifers

Raya valley basin is interconnected with the series Main Ethiopian Rift System, which is part of the Great East African Rift System. The Basin is bounded by the North Western Ethiopia plateau to the west and Afar depression to the east. The climate condition of the study area is highly controlled by topographic setup and source of moisture. The highland is relatively characterized as humid climate condition. However, the lowland graben is characterized by semiarid climate condition. The semiarid part of the study area is ecoenvironmentally fragile and scarce in surface water. The two major type of aquifers in the study area are the Quaternary alluvial and Tertiary volcanic aquifers. The alluvial deposit is most productive aquifer found in the lowland graben and the volcanic aquifer is extensively found in the western mountainous block. Groundwater, surface water and rainfall samples are collected in the summer rainy season and measured by Liquid Water Isotope Analyzer for their delta oxygen and delta hydrogen content. The analysis of the result reveals that the principal source of the groundwater is precipitation and floodwater (runoff). The recharge mechanism for the basin is understood as recharge from run off (flash flood) generated during heavy rainfall in the highland, recharge by mountain block seepage through fractured basaltic aquifer and recharge by direct rainfall in the alluvial graben. The groundwater discharge is located around Selen Wuha where the isotopic value for 18O varies between –3.11 up to 1.81‰. Surface water samples, like Lake Ashenge and Gerjale swamp area are isotopically enriched. The input signal for the lake is generally from precipitation. However, the swampy water in semiarid part of the study are is generated from the groundwater where, locally thick clay sediment deposition is occurs. Seasonal Korem flood plain is isotopically depleted compared with low land Gerjale swamp area as a result of low rate of evaporation and high humidity. The groundwater is an important hydrological component to sustain the groundwater dependent ecosystem around the Gerjale Swamp water in the semiarid part of the study area.

www.borntraeger-cramer.de
2012/0222/$ 0.25

Proceedings "Hydrogeology of Arid Environments", p. 223
Stuttgart, March 2012

Poster Abstract

Chemical and Bacteriological Quality of the Shallow Alluvial Aquifer of Guerrara (Southeast Algeria)

S. Hadj-Said, A. Zeddouri

Biogeochemistry laboratory of desert environments. University Kasdi Merbah PO 511, Ouargla 30000. Algeria,
e-mail: hadjsaidsamia@gmail.com

Key words: alluvial aquifer, groundwater, quality, arid zone

The Saharan underground contains appreciable water reserves that are however lowly renewable. Groundwater is the only water resource of the inhabitants. The Guerrara Oasis is a region of southeastern Algeria, which has a groundwater system consisting of three aquifers: the Continental Intercalary, the Terminal Complex and the alluvial aquifer. The main objectif of this work is to study chemical and bacteriological characteristics of the alluvial aquifer and determine its quality. The measurements on various water quality parameters that were carried out on groundwater samples of wells show high values of suspended load up to 13 mg/l, the values of chemical oxygen demand COD varies between 297.6 and 422.4 mg /l, electrical conductivity exceeds 1000 µS/cm. Results obtained by bacteriological analysis confirm the contamination by bacteria of faecal origin especially by faecal Coliformes, faecal Streptococci and Closridium.

www.borntraeger-cramer.de
2012/0223/$ 0.25

Proceedings "Hydrogeology of Arid Environments", p. 224
Stuttgart, March 2012

Poster Abstract

Managed Aquifer Recharge (MAR) Potential Map for Amman-Zarqa and Azraq Basins, Jordan

Ibraheem Hamdan[1], Anke Steinel[2]

[1] Federal Institute for Geosciences and Natural Resources (BGR), Amman, Jordan, ibraheem_geo@yahoo.com
[2] Federal Institute for Geosciences and Natural Resources (BGR), Hannover, Germany, anke.steinel@bgr.de

Key words: artificial groundwater recharge, stormwater runoff, infiltration, small-scale structures

Jordan is considered to be one of the water scarcest countries worldwide as the available water resources per capita is only 150 m^3/capita/year (MWI 2011) compared to international standards of 500 m^3/capita/year. Hence, competition between demands for the limited fresh water quantities is ever increasing. Groundwater abstractions exceed present-day groundwater recharge since the mid 1980s (groundwater deficit in 2009/2010: 182 MCM (MWI 2010)). This has caused groundwater level declines in most areas of Jordan. In parts, this has also led to a deterioration of groundwater quality.

For better management of existing water resources and to secure water for future generations, aquifers can be used as temporary reservoirs to store water for later use. Managed aquifer recharge (MAR) has been practiced for many years in a number of countries. Currently, runoff in ephemeral wadis is predominantly not used and represents an untapped water resource that would mostly evaporate and could instead be captured in small dams and infiltrated in the wadi beds. In this bilateral project, the German Federal Institute for Geosciences and Natural Resources (BGR) is supporting the Jordanian Ministry of Water and Irrigation (MWI) in its capability to successfully use managed aquifer recharge (MAR) via infiltration of stormwater runoff.

The one year-long study focuses on the Amman-Zarqa and Azraq basins, which are characterized by an average rainfall between 125–500 mm and 50–150 mm and a size of around 3600 km² and 12 000 km², respectively. The main task of this project comprises the development of a map for MAR potential.

In order to evaluate MAR potential for the aquifers in a detailed map, thematic layers are prepared in a GIS environment of which these four are the most important:

1. Hydrogeology: Hydrogeological classification with the spatial distribution of aquitards and aquifers for evaluating the presence or absence of a suitable aquifer to be recharged.
2. Slope: The most effective slope for infiltration is 0–5 % (Alraggad & Jasem 2010) preventing the generation of excessive runoff.
3. Land-use: In urban areas land acquisition is hindered and pollution sources as well as impervious areas are abundant.
4. Proximity to water supply sources and demand sites is considered to minimize the costs of water transport.

All input layers are overlain with ordered weighted averaging to produce the final MAR potential map encompassing four classes (low, medium, high, very high) representing the suitability for recharge.

The MAR potential map will then allow the selection of possible recharge sites for implementation. By focusing on small scale initiatives, the project provides locally suitable solutions for the improvement of the critical water situation in the semiarid regions including aspects of climate change adaptation.

References

Alraggad, M. & Jasem, H., 2010: Managed Aquifer Recharge (MAR) through Surface Infiltration in the Azraq Basin/Jordan. – Journal of Water Resource and Protection (JWARP), **2** (12): 1057–1070.

Ministry of Water and Irrigation (MWI), 2011: Water Sector Country Profile. Jordan, Draft- 1, 02/2/2011. Amman, Jordan. – Unpublished report.

Ministry of Water and Irrigation (MWI), 2010: Water Budget 2009/2010. Amman, Jordan. – Unpublished report (Arabic).

www.borntraeger-cramer.de
2012/0224/$ 0.25

Proceedings "Hydrogeology of Arid Environments", p. 225
Stuttgart, March 2012

Poster Abstract

FAD: A Computer Based System for Frequency Analysis of Droughts

Yasser Hamdi[1]

[1]Département de génie civil, Ecole Nationale d'Ingénieurs de Gabès, Tunisie. Email: yasser.hamdi@ymail.com

Key words: arid, droughts, frequency analysis, historical information, decision support system

Decision support systems are widely developed to assist water resources engineers, water managers and hydrologists in the design, operation and management of water resources structures. Unfortunately most decision support systems based on frequency analysis are not accepted by their projected users. The user's conceptual representation of certain situations does not often correspond to the hydrological frequency analysis model representation as applied by most decision support systems. The main purpose of this paper is to present a software for hydrological frequency analysis with historical information. The FAD software (acronym for Frequency Analysis of Droughts) developed in a Windows platform and compiled by MATLAB 6.5, represents a user friendly tool that can be used by practitioners for solving frequency analysis problems in the field of hydrology in arid and semi-arid regions. The software represents also a decision support system for experts to assist water resources engineers and hydrologists and a didactic tool for students who approach this kind of problem for the first time.

www.borntraeger-cramer.de
2012/0225/$ 0.25

Proceedings "Hydrogeology of Arid Environments", p. 226–227
Stuttgart, March 2012

Poster Abstract

Modeling Artificial Groundwater Recharge by Floodwater Spreading: Estimation, Effect and Enhancement

Hossein Hashemi[1], Ronny Berndtsson[2], Mazda Kompani-Zare[3]

[1,2] Center for Middle Eastern Studies and Dept. of Water Resources Engineering, Lund University, Box 118, 221 00, Lund, Sweden, email: Hossein.hashemi@tvrl.lth.se, ronny.berndtsson@tvrl.lth.se
[3] Dept. of Desert Regions Management, Shiraz University, Shiraz, Iran, email: kompani@shirazu.ac.ir

Key words: artificial recharge, groundwater modeling, MODFLOW, recharge estimate

Artificial recharge (AR) is a method to balance and recover groundwater resources. AR has been the main and parsimonious solution for water scarcity problems in the arid and semiarid Middle East for thousands of years and is being increasingly encouraged in many other arid countries. The increased interest has resulted in a renewed interest in finding new and improved methods for AR.

AR may be defined as augmenting the natural infiltration of surface water into underground formations by various techniques such as by spreading of water in infiltration basins or by artificially changing recharge conditions (Todd & Mays 2005). Unconfined aquifers can be artificially recharged by spreading of water on the ground surface. AR by water spreading is practiced in 36 multipurpose floodwater spreading stations in Iran since 1983. The systems serve as sedimentation basins and infiltration ponds for the AR of groundwater and also as experimental plots for investigation of several problems, such as sediment stabilization and afforestation (Kowsar 1992). Due to water shortage in Gareh-Bygone Plain, arid south-eastern Iran, a Floodwater Spreading System (FWS) to artificially recharge the groundwater was established between 1983 and 1987 on about 2000 ha. This system is an inexpensive method for flood mitigation and AR of aquifers that results in a large economic return for relatively small investment (Kowsar 2008).

A crucial subject in aquifer management is estimation of recharged water. A variety of techniques are available to quantify recharge. However, it is often difficult to choose the appropriate techniques (Scanlon et al. 2002). In this study, in order to estimate the recharged water for a 14-year period, a 3D conceptual model was built to be representative of the study area and to facilitate efficient estimation of hydraulic parameters. Based on monthly observed data during both steady and transient periods for the years between 1993 and 2007 groundwater flow was simulated and calibrated by MODFLOW-2000 (Harbaugh et al. 2000). The aquifer parameters such as hydraulic conductivities, specific yield, and recharge rate were determined through calibration of model during steady state, unsteady state with no recharge, and unsteady state with recharge cases, respectively. The recharge amount varied from a few hundred thousand cubic meters per month during drought periods to about 4.5 million cubic meters per month during rainy periods.

This study aimed at assessing the effects of different strategies to improve the efficiency of the FWS using the above groundwater model. The study focused on spatial distribution of the system, change in the hydraulic structures and hydraulic parameters of the aquifer, and application of different abstraction scenarios in order to increase the efficiency and management of the system. The results will be used to better manage existing and plan new FWS projects in order to achieve sustainable water resources using an economical and efficient AR system in arid areas.

www.borntraeger-cramer.de
2012/0226/$ 0.50

References

Todd, D. K. & Mays, L. W., 2005: Groundwater Hydrology. – John Wiley & Sons, Inc, 656 pp.

Scanlon, B. R., Healy, R. W.& Cook, P. G., 2002: Choosing appropriate techniques for quantifying groundwater recharge. – Hydrogeol. J. **10**: 18–39.

Harbaugh, A. W., Banta, E. R., Hill, M. C. & McDonald, M. G., 2000: MODFLOW-2000, the U.S. Geological Survey modular ground-water model-user guide to modularization concepts and the ground-water flow process. – U.S. Geological Survey 2000; open-file report: 00-92.

Kowsar, S. A., 1992: Desertification control through floodwater spreading in Iran. – Unasylva (English ed.) **43**(168): 27–30.

Kowsar, S. A., 2008: Desertification Control through Floodwater Harvesting: The Current State of Know-How. – Future of Drylands, p. 229–241.

Proceedings "Hydrogeology of Arid Environments", p. 228
Stuttgart, March 2012

Poster Abstract

Pedotransfer Functions to Estimate Annual Groundwater Recharge Rates in Countries of the Arab Region

Volker Hennings[1], Jobst Massmann[2]

[1, 2] Federal Institute for Geosciences and Natural Resources, volker.hennings@bgr.de

Key words: groundwater recharge, soil water balance, pedotransfer functions, Arab region

For quantitative water resources management in an arid environment knowledge of the groundwater recharge rate is essential. From the soil scientist's perspective percolation beyond the lower boundary of the root zone equals groundwater recharge. On a regional scale robust methods such as empirical equations or nomograms are needed, based on input variables that can be determined easily or are available from existing databases. Some soil scientists use the term "hydro-pedotransfer functions" (HPTFs) to characterize this kind of approaches.

To develop this type of HTPFs for countries of the Arab region a simulation model of the soil water balance was used. Actual evapotranspiration and percolation rates were calculated for different climatic regions, soils and land use classes. Results of all scenarios were analyzed by multiple regression statistics and equations were derived, from which reliable estimates of the target variable can be determined. Long-term means of actual evapotranspiration rates in the Arab region were estimated by models based on the FAO 56 concept, developed by the Land and Water Development Division of FAO for planning and management of irrigation. Agroclimatic data from 188 meteorological stations from eight Arabic countries (Morocco, Algeria, Tunesia, Libya, Egypt, Jordan, Lebanon, Syria) were taken from the CLIMWAT database from the FAO homepage.

Typical kinds of land use of the Mediterranean environment were taken into consideration (winter wheat, barley, small vegetables, citrus trees, olive trees, cotton, pastureland). Mean annual groundwater recharge rates can be predicted from information on mean annual precipitation, mean annual potential evapotranspiration and available water capacity of the uppermost meter of the soil profile. As proven by correlation coefficients, the accuracy of these HPTFs is generally high. Regression equations / nomograms were developed for specific crops and varying soil properties, for specific crops and varying irrigation practices, and for specific locations and varying kinds of land use. Validation studies indicate that sometimes evaporation from bare soil is overestimated, sometimes crop coefficients have to be calibrated and adjusted, but on the average HPTFs perform quite well. (Hydro-) Pedotransfer functions that require only easily available soil, crop and climate information as presented here can serve as a useful tool to provide reliable estimates of the groundwater recharge rate for most of typical land use types of the Middle East.

www.borntraeger-cramer.de
2012/0228/$ 0.50

Proceedings "Hydrogeology of Arid Environments", p. 229
Stuttgart, March 2012

Poster Abstract

Importance of Geophysical Logging in Arid Heterogeneous Unconfined Aquifers

Rolf Herrmann

Schlumberger Water Services, P.O. Box 21, Abu Dhabi, United Arab Emirates, email: rherrmann@slb.com

Key words: geophysical, logging, vadose, characterization

Advanced geophysical borehole logging has evolved tremendously over the years for groundwater investigation. Detailed hydrogeologic data are required for high resolution aquifer characterization and meaningful predictive solute-transport modeling of long-term water quality, which are often not obtainable using traditional groundwater investigation techniques.

The extensive surficial aquifers in the Emirate of Abu Dhabi, UAE, consists of a thick heterogeneous alluvial sandy aquifer. The depositional environment has been affected by the tectonic structures of the mountain range in close vicinity, which resulted in a complex unconfined aquifer system.

During an exploration phase several wells have been drilled which were subject to extensive geophysical logging to obtain high resolution information of the aquifer characteristics and properties. The logging information was processed and interpreted and was used to develop a detailed 3D hydrogeological model with high vertical resolution to represent a realistic model of the complex heterogeneous setting of the aquifer. Geostatistical modeling was applied to populate the numerical model with hydraulic properties.

The interpreted results allow to identify several key aquifer properties, such as effective porosity, hydraulic permeability, salinity. The properties can be obtained for the complete vertical thickness of the aquifer system (including the vadose zone) with a vertical resolution of less than 10 cm. In contrast to many other techniques used in the water industry, geophysical logging allows to extract hydraulic properties for the saturated and the vadose zone. This is of great importance where unconfined aquifers are used for injection creating a mound in the vadose zone or where infiltration ponds are used to recharge the aquifer.

The approach that was used allows to create a much more improved and detailed hydrogeological model of the aquifer, compared to other methods.

www.borntraeger-cramer.de
2012/0229/$ 0.25

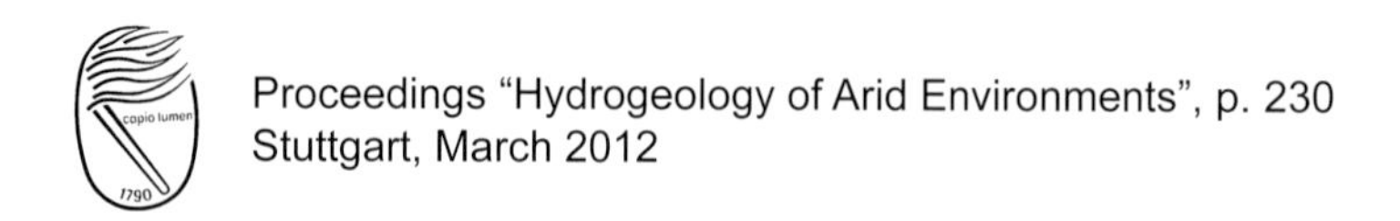

Proceedings "Hydrogeology of Arid Environments", p. 230
Stuttgart, March 2012

Poster Abstract

Hydrogeology and Contamination of the Basin of Tripoli

Fathi Hijazi, Jalal Halwani, Mirna Daye, Moumen Baroudi

Lebanese University, Water & Environment Sciences Laboratory, Tripoli – Lebanon; Email : jhalwani@ul.edu.lb; fathi-hijazi@hotmail.com

Key words: saline intrusion, groundwater, Lebanon, over pumping

Hydrogeologically, the basin of Tripoli can be divided into two adjacent basins which are not firmly separated by the fault of Tripoli with a throw of about 100 m. The west basin is lowered and forms the plain of Tripoli in the form of a small peninsula on the coast. The Miocene aquifer with a thickness of 150 to 220 m of limestone, marl and sandy marl is covered by 10–20 m of quaternary conglomerates, fossil soil and sand dunes. The Miocene is superposing the chalky marl of upper cretaceous and lower marine tertiary formation labeled with C6 on the geological map of Lebanon, which is an excellent ground for the aquifer.

Tripoli lies on the middle of a trough called Zgharta-Amioun syncline which is limited by the Tourbol anticline in the north and Qalhat anticline in the south, eastward is the aquifer limited by the so called Lebanese flexure. The limestone layers are karstified and highly fractured. The main sources of recharge to the aquifer are precipitation (800 mm/year) and the infiltration from rivers, mainly from Quadisha River. The main sources of discharge are natural outflow to the sea and withdrawal of groundwater by pumping wells.

In order to preserve the groundwater of Tripoli, 38 representative private wells, and distributed in the new residential "Dam & Farz" were selected. The extent of seawater intrusion was followed for over two years through the analysis of parameters: hardness, alkalinity, conductivity, Cl, Na, Mg, Ca and K. A division of the region in terms of these parameters into 4 was carried out which clearly shows the seriousness of the situation. In addition, the exploitation of groundwater has been studied for the first time in Lebanon through meters installed on the wells. Monthly consumption per building was regularly sampled for 2 years, which was determined by a statistical study daily consumption per capita in the area.

The evolution of the parameters shows a continuing trend of the phenomenon of saltwater intrusion in groundwater due to overexploitation of the latter due to the uncontrolled growth of private wells and pumping without any constraint of water for the purposes daily. This exploitation has led to the lowering of the piezometric level of groundwater, causing the intrusion of salty sea water; the average daily amount of water consumed per capita is 255 liters / days. This value is two times larger than the value commonly used by the Ministry of Water (120 liters / day).

Although the Ministry of Water is no longer issuing permits to drill private wells where there is public water system, we found that the construction of each new building is preceded by a private drilling. And it is clear that the continuity of such practice will lead to worsening of salinization. A formal ban on new drilling and a 5-year moratorium on the use of existing wells are necessary to allow the water to find its water balance. During this period, water will be provided only by public network.

References

Al-Abdallah, A. et al., 2010: Late Cretaceous to Cenezoic tectonic evolution of the NW Arabian platform in NW Syria. – In: Homberg, C. & Backmann, M. (eds.): Evolution of The Levant Margin and Western Arabia platform since the Mesozoic. Geological Society, London, Special Publications, **341**, pp. 305–327.

PNUD, 1972: Etude Hydrogéologique de la région de Koura-Zgharta. – Ministère des Ressources Hydrauliques et Electriques, Juin 1972, **HG19**, Beyrouth, Liban.

Walley, C. D., 1998: Some outstanding issues in the geology of Lebanon and their importance in the tectonic evolution of the Levantine region. – Tectono-physics **298**: 37–62.

Wetzel, R. & Dubertret, L., 1951: Carte Géologique au 1:50000, Feuille de tripoli avec notice explicative. – Ministère des Travaux Publics, Beyrouth, Liban.

www.borntraeger-cramer.de
2012/0230/$ 0.25

Proceedings "Hydrogeology of Arid Environments", p. 231
Stuttgart, March 2012

Poster Abstract

Exploration and Management of an Inland Fresh Water Lens in the Paraguayan Chaco

G. Houben[1], U. Noell[1], C. Grissemann[1]

[1] Bundesanstalt für Geowissenschaften und Rohstoffe (BGR), georg.houben@bgr.de

Key words: fresh water lens, Paraguay, Chaco, hydrogeophysical exploration

The town of Benjamín Aceval is located in the lower Chaco Boreal region, 40 km north of Paraguay´s capital Asuncion. The Chaco itself is a semi-arid sedimentary plain stretching from the Andes to the Rio Paraguay to the East. As evaporation exceeds precipitation, freshwater is scarce and groundwater mostly saline. Settlements are restricted to the few locations with available freshwater, e.g. the town of Benjamín Aceval. This city with its population of about 17,000 is located on some isolated hills that rise a few meters above the Chaco plain. Settlement was only possible due to the occurrence of fresh groundwater underneath these hills. Land use, except from human settlements, is mostly dedicated to sugar cane plantations and cattle farming.

Hardly anything about the groundwater resource of Benjamín Aceval was known. Overexploitation and intrusion of saline groundwater from the adjacent lowlands was a cause of concern. Therefore, in the framework of the Paraguayan-German project PAS-PY (Protection and sustainable management of groundwater in Paraguay) a hydrogeophysical exploration was performed, using geoelectrical and electromagnetic methods. We were able to delineate the geometry of the fresh water aquifer which closely resembles a Ghyben-Herzberg-lens.

Groundwater extraction and recharge rates were determined to deduce a water balance. This information is imperative to develop a sustainable water management concept for this fragile and irreplaceable water resource. The water balance is positive so far, recharge slightly exceeding consumption. Return flow of wastewater from the numerous cesspits additionally provides water for the aquifer, albeit of questionable quality. Nevertheless, the limited resources put certain constraints on the future development of the region. Large users such as industries cannot be installed in this zone.

Stable isotope data showed that most of the water present in the lens was recharged during a time with differing isotopic rainfall composition and probably different climate. Tritium data nevertheless showed that recharge is still occurring.

Water quality issues were also addressed as Benjamín Aceval lacks a sewage collection and treatment system. Most household wastewater is re-infiltrated into the subsurface by thousands of infiltration wells, often abandoned dug wells. More than 70% of all wells show indications of microbial fecal pollution.

www.borntraeger-cramer.de
2012/0231/$ 0.25

Proceedings "Hydrogeology of Arid Environments", p. 232
Stuttgart, March 2012

Poster Abstract

Hydrogeological and Hydrogeochemical Investigation of the El Tatio Hydrogeothermal Field and the Rio Salado/Rio Loa River System, Antofagasta Region, Chile

Jörg Hunger[1]

[1] Department of Geology, Section for Hydrogeology and Environmental Geology, Technische Universität Bergakademie Freiberg, Am Brauhügel 17, D-98617 Rhönblick, Germany, email: jorgohunger@aol.com

Key words: El Tatio geothermal field, Salado River, hydrochemistry, arsenic

Hydrogeological and hydrogeochemical investigation at the El Tatio geothermal field and the Rio Salado/Loa river system in the Antofagasta region in Chile were carried out during January and February 2011. The region of interest has semi arid to arid climate with annual precipitation of less than 150 mm at altitudes between 4,224 and 4,376 m. During field work a total number of 109 distinctive hot spring manifestations were examined for their on-site parameters (temperature, pH value, electrical conductivity and redox potential) to characterize the geothermal waters and to identify spatial distribution patterns. Samples were taken from 39 hot springs in the northern geothermal field and at 14 locations along the Salado and the Loa River for hydrochemical analyses with the purpose to amplify the knowledge about hot water-rock interaction in the geothermal system of El Tatio and the short- and medium term chemical behaviour of the thermal fluids after sputtering to the surface and flowing into the Salado/Loa River.

On-site parameter examined in the hot springs were spatially clustered and allocated to different hot spring groups. Distinct distribution patterns could be encountered, especially the electrical conductivity distribution suggests typical structures of a geothermal system. The hot springs revealed different water types that are enriched in sodium and chloride and thus considered to be in direct contact to the deep thermal reservoir. Measured arsenic concentrations are the highest reported for naturally occurring surface waters and occur together with other high concentrated trace elements, such as rubidium, strontium, caesium, antimony and boron. Besides the intensive interaction of the thermal water with young volcanic rocks, the dissolution of buried evaporites influences the water chemistry.

To estimate precipitation or dissolution processes in the hot spring discharge point and with increasing flow-distance to the geothermal field, saturation indexes for important mineral phases were modelled by PHREEQC. Over-saturation with respect to silicates, oxides, hydroxides and oxyhydroxides was encountered. Abrupt drop of arsenic concentration during the first flow-kilometres is due to dilution by confluents and sorption processes to the surfaces of oxides, hydroxides and oxyhydroxides.

Furthermore, the possible use of the river water as source of drinking water for the small Andean villages being inhabited by indigenous people is discussed. High concentrations of sodium, chloride, arsenic, boron, antimony, lithium, rubidium and caesium would make a complex treatment necessary, which can be hardly realized at the given conditions.

www.borntraeger-cramer.de
2012/0232/$ 0.25

Proceedings "Hydrogeology of Arid Environments", p. 233
Stuttgart, March 2012

Poster Abstract

Impact of Climate Change on the Figuig Oasis Aquifer, Morocco, Using a Numerical Model

Abdelhakim Jilali[1,2], Yassine Zarhloule[2], Alain Dassargues[1], Pascal Goderniaux[1], Samuel Wildemeersch[1]

[1] Group of Hydrogeology and Environmental Geology – Aquapôle, University of Liège, Chemin des Chevreuils, 1, Building B52/3, 4000 Liège, Belgium. email: abdelhakim.jilali@alumni.ulg.ac.be, alain.dassargues@ulg.ac.be, Pascal.Goderniaux@ulg.ac.be, swildemeersch@ulg.ac.be
[2] Laboratory of Hydrogeology & Environment, Faculty of Sciences, B.P. 717, Oujda 60000 Morocco. email: zarhloule@yahoo.fr

Key words: oasis, groundwater, modeling, climate

Numerical groundwater modeling was applied to investigate the climate change impact on the transboundary aquifer of Figuig, located at the edge of Moroccan eastern High Atlas. The stresses imposed to the model were derived from the IPCC emission scenarios and included recharge variation. The Figuig aquifer consists of carbonates, alternating marl-limestone, sandstone and alluvial deposit. The limits of the model are topographic boundaries except for the western boundary which has been truncated. The model two-dimensional, composed of a single layer representing the thickness of the aquifer, and steady state conditions were simulated. Modeling improves the understanding of the interactions between groundwater and surface water by measuring the flow of water exchanged. The decline in recharge due to climate change scenarios has a negative effect on groundwater quantity and quality and, therefore, on the Figuig oasis. This is in agreement with field observations, since observed decline of groundwater resources has affected both the water salinity, the number of springs that supply the oasis and the socio-economical aspect of the region.

www.borntraeger-cramer.de
2012/0233/$ 0.25

Proceedings "Hydrogeology of Arid Environments", p. 234–235
Stuttgart, March 2012

Poster Abstract

Ground Water Table Rise in an Arid Environment – Investigations in Greater Doha Area

Tim Kelly[1], Muhammad Riaz Akram[2], Ulrich Schott[3]

[1] Public Works Authority (Ashghal) Infrastructure Affairs, Doha, State of Qatar, email: tkelly@ashghal.gov.qa
[2] Public Works Authority (Ashghal) Infrastructure Affairs, Doha, State of Qatar, email: rmuhammad@ashghal.gov.qa,
[3] Schlumberger Water Services, Doha, State of Qatar, email: uschott@slb.com

Key words: urban recharge, groundwater table rise, hydrogeological model

Introduction

While generally, groundwater levels are declining in the arid environment of the State of Qatar, leading to coastal saltwater intrusion and abandonment of farms, in the metropolitan area around the capital city Doha, Qatar, groundwater levels in the shallow aquifers are rising. Previous studies and observations indicate that the rise in groundwater levels may be caused by leakage of the water supply system, percolation from irrigated areas, septic tanks and the drainage system.

The Public Works Authority (Ashghal) has initiated a project to assess the current groundwater situation, to identify the causes and propose remediation measures. The project forms an essential element of the Qatar National Vision 2030.

The poster presents the development of a hydrogeological model that allows for a first assessment of the groundwater system.

Objective and Scope

The main objective of the project is to undertake a comprehensive assessment of the groundwater situation for planning and design of an effective drainage network as part of an integrated storm water & groundwater management plan.

The project duration is about three years and comprises the drilling, geophysical logging and testing of approximately 1,000 monitoring wells over an area of around 3,000 km^2. The field investigations comprise groundwater level monitoring, water quality sampling and groundwater flow and transport modeling.

Methodology

To capture spatial and temporal variability of the area, the methodology accounts for drilling of a dense grid of monitoring wells accompanied with high frequency monitoring. This large amount of data and information requires the implementation of a geo-data management system. Synchronization of lithological information derived from the drilling cuttings with geophysical borehole logging interpretation reduces inaccuracies regarding depths of lithological units and provides qualitative data on the location and properties of the various formations within the exploration zone. For enhanced understanding of the hydrogeological conditions, hydrochemical analyses, hydraulic testing and frequent monitoring of water level has been undertaken based on which the hydrogeological model is setup. This hydrogeological model is a prerequisite for the development of the planned groundwater model. Cross-checking and overlaying of the collected and interpreted hydrogeological information with historical groundwater-related data and information on development of the urban infrastructure is expected to lead to an initial comprehensive understanding of the groundwater condition.

Conclusion

The initial hydrogeological model indicates that the groundwater flow in the project area is governed by a complex Tertiary shallow karstic aquifer system, heavily influenced by urban recharge and local groundwater abstractions. Initial results also seem to indicate that the apparent problems of water ta-

www.borntraeger-cramer.de
2012/0234/$ 0.50

ble rise in the urban areas may be associated with local anomalies such as topographic depressions and induced recharge on low permeability strata. In low permeability areas, significant local groundwater table rise may be caused even by relatively low recharge rates. Additional field testing and groundwater modeling will be applied to test this assumption. Identifying and mitigating the negative effects of anthropogenic urban recharge may not only reduce cost but also represents a main component of sustainable water management in urban areas. The development of a detailed hydrogeological model is of paramount importance specially in complex geological and hydrogeological settings.

References

Alsharhan, A. S., Rizk, Z. A., Nairn, A. E. M., Bakhit, D. W. & Alhajari, S. A., 2001: Hydrogeology of an arid region: the Arabian Gulf and adjoining areas. – Elsevier.

Bredehoeft, J., 2005: The conceptualization model problem – surprise. – Hydrogeology Journal **13**: 37–46.

Cavelier, C., 1970: Geological description of the Qatar peninsula (Arabian Gulf). – Government of Qatar, Department of Petroleum Affairs.

El-Kassas, I. A. & Ashour, M. M., 1990: Lineament analysis of Qatar peninsula based on Landsat imagery. – Qatar Univ. Sci. Bull. **10**: 421–442.

Garcia-Fresca, B., 2004: Urban-enhanced groundwater recharge: review and case study of Austin, Texas. – In: Howard, K. W. F. (ed.): 32nd International Geological Congress – Groundwater Management in Urban Areas, Florence.

Lerner, D. N., 1990: Groundwater recharge in urban areas. – In: Massing, H., Packman, J. & Zuidema, F. C. (eds.): Hydrological processes and water management in urban areas. – Invited lectures and selected papers of the Unesco/IHP International Symposium URBAN WATER '88, held at Duisburg, Federal Republic of Germany and at Lelystad, Amsterdam and Rotterdam, The Netherlands, 24–29 April 1988, Vol. IAHS Publ. No. **198**, Published by the International Association of Hydrological Sciences, The Netherlands.

Sadiq, A. M. & Nasir, S. J., 2002: Middle Pleistocene karst evolution in the State of Qatar, Arabian Gulf. – Journal of Cave and Karst Studies **64**: 132–139.

Proceedings “Hydrogeology of Arid Environments”, p. 236
Stuttgart, March 2012

Poster Abstract

Using Meteorological Satellite Data in a Hydrological Model to Achieve Full Space-Time Coverage in the Poorly Surveyed Awash Catchment (Central Ethiopia)

Malte Knoche[1], Peter Krause[2], Richard Gloaguen[3]

[1] UFZ – Helmholtz-Centre for Environmental Research, Department Catchment Hydrology, Theodor-Lieser-Straße 4, 06120 Halle, Germany. email (corresponding author): malte.knoche@ufz.de
[2] Friedrich-Schiller-University Jena, Department of Geoinformatics, Hydrology and Modelling, Grietgasse 6, 07743 Jena, Germany
[3] Freiberg University of Mining and Technology, Institute of Geology, Remote Sensing Group, Bernhard-von-Cotta-Straße 2, 09599 Freiberg, Germany

Key words: semi-arid, hydrological model, remote sensing, Rift Valley

The upper Awash watershed in the central Main Ethiopian Rift is tectonically highly affected. The uplift of the rift shoulders and the drop of the rift bottom force the hydrological regime into a complex topography. Climate as well as flow paths are affected by an altitude drop of about 3600 m between the highland and the lowest point in the rift. The regional climate differs from arid conditions in the lower rift to almost humid conditions at high altitudes. Within this complex terrain, the Awash River is the only larger river draining the highland into the rift. Adverse to recent hydro-chemical groundwater investigations, we provide the first hydrological model on a macro-scale in a high resolution for that area. We apply the spatially distributed conceptual hydrological model J2000 included in the Jena Adaptable Modelling System (JAMS). The highly gradient-dependent climatic setting is hardly reflected by a satisfying number of meteorological stations. Instead of using limited meteorological station data, we use satellite datasets in its full raster resolution and a daily time resolution. In this way we avoid poor interpolations and apply a new approach, which is useful for all regions having a weak meteorological infrastructure. The model outputs include groundwater recharge, surface runoff and actual evapotranspiration amongst others. Based on the model results, we assume that in the arid to semi-arid Rift Valley groundwater recharge takes place at the chain of rift valley volcanoes between Mount Bosetti and Mount Fantale mainly. This assumption is in line with hydro-chemical studies. Moreover, the magnitude of recharge at the rift valley volcanoes seems to be comparable to the recharge of the escarpments and the highland.

www.borntraeger-cramer.de
2012/0236/$ 0.25

Proceedings “Hydrogeology of Arid Environments”, p. 237–238
Stuttgart, March 2012

Poster Abstract

Erosion and Sedimentation Studies in the Wadi Al Arab Catchment/ North Jordan – Alternative Method Application and First Results in a Data Scarce Environment

Sabine Kraushaar[1], Thomas Schumann[1], Gregor Ollesch[2], Christian Siebert[1], Tino Rödiger[1]

[1] Helmholtz Centre for Environmental Research, UFZ, Germany, email: Sabine.Kraushaar@ufz.de, Thomas.Schuhmann@ufz.de, Christian.Siebert@ufz.de, Tino.Rödiger@ufz.de
[2] Landesbetrieb für Hochwasserschutz und Wasserwirtschaft Sachsen Anhalt, Germany, email: Gregor.Ollesch@lhw.mlu.sachsen-anhalt.de

Key words: Jordan, soil erosion, olive mound, erosion pin, sediment traps, fingerprinting, dam sedimentation

Background

Jordan has a quantitative and qualitative water problem in combination with a growing demand by population increase (0,98% 2011 est., CIA 2011).

In the course of water resource management it is a declared aim for Jordan to minimize the overland flow and maximize the infiltration to replenish the known water resources. The soil layer and its physical integrity is a key player to slow down runoff, safeguard infiltration, buffer and filter pollutants and important for hydrological models. Erosion processes harm this potential, fill up surface water reservoirs and contaminate water bodies in Jordan.

Different approaches of erosion estimation are presented for Wadi Al-Arab, an area that lacks on base data for representative and useful erosion modeling until now.

Research Area

The research area Wadi Al-Arab in the northwest of Jordan extends around 30km from Irbid to the west and transcends from around 400 m from the Jordan Valley Escarpment down to the Jordan River(–260 m). The climate is Mediterranean to semi-arid with <300 to 550 mm of rain in winter. Mainly marl and limestone of the Upper Cretaceous and the Tertiary make up the catchment's geology (262 km^2). It is a coherent plateau with rolling hills and agricultural plains in the East.

To the west the relief energy is higher and agricultural areas are limited to top, saddle and foot slope positions. Around 45% of the Wadi and its tributaries are used for agriculture since thousands of years.

Methods

Land use and soil conditions prime to erosion were identified and methods for first erosion estimates implemented. These are: Olive mound measurements, erosion pin fields, sediment traps, tracer deployment, and the analysis of historic sediment traps as roman cisterns as well as the dam reservoir itself. However, most applied methods give only a selective insight into a temporal limited process. Furthermore, sediment that is eroded on hill slopes is not in accordance to the actual sediment delivery ratio to the sink (Moldenhauer et al. 2008). With the help of sediment fingerprinting and the mapping of the sedimentation in the dam reservoir the relative and qualitative importance of different sediment source areas compared to the sediment sink are to be identified (Roddy 2010). Therefore the chemical characteristics of each source need to be conservative, unique and non-water soluble.

www.borntraeger-cramer.de
2012/0237/$ 0.50

Possible sources and their properties are introduced and their amount of contributed sediment to the sink discussed.

References

Moldenhauer, K.-M., Zielhofer, C. & Faust, D., 2008: Heavy Metals as indicators for Holocene sediment provenance in a semi-arid Mediterranean catchment in northern Tunisia. –- Quaternary International **189**: 129–134.

Roddy, B. P., 2010: The use of the sediment fingerprinting technique to quantify the different sediment sources entering the Whangapoua estuary, North Island, in New Zealand. – Dissertation, University of Waikato, 466 pp.

CIA, 2011: The world factbook. https://www.cia.gov/library/publications/the-world-factbook/rankorder/2002rank.html?countryName=Jordan&countryCode=jo®ionCode=mde&rank=117#jo; Zugang: 20.12.2011; 15:02.

Proceedings "Hydrogeology of Arid Environments", p. 239–240
Stuttgart, March 2012

Poster Abstract

TERENO-MED: Observation and Exploration Platform for Water Resources in the Mediterranean

Elisabeth Krueger[1], Steffen Zacharias[1], Harry Vereecken[2], Heye Bogena[2]

[1] Helmholtz Centre for Environmental Research – UFZ, Permoserstr. 15, 04318 Leipzig, email: elisabeth.krueger@ufz.de
[2] Forschungszentrum Jülich, 52425 Jülich, email: h.vereecken@fz-juelich.de

Key words: long-term monitoring, integrated research, water scarcity, Mediterranean

Background

According to the latest IPCC projections, the Circum-Mediterranean region will be particularly affected by Global and Climate Change (Kundzewicz et al. 2007, Alcamo et al. 2007). These changes include population growth, increases in food, water and energy demands, changes in land use patterns and urbanization/industrialization, while at the same time, the renewable water resources in the region are predicted to decrease by up to 50 % within the next 100 years (EC 2007). However, a profound basis for estimating and predicting the long-term effects of Global Change on the development of the quantity and quality of water resources and on ecosystems is still lacking (Lin et al. 2011). The main reason for this is that environmental monitoring, in particular in the Mediterranean region, is strongly disciplinarily oriented, and financing is usually limited to short-term periods. Yet reliable prognoses are the basis for political and structural decisions for water and environmental management as well as for infrastructure planning.

Description of the planned measure

TERENO-MED (Terrestrial Environmental Observatories in the Mediterranean) is an infrastructure measure, which aims to fill the described gaps. Together with partners in the region, TERENO-MED will establish a Circum-Mediterranian network of Global Change observatories, and will investigate the effects of anthropogenic impacts and of climate change on Mediterranean water resources and ecosystems. Within a set of representative catchments around the Circum-Mediterranean region (Southern Europe, Northern Africa, Near East), observatory sites will be established with state-of-the-art and innovative monitoring equipment, in order to measure hydrological states and fluxes on a long-term basis (minimum 15 years). Monitoring equipment will cover all scales, from the point to the regional scale, using ground-based and remote sensing technologies.

Objectives

Based on the acquired information, TERENO-MED, together with partners across the Mediterranean region will develop model scenarios that may serve as a basis for sustainable political and economical decisions. In order to gain a deep understanding of the most relevant processes and feedbacks, and to deliver reliable future scenarios for the Mediterranean region, the two initiating Helmholtz Centres, UFZ (Helmholtz Centre for Environmental Research) and FZJ (Forschungszentrum Jülich), are seeking interested German and international partners to conduct joint research within the planned monitoring network. TERENO-MED aims to make a significant contribution to solving pressing water and environmental problems in a region that is of high political and economical importance not least for Europe as a whole.

References

Alcamo, J., Moreno, J.M., Nováky, B., Bindi, M., Corobov, R., Devoy, R. J. N., Giannakopoulos, C., Martin, E., Olesen, J.E. & Shvidenko, A., 2007: Europe. – In:

www.borntraeger-cramer.de
2012/0239/$ 0.50

Parry, M.L., Canziani, O.F., Palutikof, J.P., van der Linden, P.J. & Hanson, C.E. (eds.): Climate Change 2007: Impacts, Adaptation and Vulnerability. Contribution of Working Group II to the Fourth Assessment Report of the Intergovernmental Panel on Climate Change, Cambridge University Press, Cambridge, UK, 541–580.

EC, 2007: Mediterranean Water Scarcity and Drought Report – Technical report on water scarcity and drought management in the Mediterranean and the Water Framework Directive (available at: http://www.emwis.net/topics/WaterScarcity).

Kundzewicz, Z. W., Mata, L.J., Arnell, N.W., Döll, P., Kabat, P., Jiménez, B., Miller, K.A., Oki, T., Sen Z. & Shiklomanov, I.A., 2007: Freshwater resources and their management. – In: Parry, M.L., Canziani, O.F., Palutikof, J.P., van der Linden, P.J. & Hanson, C.E. (eds.): Climate Change 2007: Impacts, Adaptation and Vulnerability. Contribution of Working Group II to the Fourth Assessment Report of the Intergovernmental Panel on Climate Change, Cambridge University Press, Cambridge, UK, 173–210.

Lin, H., Hopmans, J. W. & Richter, D., 2011: Interdisciplinary Sciences in a Global Network of Critical Zone Observatories. doi:10.2136/vzj2011.0084, Vadose Zone Journal 2011 **10**: 781–785.

Proceedings “Hydrogeology of Arid Environments”, p. 241–242
Stuttgart, March 2012

Poster Abstract

Conceptual Framework to Quantify the Influence of Soil, Land Use and Vegetation Heterogeneity on Soil Water Balances and Dynamics along the Okavango River Basin

Lars Landschreiber, Alexander Gröngröft, Annette Eschenbach

Institute of Soil Science, University of Hamburg, email: l.landschreiber@ifb.uni-hamburg.de

Key words: Okavango river basin, ecosystem functions and services, soil water balances, land use

Since 2010 “The Future Okavango” (TFO) project, funded by the German Ministry of Education and Research (BMBF), investigates with an inter- and transdisciplinary approach the quantification of ecosystem functions and services (ESF&S) at four core research sites along the Okavango river basin (Angola, Namibia, Botswana).These research sites have an extend of 100 km^2 each. Within our subproject “Interactions of soil related ESF&S with land use practise under climate change” one task is to analyse and project spatio-temporal changes in soil water balances and dynamics due to soil and vegetation heterogeneity coupled with different land use systems and changes as well as the impact of climate change on the different regions.

The scientific framework consists of three major parts including soil survey and mapping, lab analysis, field measurements and modelling approaches on different scales.

A detailed soil survey leads to a measure of the spatial distribution, extend and heterogeneity of soil types for each research site. For generalization purposes geomorphological and pedological characteristics are merged and combined with landuse types to derive landscape units. Based on this stratification representative plots are sampled for laboratory analysis to derive the spatial distribution of soil properties as well as to refine the boundaries between neighbouring landscape units. The parameters analysed in the lab describe properties according to grain size distribution, saturated and unsaturated hydraulic conductivity as well as pore space distribution.

Further on soil water contents and pressure heads are logged throughout the year with a 12 hour resolution in depth of 10 to 160 cm comparing nearly pristine plots and areas under land use but with comparable soil properties at representative test sites. Infiltration rate and surface soil water content measurements together with evaporation rate estimation methods are applied to complete the in situ characterization of soil properties. This monitoring gives information about soil water dynamics at point scale and the database is used to evaluate model outputs of soil water balances later on.

To derive point scale soil water balances for each landscape unit the one dimensional and physically based model SWAP (Kroes et al. 2008) is applied. Together with other subproject disciplines input parameters are compiled describing different vegetation properties, land use practices and meteorological data for recent times as well as for future scenarios. The output of the calibrated model quantifies the amount of runoff, groundwater recharge, interception and actual transpiration and evaporation under different circumstances. Based on a combination of several model runs with different parameter settings accounting for the spatial extent of the given landscape units it is possible to upscale from the point to the site scale representing 100 km^2. By cross checking the detailed results of the extrapolated point scale balances with the output of more generalized spatial soil water models the plausibility of the applied models can be evaluated. By looking at the relation between system water input and partitioning due to soil physics and vegetation it is

www.borntraeger-cramer.de
2012/0241/$ 0.50

possible to determine the efficiency of the ecohydrological system and its vulnerability in response to land use and climate change. That information will serve socio-economic analysis as input parameters for a valuation of overarching ESF&S.

References

Kroes, J. G., Van Dam, J. C., Groenendijk, P., Hendriks, R. F .A. & Jacobs, C. M. J., 2008: SWAP. version 3.2. Theory description and user manual. Alterra Report 1649. Alterra, Wageningen, 262 pp.

Proceedings "Hydrogeology of Arid Environments", p. 243
Stuttgart, March 2012

Poster Abstract

Preliminary Results from a Water Economy and Livelihoods Survey (WELS) in Nigeria and Mali, Sub-Saharan Africa, Investigating Water Security and Access

D. J. Lapworth[1], A. M. MacDonald[1], H. C. Bonsor[1], M. N. Tijani[2], R. C. Calow[3]

[1] British Geological Survey, United Kingdom, djla@bgs.ac.uk
[2] University of Ibandan, Nigeria
[3] Overseas Development Institute, United Kingdom

Key words: water access, water security, sub-Saharan Africa, improved supplies

It is estimated that around proportion (60%) of people in sub-Saharan Africa live without access to improved water sources for drinking water (JMP 2010). The need for sustainable development and management of water resources in sub-Saharan Africa, particularly groundwater resources, remains a major priority, especially within the context of climate variability, population growth and pressures to increase food production. In stark contrast to food scarcity, to date little systematic data collection has been done to investigate the role water scarcity has on livelihoods within rural communities in sub-Saharan Africa, particularly during droughts or periods of water stress. A water, livelihoods and economy survey (WELS) in West Africa was conducted in 2010 as part of a one year DFID-funded research programme, aimed at improving understanding of the impacts of climate change on groundwater resources and local livelihoods. The main purpose of this survey was to investigate the access to and domestic use of a range of water sources (hand pumps, wells, springs, surface water sources and rainwater harvesting) within rural communities across a rainfall transect within sub-Saharan Africa. The seasonal water use and scarcity/stress patterns were investigated for rural communities, located on both sedimentary and basement settings, using community discussions and questionnaires based on a scaled down version of the WELS methodology. The main aim of this study was to investigate seasonal access to water supplies, by gathering information on the time taken to collect water, the different sources available at different times of year (wet and dry season) and the geological and hydrogeological conditions at each community. A secondary aim was to test whether a slimmed down WELS methodology based on that described by Coulter & Calow (2011) can be effectively applied to give useful information. Preliminary results from this study are presented, focussing on the two most arid case studies in northern Nigerian and Mali.

References

Coulter, L. & Calow R.C., 2010: Assessing seasonal water access and implications for livelihoods. – RiPPLE WELS Toolkit report, RiPPLE-ODI, 11 pp.

Joint Monitoring Programme (JMP), 2010: Progress on sanitation and drinking water: 2010 update, Joint Monitoring Programme WHO/UNICEF, World Health Organization, Geneva, 55 pp.

www.borntraeger-cramer.de
2012/0243/$ 0.25

Proceedings "Hydrogeology of Arid Environments", p. 244–245
Stuttgart, March 2012

Poster Abstract

SAMIR, a Tool for Spatialized Estimates of Evapo-Transpiration and Water Budget by Remote Sensing

M. Le Page[1], V. Simonneaux[1], B. Duchemin[1], H. Kharrou[2], D. Helson[1], J. Métral[1], M. Cherkaoui[2], B. Berjamy[3], B. Mougenot[1], S. Er-Raki[4], A. Chehbouni[1]

[1] CESBIO, Centre for Spatial Studies of the Biosphere – Institut de Recherche pour le Développement, Toulouse, France
[2] ORMVAH, Regional Office of Agriculture of the Haouz, Marrakech, Morocco
[3] ABHT, Hydrologic Basin Agency of the Tensift, Marrakech, Morocco
[4] UCAM, Cadi Ayyad University, Marrakech, Morocco

Key words: evapotranspiration, remote sensing, semi-arid climate, irrigation

The SudMed project (Chehbouni et al. 2008) is aimed at developing methods for the sustainable monitoring of water resources in the Tensift basin (Marrakech, Morocco), based on ground data, remote sensing and physical modeling. The climate of this area is semi arid, characterized by low rainfall amount (240 mm on average) affected by a strong spatiotemporal irregularity. Several drought periods occurred during last years. Irrigated cultivation covers about 450000 ha and uses about 85% of the whole available water, which means that optimal use of the resources is one key of the development of the area. Irrigation optimization requires the control of all the terms of the water budget, and especially the crops water consumption, i.e. their evapotranspiration (ET). This means that at any time, estimates of their past consumption are needed for computing the water budget of the crops. Moreover, forecasting of their water requirements is necessary for a better irrigation planning. This knowledge is useful for the irrigation manager, but it is also useful for the water resources manager, i.e. the basin agency, as this flux is one major component of the water cycle in this watershed. To fulfill these objectives, we designed the SAMIR tool (Satellite Monitoring of Irrigation) dedicated to the water management of irrigated areas, making extensive use of satellite images. The tool, which is a plugin to ENVI/IDL, is designed to be versatile in different information situations: A detailed budget may be achieved in a well monitored situation (climate, irrigations, soils) while coarse maps of evapotranspiration estimates will be preferred for large areas on which local errors are assumed to compensate somehow.

SAMIR is based on the FAO-56 method (Allen et al. 1998), requiring three types of input data: climatic variables for calculation of reference evapotranspiration (ET0), land cover for computing crop coefficients (Kc), and periodical phenological information for adjusting the Kc. Although less complex than physical SVAT based methods, the model simplicity makes it well adapted for spatialization over large areas, where physical modeling would lack from the physical variables needed as input. The good trade-off between ease of use and performance makes FAO-56 the current reference method for agricultural monitoring of ET over large areas.

The Normalized Difference Vegetation Index, relying on Near Infra Red and Red Wavelength, is available from most Earth Observation Satellites. It is the basis to produce a broad classification map of the study area and derive crop coefficients and fraction cover (Simonneaux et al. 2007).

Outputs from SAMIR have been validated with different imagery sources (Formosat, Spot, Landsat, Modis), and has been used for different purposes: Local evapotranspiration of actual ET, estimation of pumping, pre-processing to a DSS system for groundwater management, Irrigation recommendation.

www.borntraeger-cramer.de
2012/0244/$ 0.50

References

Richard, A. et al., 1998: Guideline for Computing Crop Water Requirements (Livre). – [s.l.]: FAO, Vol. Irrigation and Drainage Paper n 56.

Chehbouni, A., 2008. et al.: An integrated modelling and remote sensing approach for hydrological study in arid and semi-arid regions: the SUDMED Programme [Conférence] // 2nd International Symposium on Recent Advances in Quantitative Remote Sensing. – Torrent, sp: Taylor & Francis, Vol. 29.

Simonneaux, V. et al. 2007: The use of high-resolution image time series for crop classification and evapotranspiration estimate over an irrigated area in central Morocco [Article] // International Journal of Remote Sensing, pp. 95–116.

Proceedings "Hydrogeology of Arid Environments", p. 246
Stuttgart, March 2012

Poster Abstract

Deep Fresh Groundwater Resources in the Cubango Megafan, North Namibia

F. Lindenmaier[1], H. Beukes[2], G. Christelis[2], H. G. Dill[1], J. Fenner[1], T. Himmelsbach[1], S. Kaufhold[1], R. Kringel[1], C. Lohe[1], R. Ludwig[1], R. Miller[2], A. Nick[1], M. Quinger[1], F. Schildknecht[1], A. Walzer[1], B. van Wyk[2], H. Zauter[1]

[1] Bundesanstalt für Geowissenschaften und Rohstoffe, Stilleweg 2, D-30655 Hannover, Germany, email: falk.lindenmaier@bgr.de
[2] Department of Water Affairs and Forestry, Government Office Park, Windhoek, Namibia

Key words: case study, Cuvelai-Etosha Basin, 3D-model, sedimentary basin

Namibia and Angola share the large intra-montane Cuvelai-Etosha Basin (CEB). It is Namibia's most populated area and is of increasing economical relevance. This leads to an increasing water demand that cannot be met by the existing surface water supply fed by the Angolan Kunene River. In addition, rural water supply is currently based on the exploitation of intermittent shallow groundwater resources which are developed through scattered drillholes with depths of up to 150 m. Many groundwater bodies of the CEB show high fluoride content and various grades of salinisation. The need for a better quantification of the potable water resources in the CEB is addressed through the Namibian-German technical cooperation project "Groundwater for the North of Namibia". It is linked to a research project which investigates the sedimentological and hydrogeological characteristics of the CEB.

The interdisciplinary approach to delineate the groundwater systems includes geophysical, micropaleontological, sedimentological and hydrogeological methods. Besides the need of a quantification of the basin-wide hydrogeological situation, the focus was first set to the exploration of so far unused deep groundwater resources. A geophysical exploration survey that covered large parts of the Namibian CEB used the Transient Electromagnetic Method (TEM). It was followed by the set-up of a 3D Model of electrical resistivity to delineate promising groundwater bodies.

The sedimentary history of the northern CEB is dominated by Paleo Lake Etosha as well as two sub aerial megafans of Kunene and Cubango. Several deep drillings, including some of the first core drillings within the Cubango Megafan, were conducted to explore and derive hydrogeological characteristics. Investigation of cores and cuttings indicate the existence of a complex multi-aquifer groundwater system closely linked to the deposition environment and diagenesis of the Cubango Megafan and its paleo-climatological setting. The sediment column of the almost 400 m deep boreholes consists mainly out of unstructured fine sands. Interestingly, zones with very low hydraulic conductivity are mainly sands whose pores are clogged with authigenic smectites and smectite gels. This special composition seems to be related to the deposition mechanisms of material and the influence of evaporation and salinity during the course of time. This presentation includes both the current spatial understanding of the CEB groundwater systems as well as insight to the sedimentation history.

www.borntraeger-cramer.de
2012/0246/$ 0.25

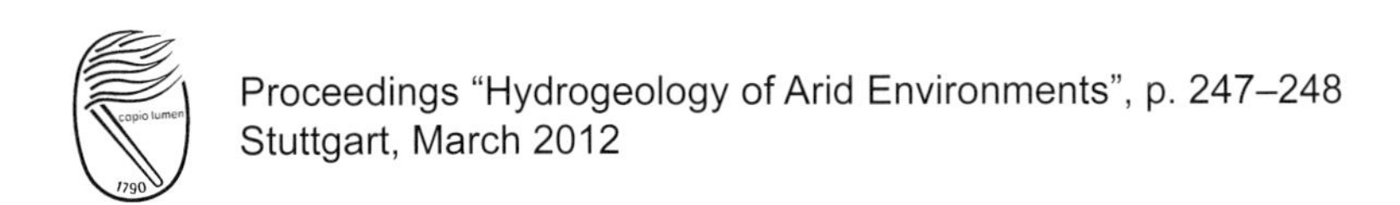

Proceedings "Hydrogeology of Arid Environments", p. 247–248
Stuttgart, March 2012

Poster Abstract

Strategic Artificial Water Storage and Recovery Project in the Liwa Desert, Abu Dhabi, United Arab Emirates

Tilman Mieseler[1]

[1] Dornier Consulting, Khalifa Street, Liberty Tower, Abu Dhabi, United Arab Emirates,
email: tilman.mieseler@dornier-consulting.com

Key words: artificial groundwater recharge (ASR), Abu Dhabi Emirate, hydrogeology, shallow aquifer system

In arid regions where natural surface water and (fresh) groundwater resources are limited, artificial recharge and storage of water in aquifers can play a major role in water resources management. In the Emirate of Abu Dhabi, this issue has become even more important with respect to securing a reliable and sustainable water supply. In many parts of the world, the principle of artificial groundwater recharge is being applied, where excess surface water from rivers, lakes or collected rainwater is being infiltrated into the underground (Peters et al. 1998). However, the recharge of groundwater with large volumes of desalinated seawater (DSW) is a new approach so far nowhere else conducted at such large scale.

At present, the storage capacity for drinking water in the capital Abu Dhabi City lasts only for 2 to 3 days (Dawoud 2009). Nevertheless, an appropriate back-up supply system in case of emergency is still missing. Therefore, the Government of Abu Dhabi requested the Consortium of the German International Cooperation (GIZ) and Dornier Consulting to carry out a feasibility study to investigate the possibility of artificially recharging existing fresh groundwater resources with DSW in the Liwa area. Based on the promising results of this feasibility study, the ASR-Pilot Project was constructed in 2002 to 2003 and was operated over one year, successfully demonstrating the efficient artificial recharge of the local aquifer system with DSW and subsequent recovery on a larger scale. During the Pilot Plant operation, a total volume of 2.2 Mio cubic meters DSW was introduced to the local dune sand aquifer. The ASR-Pilot Project impressively confirmed the suitability of the aquifer at the chosen location as well as the functioning of the applied infiltration/abstraction designs. The hydrogeological settings and relative remoteness of the desert in northern Liwa area offer huge natural storage capacity and excellent protection of natural and artificially recharged water resources from impacts from the surface. The aquifer already contains a large amount of fresh groundwater, which is still unaffected by domestic and agricultural activities.

The consultancy services for the main project "Strategic Artificial Water Storage and Recovery Project (ASR)" have been awarded in 2008 to GIZ/Dornier Consulting. After the detailed design of infrastructure components as well as the preparation and evaluation of tender documents, the construction works started in 2010. The ASR-Project in the Emirate of Abu Dhabi will consist of 3 recharge/recovery schemes with 3 infiltration basins, 315 recovery wells and 117 groundwater monitoring wells and in addition approx. 160 km of pipeline, 4 pumping stations and 5 reservoirs as well as a monitoring and process control system. At present, the drilling, construction and testing of monitoring and recovery wells is ongoing. Based on these results, the existing large-scale numerical groundwater model for the entire Emirate of Abu Dhabi is being updated.

After construction and commissioning of the ASR-Plant, during an infiltration period of 27 months, a major strategic freshwater resource of around 26 Mio cubic meters will have been created ensuring the safe water supply for more than 1 Mio inhabitants of Abu Dhabi City for a period of at least 90 days. The ASR-Project will represent a benchmark for water resources management in arid regions.

www.borntraeger-cramer.de
2012/0247/$ 0.50

References

Dawoud, M., 2009: Strategic Water Reserve: New Approach for Old Concept in GCC Countries. – Proc. of "The 5th World Water Forum, Istanbul", 16.-22.03.2009.

Peters, J. H. et al. (eds.), 1998: Artificial Recharge of Groundwater. – Balkema, Rotterdam, 474 pp.

Proceedings "Hydrogeology of Arid Environments", p. 249
Stuttgart, March 2012

Poster Abstract

Identification of Groundwater Natural Recharge Areas in the Eastern Region of UAE

Mohamed Mostafa A. Mohamed

Department of Civil and Environmental Engineering, United Arab Emirates University, Al Ain, P.O.Box 17555, UAE, email: m.mohamed@uaeu.ac.ae

Key words: Groundwater; recharge; UAE

Despite the continuous increase in water supply from desalination plants in the United Arab Emirates (UAE), groundwater remains the major source of fresh water satisfying domestic and agricultural demands. Additionally, groundwater has always been considered as a strategic water source towards groundwater security in the country. Understanding the groundwater flow system; including identification of recharge and discharge areas, is a crucial step to achieve groundwater sustainability. As such, the main aim of this research is to identify groundwater recharge regions to the shallow unconfined groundwater aquifer in the eastern part of UAE through development of groundwater age distribution maps using isotopic techniques. Generally, estimations of groundwater recharge and discharge fluxes in unconfined systems in arid regions are difficult because of lack of knowledge on how to distribute water flux and vapor discharge. Previous investigations of groundwater resources in the eastern region of UAE indicated that the surficial groundwater is one of the main sources of fresh groundwater. Most of the recent natural recharge to this surficial aquifer is associated with the high precipitation near Omani Mountains across the eastern border of the UAE. However, these studies warned that this aquifer has limited natural recharge and can be affected by pumping. Therefore, results of this study will be of great importance to water resources managers in UAE as it will help to accurately estimate sustainable extraction rates, assess groundwater availability, and identify pathways and velocity of groundwater flow as crucial information for identifying the best locations for artificial recharge.

www.borntraeger-cramer.de
2012/0249/$ 0.25

Proceedings "Hydrogeology of Arid Environments", p. 250
Stuttgart, March 2012

Poster Abstract

Domestic and Industrial Water Demand in Syria

Khaldoon A. Mourad[1] and Ronny Berndtsson[2]

[1] Corresponding Author: Department of Water Resources Engineering, Lund University, Box 118, SE-22100, Lund, Sweden, and Ministry of Environment, Damascus, Syria. email: Khaldoon.Mourad@tvrl.lth.se
[2] Center for Middle Eastern Studies and Department of Water Resources Engineering, Lund University, Lund, Sweden

Key words: WEAP, groundwater, greywater, water harvesting

Syria faces economic and physical water scarcity. This together with large population and industrial increase will effect on the water future of the country. **Water Evaluation and Planning system (**WEAP) has been used to estimate domestic and industrial water demand up to 2050. The system analyses the main seven Syrian water basins depending on the historical statistics about population and industrial growth in each basin. The results showed that population will be doubled in 2050, which will effect on the future water demand (Table 1). The paper also presents the needs for groundwater recharge and grey water reuse to tackle these coming high demands. The results show that there are reasons to be alarmed but also cautiously optimistic regarding meeting water needs in Syria. This, however, depends on the continuing progress and development within water demand management, reclaimed water, and rainwater harvesting.

Table 1. Water demands projections (MCM).

Water basin	2030				2050			
	Population	Water demands			Population	Water demands		
		Domestic	Agriculture	Industry		Domestic	Agriculture	Industry
Barada & Awaj	8029476	430	631	49	10991550	559	625	71
Al-Yarmouk	2242356	169	192	48	3447847	247	190	69
Orontes	6011108	421	2053	350	9094206	605	2033	504
Dajleh & Khabour	1922739	163	4367	16	2677222	216	4323	24
Euphrates & Aleppo	10140774	809	6550	228	16614645	1260	6485	328
Desert	569028	58	110	15	846941	82	109	21
Coastal	2325178	148	496	67	2969020	180	491	96
Total	31240660	2199	14399	772	46641431	3149	14255	1112

www.borntraeger-cramer.de
2012/0250/$ 0.25

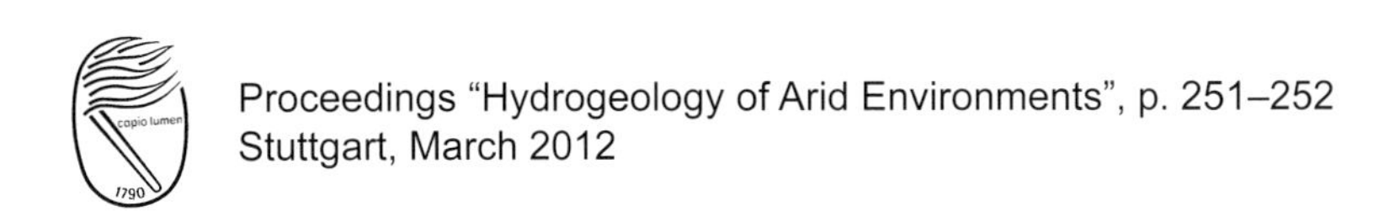
Proceedings "Hydrogeology of Arid Environments", p. 251–252
Stuttgart, March 2012

Poster Abstract

Enhancing of Environmental Quality through Groundwater Artificial Recharge in Dar es Salaam Coastal Aquifer

Yohana Mtoni[1, 2,*], Ibrahimu Chikira Mjemah[3], Kristine Walraevens[1]

[1] Laboratory for Applied Geology and Hydrogeology, Ghent University, Krijgslaan 281 S8, 9000, Ghent, Belgium.
[2] National Environment Management Council (NEMC), P.O. Box 63154, Dar es Salaam, Tanzania
[3] Sokoine University of Agriculture (SUA), P.O. Box 3038, Morogoro, Tanzania
* Email for corresponding author: yohanaenock.mtoni@ugent.be ; mtoni.yohana@gmail.com

Key words: aquifer salinization, artificial recharge

Problem setting

The current groundwater replenishment under natural infiltration in Dar es Salaam Quaternary coastal aquifer (DQCA) is about 184 mm (equivalent to 71.39 x 10^6 m^3 $year^{-1}$) indicating that only 16.5% of the long term average annual precipitation of 1114 mm ends up as groundwater recharge (Mtoni et al. 2011). The water balance of the catchment suggests for the average sustainable yield (calculated as 40% of natural groundwater recharge) a value of 28.56 x 10^6 m^3 $year^{-1}$. The current groundwater abstraction is approximately 69.3 x 10^6 m^3 $year^{-1}$ and such overexploitation of groundwater usually results in a rise of the freshwater-saltwater interface and lateral seawater intrusion in a coastal aquifer, and thus degradation of groundwater quality.

Research objective

The objective of this study was to assess the long term viability of the water supply, including water quality, and the potential to increase the yield of the aquifer through artificial groundwater recharge (AGR).

Methodology

The assessment of the potential for AGR in DQCA included the evaluation of the dynamics of groundwater flow and recharge, and consideration of the options for artificial recharge techniques that can be used. A primary concern was to understand the hydrological variability within the aquifers, as well as to identify potential sources of water for aquifer recharge. Based on meteorological data, groundwater recharge was estimated using a soil moisture water balance. Calculations with monthly meteorological data were used for the period 1971–2009.

Discussion and Conclusions

By implementing AGR, excess water in the catchment area can be collected and allowed to infiltrate, to increase groundwater storage in rainy seasons to be utilized later in dry seasons. Sources of water for AGR include the surface runoff (which accounts to about 10%) from the rainfall (that currently flows to the Indian Ocean) and grey water (from baths, kitchen, washing machines and sinks). The latter which accounts to over 50% of the outflow from homes, has less pathogens and nitrogen comparing to the wastewater from the toilets, thus it does not require expensive treatment process. AGR can be done by methods such as water spreading, recharge through pits, channels or wells depending on the local topographical, geological and soil conditions. Utilizing runoff (which otherwise drains off) and grey water (which is improperly disposed) will bring great benefits for improving groundwater levels (which have dropped due to overexploitation), providing a barrier for seawater intrusion and prevention of diseases and floods by deviating peak flows.

www.borntraeger-cramer.de
2012/0251/$ 0.50

References

Mtoni, Y., Mjemah, I. C., Van Camp, M. & Walraevens, K., 2011: Enhancing Protection of Dar es Salaam Quaternary Aquifer: Groundwater Recharge Assessment. – Springer, Environmental Earth Sciences, Advances in Research of Aquatic Environment volume 1, pp. 299-306 (DOI 10.1007/978-3-642-19902-8).

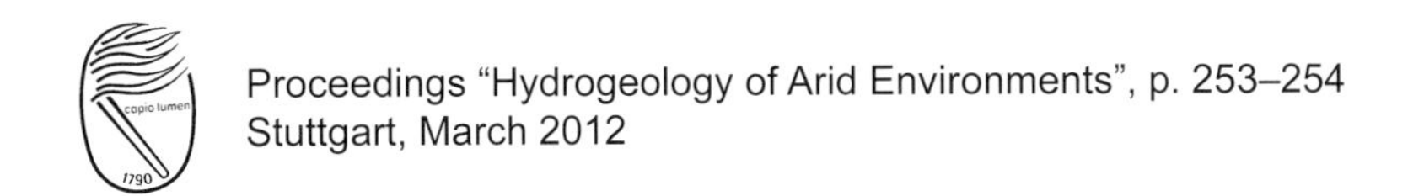
Proceedings "Hydrogeology of Arid Environments", p. 253–254
Stuttgart, March 2012

Poster Abstract

Use of Helium as a Tracer for Groundwater-Model Calibration of the Dhofar Region of Southern Oman

Thomas Müller[1], Ward Sanford[2], Karsten Osenbrück[3], Werner Aeschbach-Hertig[4], Christian Herb[3], Gerhard Strauch[1]

[1] Helmholtz Centre for Environmental Research - UFZ, Permoserstrasse 15, 04318 Leipzig, Germany, email: th.mueller@ufz.de
[2] US Geological Survey, Mail Stop 431, Reston, Virginia 20192, USA, email: wsanford@usgs.gov
[3] University of Tuebingen, WESS Department of Geosciences
[4] University of Heidelberg, Institute of Environmental Physics

Key words: groundwater flow modeling, simulation of groundwater ages

Paleoclimate studies have shown that during the last 30,000 years a change from humid to arid conditions took place over most of the Arabian Peninsula, which is today one of the most arid regions in the world. The Dhofar Mountains, which are located adjacent to the Arabian Sea along the southern edge of the Arabian Peninsula, divide the monsoon influenced coastal plain and southern mountain slopes from the hyper-arid interior Dhofar, where the mean annual precipitation is 31 mm/yr and the mean annual temperature is 26 °C. Hydrologic conditions in the interior Dhofar, the southernmost province of the Sultanate of Oman, consist of a thick unsaturated zone and a steep-to-gentle groundwater gradient in the limestone Umm Er Radhuma aquifer from the Dhofar Mountains to the discharge area at the Sabkha Umm as Sammim far to the northeast.

Previously, radiocarbon data has suggested that the bulk of the groundwater in the region is between 4,000 and 40,000 years old and would have been recharged mostly during the last glacial maximum. Recently measured dissolved-helium concentrations, however, indicate that groundwater in this aquifer is between 4,000 and 4,000,000 years old, with many of the radiocarbon values having been strongly affected by mixing during sampling. A helium production rate was calibrated using the relatively younger concordant radiocarbon-helium data. Also, unlike the radiocarbon, the observed trend of non-atmospheric helium increases in the direction of the groundwater flow, and therefore proved to be a more robust tracer of age for the calibration of a groundwater model.

A two-dimensional groundwater-flow model was developed that embodied the characteristics of the aquifer system from the recharge area in the south (Dhofar Mountains) to the discharge area in the north (Sabkha Umm as Sammim). The model reproduces the south to north gradients and the observed artesian heads in the confined aquifer. Simulation results indicate that changes between wet and dry periods caused transient responses in heads and head gradients that would last several thousand years. The model calibration also indicated that a recharge rate of <3 mm/year (similar to the present-day estimate from sporadic cyclones) is sufficient to reproduce current groundwater levels. Backward pathline tracking was used to simulate the groundwater ages. The tracking results show that a total porosity value between 10 and 20 percent is consistent with the range of the observed helium-based ages.

References

Torgersen, T. & Clarke, W. B., 1985: Helium accumulation in groundwater, i: An evaluation of sources and the continental flux of crustal 4He in the great artesian basin, Australia. – Geochimica et Cosmochimica Acta **49**(5): 1211–1218.

Clark., I. D., 1987: Groundwater resources in the Sultanate of Oman: Origin, circulation times, recharge processes and paleoclimatology. Isotopic and geochemical approaches. – Universite de Paris-Sud, Orsay, France, PhD Thesis

www.borntraeger-cramer.de
2012/0253/$ 0.50

Sanford., W., 2011: Calibration of Models using Groundwater Age. – Hydrogeology Journal **19**: 13–16.

Herb., C., 2011: Paleoclimate study based on noble gases and other environmental tracers in groundwater in Dhofar (southern Oman). – Master's thesis, Institute of Environmental Physics, University of Heidelberg.

Proceedings "Hydrogeology of Arid Environments", p. 255
Stuttgart, March 2012

Poster Abstract

Groundwater in the Dry Zones of India: Quantity and Quality Issues

K. Shadananan Nair

Nansen Environmental Research Centre, Gopal Residency II Floor, Thottekkat Road, Kochi – 682011, Kerala, India, email: nair59@yahoo.com

Key words: dry zones, India, groundwater, climate change

Quality as well as quantity of groundwater that increasingly contribute to the irrigation and domestic use in the arid and semiarid zones of India are fast declining as a result of changing climate and pressure of the fast rising population. Groundwater contributes to nearly 80% of the agricultural production and domestic water supply in rural areas and about 50% of the urban and industrial uses. Groundwater in more than one-third of the country is not fit for drinking. Increase in salinity and presence of high concentrations of fluoride, nitrate, iron, arsenic, total hardness and few toxic metal ions have been noticed in large areas. The country is in the path of rapid industrial development. Industrial overuse and misuse of groundwater is tremendous. Groundwater quality near major industries is far above safety limits. Urbanization and expansion of urban limit lead to encroachment into wetlands, affecting groundwater recharging. Mining and landuse changes have largely affected the groundwater quality. The number of open and bore wells has rapidly and indiscriminately increased. There is an increasing gap between the extraction and recharge. Groundwater resources are largely unmanaged. Remote sensing data shows that overextraction has resulted in water table depletion in north India by five fold more than expected. Groundwater across north-western and south-eastern India drops by 4 cm/year and more than 109 Km3 of groundwater disappeared in 4 years. Changing climate also has significant impact on groundwater. Increasing temperature produces more evaporation from surface water bodies making the soil dry and reducing the groundwater recharge. Rainfall is becoming highly seasonal in parts of western and southern parts, allowing wasteful runoff and reducing the time for recharging. High intensity rainfall erodes topsoil, reducing the water holding and recharging capacity of the surface. Trends in rainfall in the dry zones increase dependency of groundwater for irrigation. Climate change may result in sea levels rise and storm surges, salinating the coastal aquifers. Changes in the course of rivers as a result of flooding and sedimentation may lower the water table in the heavy rainfall regions. Melting rate of Himalayan glaciers has been accelerated in recent years, gradually leading to water crisis in entire northern parts of the country. Present study assesses the impacts of climate change and environmental degradation on the groundwater resources of India, and its reflections on different sectors of the society. A review of the existing policies and management strategies related to water and environment has been made. Better management and conservation practices can help overcoming the crisis. Since the total amount of rainfall do no show significant trends, control of runoff and adequate measures for storage and recharge can improve the groundwater condition. Reviving the traditional environment-friendly mechanisms for groundwater recharge could minimise the water scarcity in the rural areas. Existing policies and their implementation often fail because of the typical socio-economic, bureaucratic and political setup in India. India has to develop appropriate policies for water and environment and an efficient strategy for climate change adaptation. Suggestions for this have been provided.

www.borntraeger-cramer.de
2012/0255/$ 0.25

Proceedings "Hydrogeology of Arid Environments", p. 256
Stuttgart, March 2012

Poster Abstract

Groundwater Evolution of the Umm Er Radhuma Aquifer in the Rub' Al Khali Desert (Arabian Peninsula)

T. Neumann[1], W. Gossel[2], P. Wycisk[3], H. Dirks[4], R. Rausch[5], A. Al Khalifa[6]

[1, 2, 3] Martin-Luther-University Halle-Wittenberg, Institute of Geosciences and Geography, Hydrogeology and Environmental Geology, von-Seckendorff-Platz 3, 06120 Halle (Saale), Germany
[1] thomas.neumann@geo.uni-halle.de, [2] wolfgang.gossel@geo.uni-halle.de, [3] peter.wycisk@geo.uni-halle.de
[4, 5] GIZ International Services / Dornier Consulting, Riyadh, Kingdom of Saudi Arabia
[4] heiko.dirks@gizdco.com, [5] randolf.rausch@gizdco.com, krm_1403@yahoo.com
[6] Ministry of Water & Electricity, Riyadh, Kingdom of Saudi Arabia

Key words: Arabian Peninsula, Umm Er Radhuma Aquifer, groundwater chemistry, hydrochemical modeling

The study area is located in the southeastern part of the Arabian Peninsula and covers a surface area of about 887,000 km^2. The aim of the study is to conduct a hydrochemical analysis for the Tertiary Umm Er Radhuma aquifer in the Rub' Al Khali desert and to create a model of the hydrochemical evolution for this principal aquifer.

To achieve this objective in the sparsely populated region of Rub' Al Khali desert, more than 300 groundwater wells were considered. A contour map of the static groundwater level (SWL) was created from these data. Over 260 selected groundwater wells were used in drawing a contour map for the total dissolved solids (TDS) analysis. Finally, there were more than 170 groundwater samples with main ion analyses. Based on these data, the following contour maps were drawn: (1) the main ions (Ca^{2+}, Mg^{2+}, Na^+, K^+, Cl^-, SO_4^{2-}, HCO_3^-, and NO_3^-); (2) selected ion ratios; and (3) the saturation index (SI) of the main minerals (calcite, dolomite, gypsum, halite, and sylvite). The maps were drawn using GeoInformationSystem ArcView 9.3 by applying three different interpolation methods (TIN, IDW, and Ordinary Kriging).

The results show the flow of groundwater in the northern or northeastern direction to the area towards the Arabian Gulf. The distribution of TDS increases from the recharge area to the discharge area. However, lateral recharge or leakage from adjacent aquifers exists, which is indicated by differences in the TDS distribution. Further causes have been investigated in this work. When applicable, the enlargements of TDS bearing data points have been achieved by the use of the hydrochemical examinations of the main ions or the data from electric conductivity (EC) using the relation: factor = TDS [ppm] / $EC_{(adjacent\ wells)}$ [µS/cm].

Pie-, Piper- and Schoeller-diagrams of the hydrochemical characteristics of the groundwater from each well were plotted in AquaChem 5.0 software to obtain further information of the type of water. In general, groundwater changes from calcium-sulphate dominated water in the recharge areas to sodium-chloride water in the discharge areas. In combination with the increasing TDS values, this evolution indicates a part of the Chebotarev-Sequence.

The final step of this work involved the hydrochemical modeling along 9 representative flow paths to investigate the hydrochemical evolution. This was done using PHREEQC 2.18 software and with an in-built inverse modeling tool. Furthermore, the SI of the minerals of importance (calcite, dolomite, gypsum, halite, and sylvite) was modeled. Due to analysis of the data available (SI, TDS, inverse modeling, and ion ratios), following hydrochemical processes have been identified: (1) sulphates dissolution; (2) dedolomitization; (3) ion exchange (alkalization, earth alkalization); (4) dissolution of chlorides; and (5) mixing water.

The modeling results suggest upward leakage from underlying Aruma aquifer into the UER aquifer (cross formation flow), for example in the Ghawar anticline structure.

www.borntraeger-cramer.de
2012/0256/$ 0.25

Proceedings “Hydrogeology of Arid Environments”, p. 257
Stuttgart, March 2012

Poster Abstract

Monitoring of the Mineralization and Chemical Composition Regime Changes of Underground Waters in Vegetation Periods

Inom Normatov[1], Surae Buranova[2], Muslima Kholmirzoeva[3]

[1,2,3] Institute of Water problems, Hydropower and Ecology Academy of Sciences Republic of Tajikistan, 12 Parvin Str., Dushanbe, 734002, Tajikistan, email: inomnor@gmail.com

Key words: mineralization, irrigation, drainage, pollution

The surface and underground waters in Tajikistan are characterized by the considerable variety in mineralization degree and chemical composition. The least mineralized groundwater is observed near the filtration river sites and irrigation cannels. The most marked difference in chemical composition of republican irrigation waters is observed in chloral ion contents, on the base of which the evaluation of ground salinization degree is given for the areas with sulfate and sulfate chloride salinization. It is possible to establish some laws in changing of mineralization and chemical composition of underground waters in quarter deposits on irrigated lands, taking into account the chemical composition of irrigation waters of Vakhsh River and river waters in Hissar Valley. Irrigation waters of Vakhsh River contain up to 0.1–0.15 g/l of chloral ion. Compared to Vakhsh the essential change in mineralization and chemical composition of groundwater should be expected on the developed virgin lands with thick surface shallow grounds. At the first stage of developing these lands the descending waters currents will wash the salts out of the upper layers and transmit them into deeper horizons. When at the next stage the level approaches the day surface, the groundwater may have already somewhat less mineralization than at the initial state. At this moment the artificial drainage must begin functioning, gradually throwing the water and salts out of irrigated lands. The drainage flowing may be up to 30-40% of the total water-intake into irrigation systems. Due to this correlation between water-intake and the surface throw of drainage and thrown waters, the groundwater, already suitable for irrigation, may gradually being formed in the zone of active influence of drainage. Mineralization and chemical composition of subterranean and forceful waters on irrigated lands in Tajikistan are also subject to regime changes. According to year seasons the most marked changes in this respect are observed on greatly salinized lands. There in the hot season of the year the groundwater of the surface shallow grounds with sharply increased mineralization is spent on the total evaporation from the ground surface, thus increasing temporarily the salt store in the soil grounds of aeration zone and on the land surface. The waters with lesser mineralization come to their place from the bedding pebbles. By the next vegetation period the adverse effect takes place – salts dissolve in the groundwater due to precipitation. Mineralization and chemical compound of pressure waters of second, third and more deep horizon (in any care, in limit of high 100–150 m) at action of vertical drainage in upper of them (first) will not subject to essential changes. In them contour with fresh waters in compare with of first horizon town to flow goes further. The water of these horizons a less subject to pollution, expediently to use for centralized water – pipe economical – drinkable water – supply.

www.borntraeger-cramer.de
2012/0257/$ 0.25

Proceedings "Hydrogeology of Arid Environments", p. 258
Stuttgart, March 2012

Poster Abstract

Decision Support System (DSS) to Manage the Zeuss Koutine Groundwater (Tunisia) Using the WEAP-MODFLOW Framework

I. Nouiri[1,*], R. Haddad[1], J. Maßmann[2], M. Huber[3], A. Laghouane[4], H.W. Müller[2], H. Yahiaoui[4], J. Wolfer[2], O. Alshihabi[5], J. Al-Mahamid[5], J. Tarhouni[1]

[1] National Institute of Agronomy of Tunisia (INAT)
[2] Federal Institute for Geosciences and Natural Resources, Germany (BGR)
[3] Geo:Tools, Brünnsteinstr. 10 81541 Muenchen, Germany
[4] Ministry of Agriculture and Environment of Tunisia, Regional administration of Medenine,
[5] Arab Centre for the Studies of Arid Zones and Dry Lands, Syria (ACSAD)
* Corresponding author email: inouiri@yahoo.fr

Key words: Groundwater management, DSS, MODFLOW, WEAP, Zeuss Koutine, Tunisia

This paper describes the build of a Decision Support System (DSS) for monthly groundwater management of the "Zeuss Koutine" aquifer, in the Eastern South of Tunisia. The WEAP-MODFLOW framework is used to lead to this goal. A monthly MODFLOW model is developed first to simulate the behaviour of the studied aquifer. The WEAP schematic is built to represent the real physical system. A shape file ensures the link of each of the WEAP nodes to the MODFLOW model cells. Other available water sources in the region, as desalinization plants and groundwater, are considered in the present DSS and modelled as "fictive" groundwater nodes. The inputs to the hydro geological model are mainly the natural recharge and the drainage from higher neighbour aquifers. The outputs are mainly the agricultural, industrial and the urban water consumptions. The elaborated DSS is able to evaluate water management scenarios up to 2030. In particular, future water consumptions, flows in transmission links and heads in the active cells of the MODFLOW model are computed for each time step.

Results prove that the created database and the elaborated WEAP model allow an easy input modification and output displaying. Scenarios comparison can be cited as a second advantage of the DSS developed in this work. Results related to the Zeuss Koutine aquifer demonstrates that the desalinization plants already build in Jerba and Zarzis cities have contributed to the decrease of the continue drawdown observed before 1999. The use of sea water desalinization plant, to supply Jerba and Zarzis cities in the future, is a solution to stop the Zeuss Koutine aquifer drawdown. The definition of its optimal capacity and its starting year constitute the main research questions proposed by decision makers.

www.borntraeger-cramer.de
2012/0258/$ 0.25

Proceedings "Hydrogeology of Arid Environments", p. 259–260
Stuttgart, March 2012

Poster Abstract

Interdisciplinary of Structural Geology and Hydrogeology: The Case Study of Wadi Zerka Ma'in Catchment Area, Dead Sea, Jordan

Taleb Odeh[1], Stefan Geyer[1], Tino Rödiger[1], Christian Siebert[1], Mario Schirmer[2]

[1] Helmholtz Centre for Environmental Research – UFZ, Catchment Hydrology Department, Theodor-Lieser-Strasse 4/ 06120 Halle, Germany, email: taleb.Odeh@Ufz.de
[2] Eawag – Swiss Federal Institute of Aquatic Science and Technology, Department Water Resources and Drinking Water, Überlandstrasse 133, 8600 Dübendorf, Switzerland

Key words: groundwater mixing, spatial distribution, heterogeneous, strike slip fault

Abstract

Wadi Zerka Ma'in catchment area is located at the eastern side of the Dead Sea. A major strike slip fault passes perpendicularly through the two aquifers and the aquiclude layer with embedded normal faults. Theses faults form conduits that allow groundwater to flow from the lower aquifer to the upper aquifer, resulting in a mixed groundwater. The topographic profile of the Zerka Ma'in River exhibits two knickpoints because of those normal faults. Climatically and gemorphologically, Wadi Zerka Ma'in catchment area is considered heterogeneous as a result of the major strike slip fault. The catchment area has a spatial distribution of groundwater recharge as a result of that heterogeneity .However, the major strike slip fault affects also groundwater flow by generating a high permeability zones.

Introduction

The Wadi Zerka Ma'in catchment area is located at the north-eastern side of the Dead Sea, in Jordan, and spans an area of 272 km². It is the smallest catchment area and contains the largest city, Madaba, at the eastern side of the Dead Sea. Therefore, understanding the groundwater system of Wadi Zerka Ma'in is of a high importance for Jordan. Wadi Zerka Ma'in catchment area has two aquifers: an upper unconfined limestone aquifer and a lower confined sandstone aquifer. A major strike slip fault bound the lower and middle parts of that catchment area (Odeh et al. 2009). Groundwater systems of Wadis have interactions between their structural geology and groundwater flow, recharge and chemistry (Şen 2008). Therefore the objective of the research is to understand the interdisciplinary between Wadi zerka Ma'in structural geology and its hydrogeological system.

Methodology

The hydrologic response unit (HRU) method was used for the hydrological model while finite elements method was used for the groundwater flow steady state modeling. The both methods consider the heterogeneity of the catchment area. An integrated approach of remote sensing (RS) and geographic information systems (GIS) was used to supply the models with the land surface spatial data. The structural origin of the Wadi was investigated by geomorphologic indices to understand the types and the distributions of the faults. The geomorphologic indices were extracted by analyzing high resolution DEMs that were extracted and analyzed by a remote sensing and GIS approach too.

Results and Discussions

The regional strike–slip fault of Wadi Zerka Ma'in cuts through the middle and lower parts of the catchment area and is associated with normal faults that caused two kinck points. However, that

www.borntraeger-cramer.de
2012/0259/$ 0.50

normal faults form a connection between the two aquifers. The connection, through the faults, mixes the groundwater of the two aquifers and produces a new groundwater type. Furthermore, the HRU model of Wadi Zerka Ma'in shows that there a zone of high groundwater recharges and permeability within the zone of the regional strike slip fault.

Conclusions

Faults control the groundwater systems of the Wadis. These faults lead for connections between the aquifers and generate zones of high groundwater recharge and maximum permeability.

References

Odeh, T., Gloaguen, R., Schirmer, M., Geyer, S., Rödiger, T. & Siebert, C., 2009: Exploration of Wadi Zerka Ma'in rotational fault and its drainage pattern, Eastern of Dead Sea, by means of remote sensing, GIS and 3D geological modeling, Proceeding of SPIE Europe's International Symposium on Remote Sensing (ERS09), Vol. 7478, Berlin.

Şen, Z., 2008: Wadi Hydrology. CRC Press, New York.

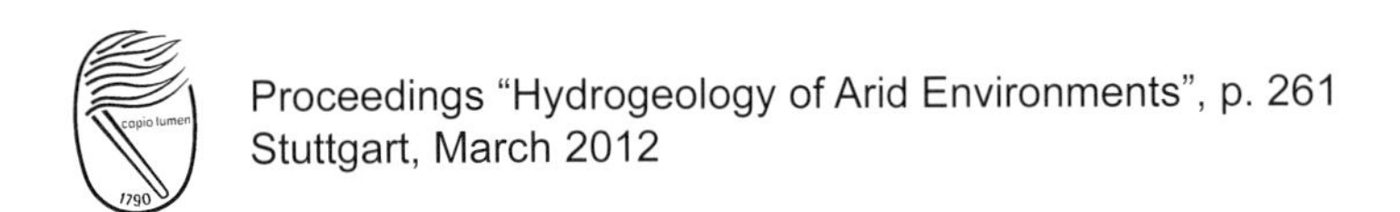
Proceedings "Hydrogeology of Arid Environments", p. 261
Stuttgart, March 2012

Poster Abstract

The SASS Water Management in the Southern Tunisia Oases: Water Shortage or Inadequate Approach?

Nizar Omrani[1], Dieter Burger[1]

[1] Karlsruhe Institute of Technology, Germany. Omrani_nizar@yahoo.fr ; Dieter.Burger@kit.de

Key words: Tunisia, oases, water management, sustainability.

In the southern Tunisia, although they are located under an extreme arid climate, the oases still represent one of its most relevant development pillars. The irrigation agriculture plays a key role in the promotion of these arid lands and contributed since many decades to fix their population. The main incomes of the oases production provide from the date palm sector which has the highest added value and mainly destined to the worldwide exportation.

The technological advances in the drilling techniques allowed the access to the deep aquifers such as the resources of the Saharian Septentrional aquifers System (SASS). It deals with sedimentary basin holding very important underground water volumes and extended over an area of one million km^2 across three countries; Algeria (700 000 km^2), Lybia (250 000 km^2) and Tunisia (80 000 km^2). During the previous decades, the water mobilization from this basin has considerably enhanced from 0,6 million km^3 in 1970 to a current ratio of 2,5 million km^3.

This intensive Tempo contributed to double the irrigated area in the southern Tunisia and its permanent extension became out of sight to the agriculture development authorities. Indeed the illegal private parcels multiplication as well as the high water consumption of the cultivation systems brought to light the inability of the irrigation network to overcome the growing water demand. The water shortage became severe in the summer period when farmer's plots could be only partially irrigated.

Furthermore, the downscaling of the climate change impacts for Tunisia revealed an expected rise in the mean temperature and more frequent droughts sequences. The Tunisia oases had been also identified as strongly vulnerable ecosystems to such global changes. Their tributary to the underground water resources will be definitely stronger.

The main issue that determines the survival of the irrigated agriculture in such conditions will definitely pass through the promotion of optimal water management solutions.

This paper attempts to highlight the technical feasibility of the irrigation practices conversion in these arid lands into higher efficient systems. The adaptation of the surface irrigation to the drip and sprinkler systems for the date palm cultivation still a key challenge for the promotion of the water saving in the oases. A comparative study between those irrigation systems had been notably led to evaluate the suitable method to be implemented for wider scale.

It targets also to gather technical irrigation references that had been mistaken for the irrigation optimization under the specific arid context of the oases such as the ongoing cultivation systems water consumption.

In a rapidly changing and ever more complex water management context, a better understanding of the irrigation problems, coupled with the technology transfer assessment, can bring to bear more efficient solutions to achieve the sustainable management of the SASS Aquifer water resources.

www.borntraeger-cramer.de
2012/0261/$ 0.25

Proceedings "Hydrogeology of Arid Environments", p. 262
Stuttgart, March 2012

Poster Abstract

Evaluation of Water, Vapor and Heat Fluxes under Arid Conditions by Inverse Modeling

H. Pfletschinger*, I. Engelhardt, C. Schüth

*Technische Universität Darmstadt, Institute of Applied Geosciences, email: pfletsch@geo.tu-darmstadt.de

Key words: Hydrus-1D, vadose zone, hydraulic parameters, PEST

Quantification of groundwater recharge in arid regions poses many difficulties, especially when dealing with deep vadose zones which determine amounts of evaporation, infiltration, redistribution and percolation. As the complex and non-linear processes involved in unsaturated water flow under non-isothermal and low water content conditions are yet not fully understood, the influence of changing initial and boundary conditions on these fluxes should be studied in detail.

In general, vertical unsaturated water fluxes are imposed by gradients in total soil water potential and in temperature. To better understand and quantify the interaction of the physical water flux processes, soil column experiments were developed that can mimic soil water states, temperature profiles and soil-atmospheric conditions as they can be expected in arid regions (Pfletschinger et al., in press). Within the experiments, the processes of water infiltration, evaporation, redistribution and percolation are analyzed. For this, measurements are performed in high spatial and temporal resolution on water content and temperature in the column as well as on water and vapor fluxes into and out of the column. By running the experiments with defined initial and boundary conditions, a model could be set up according to these conditions in Hydrus 1-D (Šimůnek et al. 2009). The model was calibrated by inverse modeling with the parameter estimation software PEST (Doherty 2010). Calibration was performed on hydraulic and thermal properties whereas experimental measurements for the calibration included transient water outflow at the bottom of the column, evaporation rates at the top of the column, saturation profiles and the temperature gradient along the axis of the column.

A sensitivity analysis was performed during calibration showing a high impact of the air entry pressure and the residual water content on model predictions whereas hydraulic conductivity was less controlling. Thus, under arid settings, care must be given for hydraulic parameters influencing evaporation rather than pore-filled water flow. Based on the calibrated model further model predictions that simulate recharge for critical precipitation rates or extreme temperature events are feasible without any further experimental investigations. Therefore, it offers a valuable tool for groundwater management issues with respect to expected future hydrological and microclimatological changes.

References

Doherty, J., 2010: PEST Surface Water Utilities User`s Manual. Watermark Numerical Computing. Brisbane, Australia. – University of Idaho. Idaho Falls, Idaho.

Pfletschinger, H., Engelhardt, I., Piepenbrink, M., Königer F. & Schüth, C., in press: Soil column experiments to quantify vadose zone water fluxes in arid settings. – Environmental Earth Sciences.

Šimůnek, J., Šejna, M., Saito, H., Sakai, M. & van Genuchten, M. Th., 2009: The Hydrus-1D Software Package for Simulating the Movement of Water, Heat, and Multiple Solutes in Variably Saturated Media. Version 4.08. HYDRUS Software Series 3. Department of Environmental Sciences, University of California Riverside, Riverside, California, USA.

www.borntraeger-cramer.de
2012/0262/$ 0.25

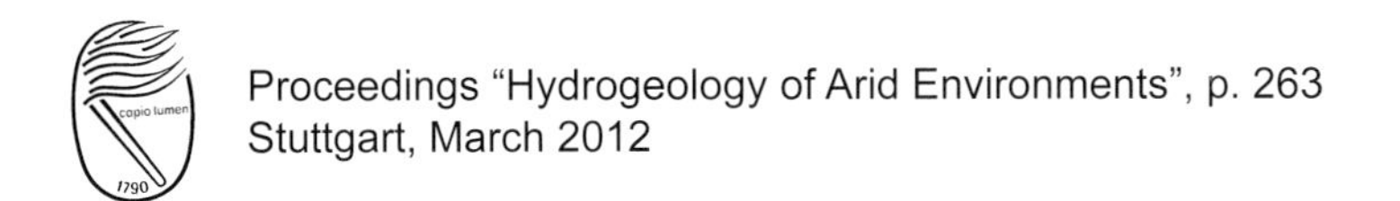

Proceedings "Hydrogeology of Arid Environments", p. 263
Stuttgart, March 2012

Poster Abstract

Infiltration During Wadi Runoff – A Modelling Tool for Improving Groundwater Recharge Assessment

Andy Philipp and Jens Grundmann

Institute of Hydrology and Meteorology, Dresden University of Technology, email: Andy.philipp@tu-dresden.de, Jens.grundmann@tu-dresden.de

Key words: transmission losses, ephemeral river flow, groundwater recharge dams

The coastal aquifer in Oman's Batinah plain is affected by saltwater intrusion due to excessive groundwater withdrawal for irrigated agriculture. To account for a more sustainable water resources management, a sound groundwater recharge assessment is needed. The infiltration of water during flood events in ephemeral streambeds (or wadis) is important for quantifying groundwater recharge. However, flash floods caused by convective rainstorms lead to substantial freshwater losses to the sea. In order to minimize these losses and to promote artificial recharge of the coastal aquifer system, several dams have recently been constructed in the wadi beds. These dams help to retain flood flow, allow for sedimentation and establish a decelerated release to the downstream section which increases infiltration opportunity times.

The investigations focus on the surface runoff processes in ephemeral streambeds and their impact on groundwater recharge. We present a comprehensive model chain capable of soundly portraying the involved surface flow and infiltration processes. Along these lines, the accurate modelling of the weak process dynamics as a consequence of dam release to an initially dry wadi bed – which can lead to standing wave effects – is focused. Furthermore, the pronounced flow dynamics upstream of the dam and the retention effects of the recharge dam itself are taken into account.

Besides a detailed sensitivity and errors analysis, the model is applied for several historical flood events in the study area of the southern Batinah region in the Sultanate of Oman. The presented modelling system allows for quantifying the temporal and spatial distribution of transmission losses (i.e. infiltration during wadi runoff), which serves as a potential contribution to groundwater recharge of the coastal aquifer. In addition, the impact of different flood characteristics on groundwater recharge will be discussed.

www.borntraeger-cramer.de
2012/0263/$ 0.25

Proceedings "Hydrogeology of Arid Environments", p. 264
Stuttgart, March 2012

Poster Abstract

Sustainable Use of Groundwater in the Coast Affected by the Tsunami – Case Study of Madiha East, Sri Lanka

Ranjana U. K. Piyadasa[1], K. D. N. Weerahinghe[2]

[1] Department of Geography, University of Colombo, Colombo, email: ranjana@geo.cmb.ac.lk
[2] Department of Agric. engineering, Faculty of Agriculture, University of Ruhuna, Kamburupytiya

Key words: aquifer, Electrical conductivity, alluvium, pH

The Tsunami of December 26, 2004 was the worst ever natural disaster to strike Sri Lanka and south Asia. Costal area have been toughly affected: destroying habitations, roads, wells, and affecting socially and physically survivors. People in the coastal areas of the Sri Lanka traditionally use groundwater for water drinking and other domestic purposes. This case study becomes rooted, in this context post Tsunami, for Rehabilitation and enhancement of sustainable use of affected wells in order to challenge public water supply which would be more and more expensive in the next years.

The study was conducted in Madiha East, south of Sri Lanka, District of Matara. The study area, falls within the WL2a agro-climatic region which is defined by a 75 % expectancy of an annual rainfall exceeding 2,400 mm. Annual total of rainfall varies over space between 1,875–2,500 mm. The mean annual temperature is 25 °C. Topography of the western border of the study area is relatively flat or gently seaward sloping. Elevation and topographic relief generally increases towards inland from the coastal line. This contrasts with typical elevations in the outer coastal plain setting of 1.5–7m above sea level. The top unconfined alluvium aquifer is distributed along the study area and water-bearing sand in the top section is more often fine while lower sections usually have coarse sand with small portions of gravel. Fifty ordinary dug wells have been studied with respect to physical parameters as Electrical Conductivity (EC), pH, Salinity, total dissolved solids (TDS) are tacking into account to evaluate the quality of water. Chemical analysis was conducted to identify type of aquifer type using the piper diagram.

The study reveled that aquifer in the study are mainly consist with sodium/ potassium chloride type. Electrical conductivity varied in the range of 500 to 1900 micro Siemens/ cm. The study has been identified relationship between groundwater Quality and Consumption with relation to availability of water resource maintenance, hygienic condition, disease, subsidies for rehabilitation and Tsunami affection have been the main point of interest. Pumping tests have been done to understand groundwater dynamics in order to quantify water potential resources in the area. This study highlight the correlation between the bad quality of groundwater and their given up by habitants. Moreover it shows that even at the sea shore, the area of Madiha East has medium quality resource, enough for common purpose as garden irrigation, washing, bathing which in a context of water pressure should be used. This study shows also that water potential resource is also sufficient for such use. Groundwater quality in the tsunami affected wells could be increase water quality under WHO standards and allow its use, increasing drinking water saving.

Reference

Weerasinghe, K. D. N., Piyadasa, R. U. K., Kariyawasam, I. S., Pushpitha, N. P. G., Vithana, S. B., Kumara, D. S. E., Wijayawardhana, L. M. J. R., Schwinn, P., Gruber, M., Maier, M. & Maier, D., 2006: Well water quality in tsunami affected areas and their purification by reverse osmosis – a case study, after the tsunami rehabilitation in Sri Lanka. – Mosaic Books, New Delhi, India.

Panaboke, C. R., 1996: Our Engineering Technology. – The Open University Review of Engineering Technology, Vol. 2, Number 1, pp. 17–19.

Panabokke, C. R. & Perera, A. P. G. R. L., 2005: Groundwater Resources of Sri Lanka. – Water Resources Board, Colombo. Sri Lanka.

www.borntraeger-cramer.de
2012/0264/$ 0.25

Proceedings "Hydrogeology of Arid Environments", p. 265
Stuttgart, March 2012

Poster Abstract

Long-Term Environmental Impact Assessment of a Managed Aquifer Recharge Project in the Northern Gaza Strip by Using Mathematical Modeling Technique

Mohammad Azizur Rahman[1], Mohammad Salah Uddin[1], Muath Abu Sadaa[2], Bernd Rusteberg[1]

[1] Geosciences Center, George-August Universität Göttingen, D 37077, Göttingen, Germany, email: mrahman1@gwdg.de, sala_135@yahoo.com, Bernd.Rusteberg@geo.uni-goettingen.de
[2] Palestinian Hydrology Group, Al- Nadha Street , Ramallah, Palestine, email: muath@phg.org

Key words: managed aquifer recharge, impact assessment, mathematical modeling, Gaza strip

The Beit Lahia Wastewater Treatment Plant (BLWWTP), located at the Northern Gaza Strip, is being dysfunctional for some time now and creating severe problems for the public health and the environment. The poor management of the BLWWTP and the incomplete treatment and improper disposal of its effluent are causing serious environmental and socio-economic impacts for people of Gaza. In response to this water resources problem, a three-phase 20-year project, involving the construction of a new WWTP further to the south of the northern Gaza strip, near the Israel boarder and a pipeline connecting the effluent to the new proposed infiltration basin is in progress. The new WWTP will involve Managed Aquifer Recharge (MAR) of treated effluents by using infiltration pond technique. Based on the water resources problem analysis and considering the water resources management plans for the years 2005 to 2025, four MAR strategies were established in this study. The water management strategies consider 3 phases in terms of wastewater resources development at the case study area. Strategy no.1 (Sc-1) represents the so-called "Do-Nothing-Approach". Strategy no. 2 (Sc-2) considers the diversion of the water from the BLWWTP to the newly constructed infiltration basin, which is located close to the foreseen position of the new North Gaza Wastewater Treatment Plant (NGWWTP) at the Israelian border. Strategy no.3 (Sc-3) considers additional wastewater treatment in the NGWWTP before recharge to the aquifer. Sc-3 considers improved effluent quality. Strategy no.4 (Sc-4) considers the infiltration of the extra volume of treated effluents, considering the extension of the NGWWTP capacity. The water management strategies Sc-2, Sc-3, and Sc-4 rely on MAR.

The present study performed a long term (year 2005 to year 2040) environmental impact assessment of the above mentioned four strategies by applying the three-dimensional groundwater flow and transport model 'Visual MODFLOW'. Change of ground water level and chloride and nitrate concentration within the aquifer have been considered for the impact assessment.

All strategies except Sc-1 show that MAR will definitely halt the declining trend of groundwater level and store water in the aquifer for further use. Implementation of Sc-4 will offer storage in the aquifer with a maximum value of 23 MCM per year after the full implementation of NGWWTP, phase–III (year 2025). Fresh water flow from Israel will be reduced due to project implementation. The results show that infiltrated water will improve significantly the groundwater quality, in terms of chloride, in all MAR implementation strategies. Regarding nitrate, the relatively high concentration in the partially treated water will have a negative impact on aquifer water quality. Thus fully treated water is desired. The environmental impact analysis of the developed water resources planning and management strategies clearly shows that managed aquifer recharge by means of infiltration ponds with proper waste water treatment is a viable response to the increasing water resources problems of the region.

www.borntraeger-cramer.de
2012/0265/$ 0.25

Proceedings "Hydrogeology of Arid Environments", p. 266
Stuttgart, March 2012

Poster Abstract

Identification of Multiple Nitrate Sources in Selected Saudi Arabian Aquifers Using a Multi-Isotope Approach

Mustefa Yasin Reshid[1], Nils Michelsen[1], Christoph Schüth[1], Susanne Stadler[2], Randolf Rausch[3], Stephan Weise[4], Mohammed Al-Saud[5]

[1] TU Darmstadt, Institute for Applied Geosciences, Schnittspahnstraße 9, 64287 Darmstadt, Germany, email: reshid@geo.tu-darmstadt.de, michelsen@geo.tu-darmstadt.de, schueth@geo.tu-darmstadt.de
[2] Federal Institute for Geosciences and Natural Resources (BGR), Stilleweg 2, 30655 Hannover, Germany, email:Susanne.Stadler@bgr.de
[3] GIZ IS, P.O. Box 2730, Riyadh 11461, Kingdom of Saudi Arabia, email: randof.rausch@gizdco.com
[4] Helmholtz-Centre for Environmental Research (UFZ), Dept. Isotopenhydrologie, Halle, Germany, email: stephan.weise@ufz.de
[5] Ministry of Water & Electricity (MoWE), Saud Mall Center, Riyadh 11233, Kingdom of Saudi Arabia, email:malsaud@mowe.gov.sa

Key words: groundwater, nitrate, Saudi Arabia, ^{15}N

Nitrate is one of the principal contaminants of drinking water resources (Widory et al. 2004). Base-level concentrations in the order of a few mg/L generally occur naturally in groundwater. Elevated concentrations, however, are mostly assumed to be caused by anthropogenic impact, e.g. fertilizer application or waste water infiltration. Yet, there are also cases in which human activity can be ruled out as a relevant nitrate source (Stadler et al. 2008, Stadler et al., accepted). Partly, this seems to apply to Saudi Arabian groundwaters as well. In the course of several studies conducted in the Kingdom, nitrate was detected in groundwaters for which non-anthropogenic origins are likely, particularly in case of deep wells or remote sampling locations (e.g. Alabdula'aly et al. 2010).

Aiming at an identification of the corresponding nitrate sources, groundwater samples from selected aquifers (mainly Biyadh, Wasia, Aruma) were analysed for their hydrochemical composition and the isotopic signature of the encountered nitrate ($\delta^{15}N\text{-}NO_3$, $\delta^{18}O\text{-}NO_3$). Moreover, the parameters $\delta^{18}O$, δD, ^{14}C-DIC, and $\delta^{13}C$-DIC were determined.

Due to the different settings of the sampled wells (agricultural, residential, and remote areas) the isotopic signatures of the nitrate show a relatively broad spectrum of 4.4–11.5 ‰ AIR ($\delta^{15}N\text{-}NO_3$) and 7.2–23.8 ‰ VSMOW ($\delta^{18}O\text{-}NO_3$). Based on the obtained values and considering other isotopic and hydrochemical data, it can be concluded that within the study area both anthropogenic and natural nitrate sources occur and deteriorate the water quality. Some samples show isotopic signatures typical for fertilizer-derived nitrate and some seem to be affected by infiltrating waste water. As for the remaining samples, the isotopic fingerprints suggest that the encountered nitrate originates to a large extent from nitrogen that has taken part in the soil-N-cycle. Atmospheric deposition is suspected to contribute to the nitrogen input. However, evidence for this is lacking.

References

Alabdula'aly, A. I., Al-Rehaili, A. M., Al-Zarah, A. I. & Khan, M. A., 2010: Assessment of nitrate concentration in groundwater in Saudi Arabia. – Environ. Monit. Assess. **161**: 1–9.

Stadler, S., Osenbrück, K., Knöller, K., Suckow, A., Sültenfuß, J., Oster, H., Himmelsbach, T. & Hötzl, H., 2008: Understanding the origin and fate of nitrate in groundwater of semi-arid environments. – J. Arid Environ. **72**: 1830–1842.

Stadler, S., Talma, S., Tredoux, G. & Wrabel, J., accepted: Identification of sources and infiltration regimes of nitrate in the semi-arid Kalahari and implications for groundwater management. – Water SA.

Widory, D., Kloppmann, W., Chery, L., Bonnin, J., Rochdi, H. & Guinamant, J. L., 2004: Nitrate in groundwater: an isotopic multi-tracer approach. – J. Cont. Hydrol. **72**: 165–188.

www.borntraeger-cramer.de
2012/0266/$ 0.25

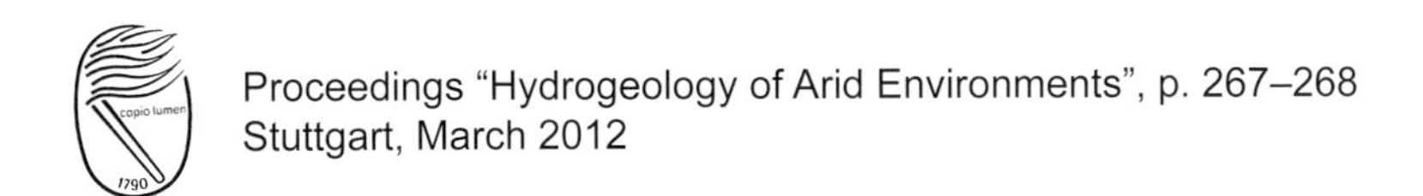

Proceedings "Hydrogeology of Arid Environments", p. 267–268
Stuttgart, March 2012

Poster Abstract

Sources of Nitrate and Variability of its Concentrations in Arid Basin Groundwaters; Implications for Flow Processes and Water Quality

Wendy Marie Robertson, John M. Sharp, Jr.

The University of Texas at Austin Jackson School of Geosciences, wendyr@mail.utexas.edu and jmsharp@jsg.utexas.edu

Key words: nitrate, arid basins, groundwater, rapid recharge

Since the 1940's, nitrate concentrations in four basins of the Trans-Pecos region of West Texas have been changing on relatively short (decadal) time scales. We have documented variability in nitrate concentrations of >1 mg/L (as nitrate) in 77 of 93 wells (82.8%) and increases in nitrate concentrations from 1 to 40+ mg/L (as nitrate) in 65 of 93 wells (69.8%) examined in this study. The observation of relatively rapid changes in nitrate concentrations is at odds with the conceptual models of flow and estimated groundwater ages in these basin systems. The current models assert that 1) very little recharge occurs in these systems (<0.5 cm/yr) (Beach et al. 2004, Beach et al. 2008) and 2) no recharge is occurring on the basin floors. Previous estimates of groundwater age are around 10,000 years old (Darling et al. 1995, Darling et al. 1998). The temporal variability in nitrate concentrations we have documented can not be reconciled with these assumptions because of the decadal times scales on which they are observed and the observation that all of the likely sources of nitrate to the basin groundwater exist only on the basin floors and not in the adjacent mountain blocks. Additionally, the presence and change in concentration over time raises questions about the sources and mechanisms of transport of nitrate to the basin groundwater. During the summer of 2011, we sampled 80 wells for nitrate concentrations and N and O isotopes of nitrate dissolved in the groundwater. Five wells were also sampled for CFCs and SF6 to determine if young (<70 year old) water was present in the basins. Evidence of young water was found in all of the wells sampled. We infer that the temporal changes in nitrate concentration and isotopic compositions of the nitrate sampled as well as the presence of CFCs in basing groundwater indicate that relatively rapid recharge is occurring on the floors of these basins as a result of both anthropogenic and natural processes. Irrigation return flow is likely occurring beneath fields and orchards in the basins and that large magnitude, short duration rainfall events are also recharging the deep basin groundwater. These mechanisms are both spatially and temporally variable; they are also important to the water budget and water quality in the basins. We also infer from reports of borehole cameras, the geologic setting, and from time-variability of salinity in irrigation wells that that channels of high permeability occur in basin sediments. Current models estimating sustainable yield in these systems do not consider basin floor recharge and oversimplify hydraulic characteristics. The occurrence of rapid recharge on basin floors is significant because of potential deterioration of groundwater quality, especially in agricultural regions, and for estimation of sustainable yields. These processes are likely occurring in similar systems around the globe and in light of this, we assert that the fundamental assumptions about groundwater flow in arid systems may need to be reexamined.

References

Beach, J. A., Ashworth, J. B., Finch, S. T., Chastain-Howley, A., Calhoun, K., Urbancyzk, K. M., Sharp, J. M., & Olson, J., 2004: Final Report – Groundwater Availability Model for the Igneous and Parts of the West Texas Bolsons (Wild Horse Flat, Michigan Flat, Ryan Flat,

www.borntraeger-cramer.de
2012/0267/$ 0.50

and Lobo Flat) Aquifers, June 2004. – Texas Water Development Board Report, 407 p.

Beach, J. A., Symank, L., Huang, Y., Ashworth, J. B., Davidson, T., Collins, E. W., Hibbs, B. J., Darling, B. K., Urbanczyk, K., Calhoun, K. & Finch, S., 2008: Final Report – Groundwater Availability Model for the West Texas Bolsons (Red Light Draw, Green River Valley, and Eagle Flat) Aquifer in Texas, November 2008. – Texas Water Development Board Report, 320 ps.

Darling, B. K., Hibbs, B. J., Dutton, A. R. & Sharp, J. M., 1995: Isotope Hydrology of the Eagle Mountains Area, Hudspeth County, Texas: Implications for Development of Ground-water Resources. – In: Proceedings, Water Resources At Risk, Denver, Colorado, American Institute of Hydrology, p. SL-12-SL-23.

Darling, B. K., Hibbs, B. J. & Sharp, J. M., 1998: Environmental Isotopes as Indicators of the Residence Time of Ground Waters in the Eagle Flat and Red Light Draw Basins of Trans-Pecos, Texas, West Texas. – Geological Society Publication #98-15, p. 259–270.

Proceedings "Hydrogeology of Arid Environments", p. 269
Stuttgart, March 2012

Poster Abstract

Accessing the Genesis of Lithium-Rich Salt Pan Brines by the Study of Chemical and Isotopic Compositions

Nadja Schmidt[1], Broder Merkel[2]

[1] TU Bergakademie Freiberg, Institute of Geology, email: Nadja.Schmidt@geo.tu-freiberg.de
[2] TU Bergakademie Freiberg, Institute of Geology, email: merkel@geo.tu-freiberg.de

Key words: salt pan, lithium, hydrochemistry, evaporation

Salt pans, which form under specific geologic and climatic conditions, are composed of evaporite layers, which can reach several 100 meters in thickness. Some of them are characterized by the existence of pores and cavities in the sediment, which are filled with a highly saline brine significantly enriched in lithium. The formation of salt pans with such characteristics is controlled by various processes. These include the leaching of surrounding rocks and older underlying evaporites as well as the rise of volcanic fluids and in most cases arid or semi arid climate (Risacher & Fritz 1991).

The focus of this study was to investigate enrichment mechanisms leading to significant amounts of lithium in the pore solution of salt lake sediments. For this purpose, drillings were performed with depths of 2 to 12 m at 11 different locations on the Salar de Uyuni, Bolivia, which is located on the Altiplano between the Andean cordilleras, and the salt pan Tuz Gölü in central Anatolia, Turkey. The Salar de Uyuni is considered as the largest in the world, having a size of 10,000 km^2. Bore holes were completed as monitoring wells and served for depth-dependent sampling of brines during the beginning until the end of the dry season in 2009–2011. Water samples were analyzed for main ions and trace elements by IC and ICP-MS. Additionally, ratios of stable isotopes for H, O and SO_4^{2-} were determined. Characterized by a mineralization of more than 300 mg/l, which is in the range of halite saturation, the brines show high concentrations of lithium, boron and bromine. It is unlikely, that lithium enrichment is solely the result from the evaporative concentration of the inflowing water from streams, rivers and rainwater. Other mechanisms, such as the influence of mud layers containing clay minerals, which occur between the halite layers and especially in the delta area of the main inflow, must as well be discussed. Seasonal fluctuations in the brine composition support the assumption of an annual recycling process and the influence of inflowing subsurface waters and rain water down to depths of at least 10 m. A general trend is observed by decreasing concentrations of sodium and calcium with increasing content of total dissolved solids caused by the precipitation of halite and gypsum. In contrast, conservative compounds as lithium and bromine stay in solution until a high grade of brine concentration which becomes apparent by rising contents with increasing TDS. The study of stable isotopes in the brine shows a linear correlation between ^2H and ^{18}O, the points are plotting on the local evaporation line.

A drilling and sampling campaign was also performed at the salt pan Tuz Gölü in central Anatolia, Turkey. It is located in a 16,000 km^2 large basin, which is surrounded by six major highlands mainly composed of metamorphic rocks and ignimbrites (Camur & Mutlu 1996). Brine samples from boreholes in the north eastern part of the salt lake show a trend with depth, as the concentration of lithium and other ions increases with depths. Continuing work includes sampling of brines on different parts of the salt lake to investigate the composition of the interstitial solution on a regional scale.

References

Camur, M. Z. & Mutlu, H., 1996: Major-ion geochemistry and mineralogy of the Salt Lake (Tuz Gölü) basin, Turkey. – Chem. Geol. **127**: 313–329.

Risacher, F. & Fritz, B., 1991: Quaternary geochemical evolution of the salars of Uyuni and Coipasa, Central Altiplano, Bolivia. – Chem. Geol. **90**: 211–231.

www.borntraeger-cramer.de
2012/0269/$ 0.25

Proceedings "Hydrogeology of Arid Environments", p. 270
Stuttgart, March 2012

Poster Abstract

Groundwater Recharge Estimation Using the Hydrological Model J2000g of the Zarqa River Catchment, NW-Jordan

Stephan Schulz[1], Tino Rödiger[1], Christian Siebert[1], Peter Krause[2]

[1] Helmholtz-Centre for Environmental Research – UFZ, email: stephan.schulz@ufz.de, tino.roediger@ufz.de, christian.siebert@ufz.de
[2] Friedrich-Schiller-Universität Jena, email: p.krause@uni-jena.de

Key words: groundwater recharge, J2000g, hydrological model, Zarqa River

Pollution and overexploitation of the groundwater resources are serious problems in the Zarqa River catchment, Jordan. This makes it necessary to quantify the amount and the spatial distribution of groundwater recharge.

For this purpose, the hydrological model J2000g is used. The model is run for 30 years (01.07.1977 to 30.06.2007) with a temporal resolution of one day. For the model setup a land use classification using Landsat 7 ETM+ images is carried out. Hydrological processes are strongly influenced by soil physical properties. Due to the fact that these properties vary along toposequences (Ziadat et al. 2010), a new soil map with a topological approach is designed. For this purpose main soil classes (according to USDA soil taxonomy) are separated into upslope, colluvial and alluvial (Wadi channels) areas. The calibration of the hydrological model J2000g is realized by matching the simulated discharge with the observed discharge at gauging station New Jarash Bridge. The first 15 years of the time series are used for calibration while the second 15 years are used for validation. Resulting coefficient of determination R-square of monthly averaged values is 0.71 for the calibration and 0.65 for the validation. Relative percentage volume error is 0.02% for the calibration and -19% for the validation. Despite of corrections, this is probably be caused by additional water from domestic sewer pipes in the observed discharge. Estimated groundwater recharge of the Zarqa River catchment is 106 Mio. m^3 (21 mm) per year. This is 17% higher than assumed by the authorities (Talozi 2007, MWI 2000). Additionally, a high sensitivity of groundwater recharge to precipitation becomes apparent. The higher the amount of precipitation in a rainy season, the larger the share of precipitation to groundwater recharges. The average share of precipitation to groundwater recharge is 11%. For direct validation of groundwater recharge two independent methods are applied. The first method is the analysis of the stable isotopes ^{18}O and ^{2}H and the second one is the Chloride Mass Balance. Both approaches show slightly higher recharge rates than the model J2000g. Recharge rates determined by isotopic investigations are 24% higher and recharge rates determined by Chloride Mass Balance are 19% higher than the modeled results. Basis for these methods are hydrochemical parameters from spring samples. Their results are related to the corresponding above ground spring catchments.

References

Ministry of Water and Irrigation (MWI), 2000: Outline Hydrogeology of the Amman-Zarqa Basin. – In: WRPS Hydrogeology Report.

Talozi, S., 2007: Water and Security in Jordan. – In: Lipchin, C., Pallant, E., Saranga, D. & Amster, A. (eds.): Integrated Water Resource Management and Security in the Middle East. – Springer Netherlands, pp. 73–98.

Ziadat, F., Taimeh, A. & Hattar, B., 2010: Variation of Soil Physical Properties and Moisture Content Along Toposequences in the Arid to Semiarid. – Arid Land Research and Management **24**: 81–97.

www.borntraeger-cramer.de
2012/0270/$ 0.25

Proceedings "Hydrogeology of Arid Environments", p. 271
Stuttgart, March 2012

Poster Abstract

Groundwater Contamination by Agricultural Activities in Arid Environment: Evidence from Nitrogen and Oxygen Isotopic Composition, Arava Valley, Israel

N. Shalev[1,2], I. Gavrieli[1], B. Lazar[2], A. Burg[1*]

[1] Geological Survey of Israel, 30 Malkhe Israel St., Jerusalem, 95501, Israel
[2] The Institute of Earth Science, The Hebrew University, Jerusalem, 91904, Israel
* burg@gsi.gov.il

Key words: Arava, contamination, nitrogen, isotopes

The Negev desert, located in the southern region of Israel, is part of the arid belt of the northern hemisphere. Its eastern part is an elongated north to south low topography valley, named the Arava valley, which is shared by Israel and Jordan. It is characterized by moderate winters with very low precipitation (25 to 50 mm/yr), followed by extremely hot summers. The valley is part of the Dead Sea transform and serves as a drainage basin for occasional flash floods from both east and west. Some of these flood waters recharge the shallow unconsolidated alluvial sediments in the valley. Moreover, water from regional fossil aquifers that exist on both sides of the valley leak into the Arava valley sediments. Because of its arid climate and the extremely slow flushing of the upper soil, as well as the existence of some ancient and deeper brines, all waters in the valley are brackish or even saline.

The Israeli part of the Arava valley is sparsely populated by small rural settlements, which are based on modern 'water saving' agriculture technology. The water supply for the agriculture demands are fulfilled mainly by wells drilled into the Arava valley sediments. Our current research is focused on the groundwater contamination processes following the development of this agriculture in the Central Arava valley during the last decades.

The main water resources in the research area are three shallow aquifers in the valley, namely the Hazeva Gr., the Arava Fm. and the younger alluvium which exists mainly along few dry wadis. These aquifers are experiencing ongoing nitrate contamination due to intensive agriculture over the last decades. The most likely contamination sources are synthetic NO_3^- fertilizers, which are applied to the cultivated fields. However $\delta^{15}N_{NO3}$ and $\delta^{18}O_{NO3}$ values in the nitrate contaminated groundwater cannot be explained solely by simple mixing between natural occurring nitrate and the synthetic NO_3^- fertilizers. Other possible sources of anthropogenic nitrogen are synthetic NH_4^+ fertilizers and manure, which is used as organic fertilizer in the cultivated fields.

A mixing/reaction model was implemented in order to explain the measured $\delta^{15}N_{NO3}$ and $\delta^{18}O_{NO3}$ values in the nitrate contaminated groundwater. The model considered the following nitrogen sources: 1. natural nitrate in non contaminated groundwater; 2. synthetic fertilizers (NO_3^- and NH_4^+ fertilizers); 3. manure. The following biogeochemical processes that fractionate the stable isotopes of nitrogen and oxygen were considered: 1. ammonia volatilization from manure or NH_4^+ synthetic fertilizers which leads to a large ^{15}N enrichment in the remaining nitrogen; 2. evaporation of the irrigation water which may influence the oxygen isotopic composition of the nitrification products; 3. nitrate or ammonium uptake by plants which results in a small isotopic fractionation. Complete nitrification of all the NH_4^+ based fertilizers either in the soil or in the unsaturated zone was assumed, following by no isotopic fractionation.

The calculations suggest that the nitrate in most of the contaminated wells was derived from equal proportions of synthetic NO_3^- fertilizers, synthetic NH_4^+ fertilizers and manure.

The alluvial aquifer showed different isotopic values (higher $\delta^{18}O_{NO3}$ values), suggest that it derived its N from synthetic fertilizers only.

www.borntraeger-cramer.de
2012/0271/$ 0.25

Proceedings "Hydrogeology of Arid Environments", p. 272
Stuttgart, March 2012

Poster Abstract

Insights in the Hydrodynamics of a Lithium-Rich Brine Reservoir in an Arid Region in South-Western Bolivia: Exploration Study at the Salar De Uyuni

Robert Sieland[1], Broder Merkel[1]

[1] TU Bergakademie Freiberg, Institute of Geology, Chair of Hydrogeology, Gustav-Zeuner-Str. 12, 09599 Freiberg, email: Robert.Sieland@geo.tu-freiberg.de

Key words: Salar de Uyuni, porosity, pumping test, X-ray tomography

The Salar de Uyuni is a giant salt flat (10,000 km^2, 3653 m a.s.l.) occupying the southern Altiplano of the Bolivian Andes. A sequence of alternating salt and lacustrine sediments was formed during extended wet and dry periods in Quaternary times. Currently, the area is influenced by arid climatic conditions with annual rain fall around 150 mm (Fritz et al. 2004) and potential evapotranspiration in the order of 1600 mm/a (Carpio 2007).

The upper salt crust of the salt flat is mainly composed of porous halite. The pore space of the salt crust is filled with a high-mineralized solution (brine) rich in lithium, magnesium, boron and other elements (Risacher & Fritz 1991). Thus, the Salar de Uyuni is an important geo-resource with high economical value. However, hydrogeological information concerning permeability and porosity of the salt as well as brine flow behaviour are still lacking.

Comprehensive field studies in 2009 and 2010 dealt with the exploration of the statigraphy and main hydraulic parameters of the salt flat. Core drillings between 2 and 13 m depths were conducted at 11 different locations over the salt flat. The maximum thickness of the first halite layer reached 11 m in the central-eastern part of the Salar de Uyuni. Pumping tests were performed at 6 locations. The evaluation of the hydraulic tests considering certain density and viscosity of the brine revealed very high permeabilities in the magnitude of 10^{-10} m^2. Field observations also showed the existence of preferential flow paths and cavities of different sizes (several cm up to meter) strongly influencing the movement of brine in the salt crust.

The porosity of the salt was determined by means of X-ray tomography on the basis of core samples. Also flow-through experiments (permeameter tests) were done using an organic fluid to avoid dissolution of salt during the experiments. Resulting data give hints on the pore structure and therewith on the effective porosity and permeability of the salt cores.

In combination with permeability values from the pumping tests and in consideration of the physical properties of the brine (density, viscosity), new insights into the hydraulic properties of the salt crust were gained. Based on spatial distribution of porosity and thickness of the upper salt crust from the new drill cores the total volume of the lithium-rich brine in this part of the Salar de Uyuni was re-estimated.

References

Carpio, J. M., 2007: Agua y recurso hídrico en el Sudoeste de Potosí. La Paz, Bolivia, FOBOMADE - Foro Boliviano sobre Medio Ambiente y Desarrollo.

Fritz, S. C., Baker, P. A., Lowenstein, T. K., Seltzer, G. O., Rigsby, C. A., Dwyer, G. S., Tapia, P. M., Arnold, K. K., Ku, T.-L. & Luo, S., 2004: Hydrologic variation during the last 170,000 years in the southern hemisphere tropics of South America. – Quat. Res. **61**: 95–104.

Risacher, F. & Fritz, B., 1991: Quaternary geochemical evolution of the salars of Uyuni and Coipasa, Central Altiplano, Bolivia. – Chem. Geol. **90**: 211–231.

www.borntraeger-cramer.de
2012/0272/$ 0.25

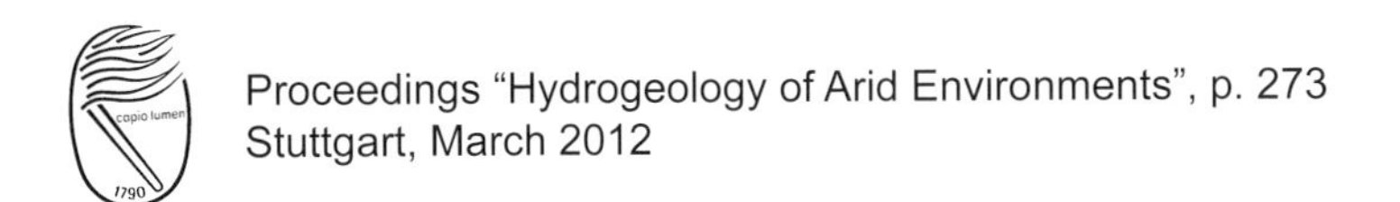

Proceedings "Hydrogeology of Arid Environments", p. 273
Stuttgart, March 2012

Poster Abstract

Using WEAP – MODFLOW to Assess Water Resource Management Options in the Azraq Basin in Jordan

Hadeel Smadi[1], Ali Hayajneh[2], Ali Subah[3], Oliver Priestley-Leach[4], Johannes Stork[5], Markus Huber[6], Jobst Maßmann[7]

[1] Ministry of Water & Irrigation, email: hadeel_smadi@mwi.gov.jo, Jordan
[2] Ministry of Water & Irrigation, email: hayajneh78@gmail.com, Amman, Jordan
[3] Ministry of Water & Irrigation, email: Ali_Subah@mwi.gov.jo, Amman, Jordan
[4] GIZ Amman, email: oliver.priestley-leach@giz.de, Amman, Jordan
[5] Ministry of Water & Irrigation/CIM, email: Johannes_stork@gmx.de, Amman, Jordan
[6] Geo tools, email: markus.huber@geo-tools.de, Munich, Germany
[7] Federal Institute for Geosciences and Natural Resources (BGR), Hannover, Germany

Key words: National Water Master Plan, WEAP modeling, national water strategy

Jordan is among the water poorest countries in the world. Currently annual demand exceeds sustainable supply by about 150 MCM. Of the 12 groundwater basins, 6 are being over-extracted, 4 are balanced with respect to abstraction and 2 are under-exploited. The problem will get worse as population expands, industrial demand increases and climate change reduces the renewable supply.

The Ministry of Water and Irrigation's National Water Master Planning Directorate is in the process of revising and updating the National Water Master Plan with a 20 year time frame. The plan will to a large extent be based on the outcome of modelled supply and demand scenarios using WEAP and MODFLOW models. One of the 11 surface water basins being modelled is the Azraq basin.

The Azraq water basin is situated in the eastern region of Jordan and is recharged mainly by water from aquifers under Syria. There are three water bearing layers – two limestone and one basalt. Approximately 50 MCM of water per year are pumped. The groundwater resources are suffering from severe depletion due to over-abstraction (legal and illegal) for irrigation and for drinking water supply for the capital city of Amman. The groundwater level is dropping by up to 2 m per year leading to an increase in salinisation and the depletion of the Azraq oasis, once an important site for migrating birds.

Separate WEAP and MODFLOW models have been linked in order to provide a dynamic overview of groundwater resources and assess the impact of different management and demand scenarios. The modelling has explored options for groundwater recharge to restore and protect the Azraq oasis, agricultural policy to restrict irrigated area, change crop types or land use and restrict legal extraction, legislation to reduce illegal abstraction and for options to reduce abstraction for drinking water through the provision of alternative supplies for Amman. Typically results show changes in unmet demand and groundwater levels over time and space.

The scenarios provide the basis for the formulation of strategies to be adopted in the new national water strategy and subsequent action plan. For example the modelling has shown that changes to agricultural policies by restricting irrigated areas is more effective in reducing groundwater drawdown than increasing irrigation efficiency.

The presentation will give an overview of the main results and conclusions of the scenario modelling exercise and how they have been translated into the National Water Master Plan.

www.borntraeger-cramer.de
2012/0273/$ 0.25

Proceedings "Hydrogeology of Arid Environments", p. 274
Stuttgart, March 2012

Poster Abstract

Freshwater Lens Investigations (FLIN): Visualizing Age Stratification and Internal Dynamics on a Laboratory Scale

Leonard Stoeckl[1], **Georg Houben**[1]

[1] Bundesanstalt für Geowissenschaften und Rohstoffe, Germany, email: leonard.stoeckl@bgr.de

Key words: Fresh water lens, Ghyben-Herzberg lens, sand-box model, age stratification

So far the internal dynamics and age stratification of freshwater lenses have not been considered sufficiently. Especially in arid and semi-arid regions a sustainable management of such lenses is of major importance due to the general scarcity of water there. Because of low groundwater recharge in those regions, an overexploitation will quickly lead to salt water intrusion (up-coning), damaging a lens in the long term. We have performed a series of experiments on the laboratory scale to investigate and visualize processes of freshwater lenses under different external conditions and the influence of anthropogenic interference.

A transparent box of 200 cm in length, 50 cm in height and 5 cm in thickness was used for physical modeling. A cross section of an island was simulated by filling coarse sand into the middle of the box, forming a homogeneous sand cone. Degassed salt water with a density of 1,025 kg l^{-1} was then injected, simulating ocean water. By using peristaltic pumps, individual drips of fresh water were installed, recharging the sand cone above the salt water level. Adding artificial tracer dyes to recharge water allowed us to visualize flow paths, residence times and dispersion processes in the lens. With this setup a freshwater lens developed. By video recordings a detailed analysis of the experiments was possible.

We used a time-dependent application of different tracers like uranine and eosine blue and yellow, representing different recharge events. Lens dynamics are driven by density contrasts of saline and fresh water only. At steady state conditions with constant recharge rates, an image of the age stratification became visible. All layers remain in contact with the discharge zone at all times. Therefore, the oldest layers near the interface of saline and fresh water were found to be the thinnest due to a constant outflow into the ocean. This has some interesting implications for the interpretation of groundwater ages, e.g. by isotopic analyses. The position of the screen of a sampled well can determine the percentage of "old" water in the sample. A short screen at greater depth may contain only old water while a longer screen at lesser depth may contain only "young" water.

To upscale results from physical experiments, numerical models are used. Different model scenarios are compared using Feflow and SC_25. It is intended to generate benchmark experiments which will be used to validate different models. Vulnerability and resilience of fresh water lenses are important issues when it comes to climate change. The effects of rising sea level, increasing temperature, and changing precipitation patterns on the dynamics of freshwater lenses are not fully understood and will be investigated in the FLIN project.

www.borntraeger-cramer.de
2012/0274/$ 0.25

Proceedings "Hydrogeology of Arid Environments", p. 275
Stuttgart, March 2012

Poster Abstract

Comparison of Various Numerical Schemes for Simulating Fluid Flow in Variably Saturated Porous Media

Heejun Suk[1], Kang-gun Lee[2]

[1] Korea Institute of Geoscience and Mineral Resources, email: sxh60@kigam.re.kr
[2] Seoul National University, email: kklee@snu.ac.kr

Key words: Kirchhoff transformation, unsaturated flow, nonlinear Richards equation

Prediction of fluid movement in unsaturated soil is an important problem in many branches if science and engineering. In virtually all studies of the unsaturated zone, the fluid motion is assumed to obey the classical nonlinear Richards equation. The nonlinear Richards equation in variably saturated porous media needs to be solved numerically with constitutive relations, heterogeneities, irregular geometries and complex boundary conditions. Many numerical solution strategies have been suggested to efficiently solve the equation such as: Picard or Newton-Raphson iterative techniques together with finite difference or finite element approximations, primary variable-switching between pressure head and water content, upstream weighting of relative permeability, and transformation techniques (Celia et al. 1990, Kirkland et al. 1992). The difficulty in solving the nonlinear Richards equation originates from its hyperbolic characteristic, despite its parabolic form, depending on the degree of saturation in the solution domain. The purpose of this study is to demonstrate a full linearization of the Richards equation with Gardner constitutive relations by using Kirchhoff integral transformation in a transient variably saturated flow and show that the integral transformation approach is not only more computationally efficient but also more robustness than other existing numerical methods: h-based model, Celia model (Celia et al. 1990), Kirkland model (Kirkland et al. 1992), and 3DFEMWATER.

Acknowledgment

This study was supported by Korea Ministry of Environment as The GAIA Project (Grant no. 173-092-009).

References

Celia, M. A., Bouloutas, E. T. & Zarba, R. L., 1990: A general mass-conservative numerical solution for the unsaturated flow equation. – Water Resour. Res. **26**: 1483–1496.

Kirkland, M. R., Hills, R. G. & Wierenga, P. J., 1992: Algorithms for solving Richards' equation for variably saturated soils. – Water Resour. Res. **28**: 2049–2058.

www.borntraeger-cramer.de
2012/0275/$ 0.25

Proceedings "Hydrogeology of Arid Environments", p. 276
Stuttgart, March 2012

Poster Abstract

Regional Droughts and Irrigation Water Demand under Current and Future Climates in the Jordan River Region

Tobias Törnros, Lucas Menzel

Institute of Geography, Heidelberg University, Im Neuenheimer Feld 348, 69120 Heidelberg, Germany,
email: tobias.toernros@geog.uni-heidelberg.de; lucas.menzel@geog.uni-heidelberg.de

Key words: Drought; Hydrological modelling; Jordan River region

The arid to semi-arid Jordan River region counts to one of the most water scarce regions in the world. During droughts, the pressure on the water resources becomes even more evident. The Standardized Precipitation Index (SPI) was applied in the region to assess the characteristics of current (1961–1990) and future (2031–2060) agricultural droughts. Thereafter, the regional irrigation water demand was simulated with a hydrological model.

The drought index SPI was applied on spatially interpolated observed precipitation (1x1 km). The index can be computed on several time-scales to assess different kinds of droughts (McKee et al. 1993). As an example, the 3-month SPI sums the precipitation from the last three months during the calculation of the monthly SPI-value. To assess the performance of the drought index and to identify the most appropriate SPI time-scale, correlation analyses were conducted between multiple timescales of SPI and the Normalized Difference Vegetation Index (NDVI) obtained from NOAA´s Advanced Very High Resolution Radiometer (AVHHR). A spatiotemporal correlation was found between the two variables and the overlapping years 1982-2001. In overall, the 6-month SPI best explained the interannual variation of NDVI.

If a drought index performs well for current conditions, its suitability can also be assumed for future conditions. To investigate the characteristics of current and future droughts, the 6-month SPI was applied on three climate change projections based on the IPCC emission scenario A1B. When comparing the future time period 2031–2060 with the reference period 1961–1990, it was shown that the mean drought length is expected to be prolonged with about 1.5 months. The future droughts are further expected to become more severe when a shift from moderate, to severe and extreme droughts is taking place (Törnros & Menzel 2011).

A model approach was used to further assess the hydrological impacts of the increased drought severity. The physically based hydrological model TRAIN has among others been applied to assess the hydrological impacts of land use change within the Jordan River region (Menzel et al. 2009). Within this study, drought periods were identified with SPI. Thereafter, TRAIN was applied to simulate the annual irrigation water demand (IWD) during average conditions as well as during the longest current and future drought. For the years 1961–1990, the annual IWD was 80 mm. This corresponds to about 1810 million m^3 of water for the whole study region. The annual IWD during the longest current and future drought was 122 mm (2770 million m^3) and 174 mm (3950 million m^3), respectively. When comparing to the average conditions, this implies to an increased demand of 53 and 118%, respectively.

References

McKee, T., Nolan, J. & Kleist, J., 1993: The relationship of drought frequency and duration to time scales. – In: Proceedings of the 8th Conference of Applied Climatology, 17-22 January, Anaheim, CA. American Meteorological Society, Boston, MA. 179–184.

Menzel, L., Koch, J., Onigkeit, J. & Schaldach, R., 2009: Modelling the effects of land-use and land-cover change on water availability in the Jordan River region. – Adv. Geosci. **21:** 73–80.

Törnros, T. & Menzel, L., 2011: Applying a regional drought index in the Jordan River region. – Manuscript submitted for publication.

Proceedings “Hydrogeology of Arid Environments”, p. 277
Stuttgart, March 2012

Poster Abstract

Hydrogeological and Hydrogeochemical Investigations in the Copiapó Valley, Región de Atacama, Chile

Martina Ueckert[1], Lisa Brückner[1], Broder Merkel[1]

[1] Technische Universität Bergakademie Freiberg, email: tine.ueckert@gmx.de, lissa-b@web.de, merkel@geo.tu-freiberg.de

Key words: water shortage, overexploitation of groundwater, drawdown, Copiapó

Copiapó Valley is located in the north of Chile, more precisely in the southern part of the “Región de Atacama” and covers an area of 18.400 km^2 in total. Although, the climate is extremely arid in this region, the Copiapó Valley is also influenced by the high mountains of the Andes rising up to 6000 m a.s.l. Precipitation rates vary from 0.1 mm/year to 6 mm/year in the lower Copiapó Valley up to averagely 150 mm/year to 200 mm/year in the upper Copiapó Valley. Therefore origin the main tributaries, Manflas, Jorquera and Pulido, of the Copiapó River in the higher Cordillera. These three rivers confluence together at “Las Juntas” (1230 m a.s.l.) and form the Copiapó River which discharges after 120 km into the Pacific Ocean.

The area is one of the primary mining basins in Chile, because the surrounding mountains are rich in copper, iron, silver and gold ores. According to the public water office, 115,150 tons of copper and 53,914 tons of gold were produced in 2006. Additionally, Copiapó Valley is a maior agricultural resource. About 10,658 ha are used as farmland, thereby is the upper part of the Valley mainly used for growing grapes. In the lower part of the Valley soils are affected by salinization and thus olives are grown.

Both kinds of land use require rather big volumes of water. To satisfy the water demand both groundwater resources and superficial waters are utilized. This caused a desiccation of the Copiapó River, prior to reaching the City of Copiapó. Furthermore, groundwater levels decreased up to 40 m over the last 25 years.

These days, the authorities are aware of the problem and some fundamental hydrologic and hydrogeological studies of the Copiapó Valley have been performed respectively are on the way. But still, overexploitation is going on. Simultaneously a lack of frequently taken data regarding hydrogeological parameters and water chemistry data in the Copiapó Valley and the adjacent catchments can be stated.

The concern of this project was taking superficial and groundwater samples as well as measuring on-site-parameters (pH, Eh, EC, T and O_2). By means of ion chromatography (IC) and inductive coupled plasma mass spectrometry (ICP-MS) major and minor constituents were determined to reveal impacts on ground- and superficial water quality caused by mining and farming. Also organic compounds were detected by gas chromatograph (GC). PHREEQC was used to evaluate the data. Furthermore, the aquifer of the lower Copiapó Valley was modeled by means of Visual MODFLOW, using data from the local water authorities respectively own readings. Based on this model it is possible to simulate scenarios of future water consumption in the region and the impact on the aquifer. This may help politicians and decision makers to establish a proper and sustainable water management to secure water supply and help to protect the water resources.

www.borntraeger-cramer.de
2012/0277/$ 0.25

Proceedings "Hydrogeology of Arid Environments", p. 278
Stuttgart, March 2012

Poster Abstract

A Three Dimensional Model of the Omdel Aquifer in the Namib Desert of Namibia

A. E. van Wyk[1], W. Kambinda, M. Amukwaya, A. Mwetulundila, R. Nuujoma, L. Menge, P. Shidute

[1] Department of Water Affairs and Forestry, Windhoek, Namibia. email: vanwyka@growas.org.na

Key words: Omdel Aquifer, three-dimensional model, groundwater quality, seawater interface

The Omaruru Delta Aquifer (Omdel) located in the Namib Desert is an important aquifer utilized for bulk water supply purposes to coastal towns and mines, that are mainly mining uranium, within the Erongo Region of Namibia. The Omdel consists of the so-called "Main Channel" which is deeply incised into bedrock and contains an alluvial aquifer with fresh water reserves. The aquifer is currently utilized at a rate of 9 Mm^3/a through a well field consisting of 45 production boreholes. Adjacent to the Main Channel, are shallower side channels, termed the "Northern Elevated Channel" and "Southern Elevated Channel", that contain a variety of fresh, brackish and saline groundwater. During 1994 a managed aquifer recharge scheme that captures flood water from the Omaruru River was commissioned. The water from the dam is released into the aquifer through infiltration beds. Due to a significant increase in the mining activities, through the so-called "uranium rush" of Namibia, that resulted in economic growth of the coastal towns, the aquifer is currently utilized beyond its sustainable levels. To cope with the higher water demand at the coast, a seawater desalination plant was built and a second one is planned for the future which will alleviate the Omdel from its current over-abstraction. Seasonal flooding of the Omaruru River, that is an important component for sustainable yield calculations, has been less than predicted for the past 15 years. The most recent review of the sustainable yield set the limits between 3.5 to 4 Mm^3/a while the current stored fresh groundwater reserves are estimated at 165 Mm^3. The combination of over-abstraction and "less than expected flooding" from the Omaruru River resulted in a decline in the water levels of up to 20 meters in some parts of the aquifer. This poster illustrates a 3 dimensional geological model of the aquifer that in particular demonstrates the hydraulic interconnection between the main aquifer and the side channels, the postulated interface between the groundwater and the seawater, the distribution of groundwater quality and aquifer water levels.

www.borntraeger-cramer.de
2012/0278/$ 0.25

Proceedings "Hydrogeology of Arid Environments", p. 279
Stuttgart, March 2012

Poster Abstract

Safeguarding Lake Chad Basin Transboundary Aquifers under Uncertain Climatic Regime

Sara Vassolo[1], Roland Geerken[2] Mohammed Bila[3]

[1] GIZ/BGR/Lake Chad Basin Commission Sustainable Water Management Project, Ndjamena, email: svassolo@yahoo.com
[2] GIZ/BGR/Lake Chad Basin Commission Sustainable Water Management Project, Skype, email: rgeerken.yale
[3] Lake Chad Basin Observatory, Lake Chad Basin Commission, Ndjamena, email: mdbila@yahoo.com

Key words: Lake Chad, drought, transboundary aquifer, conjunctive use

The changes affecting the Lake Chad are a manifestation of the regional climatic and hydrological variability. These changes impact all components of the hydrological system leading to an increasing dependence on the groundwater resources by an increasing population. The decade long shrinking in the surface area of Lake Chad is influenced by droughts, population growth, reservoir construction, water diversion, and groundwater pumping. Seasonal and inter-annual fluctuations of the Lake Chad levels are primarily controlled by precipitation and run-off but longer term variability is likely dependent on regional groundwater levels. Under an assumption of a simplified climate change scenario, a simulation of the impact of 1° C temperature increase on surface run-off indicated that areas creating surface run-off that recharges the Lake Chad may lead to a run-off reduction of about 2.3 km^2. The linkage between surface water and ground water in the Lake Chad shows a difference in behavior between pre-drought and post-drought conditions. Pre-drought lake levels and precipitation show a strong correlation while there is little relationship after post-drought condition but instead an increased variation in the level of the lake. In the pre-drought period, groundwater was recharging the Lake Chad during the annual dry season thereby compensating for any evaporative losses, while in the post-drought period the Lake Chad is losing water to the surrounding aquifers in addition to the evaporative losses. In the Lake Chad Basin, the wetlands traditionally store substantial amounts of water that are believed to play a particular role in groundwater renewal. Human interference in the form of water diversions and agricultural practices are threatening this vital function by impacting on river discharge and groundwater renewal. Understanding the mechanism and behavior of this exchange between the Yaere and the River Logone during extreme events is crucial for the understanding of lake level fluctuations and for forecasting the impacts of further climate change. Recent knowledge synthesised since the beginning of the millennium highlighted the need to urgently establish a transboundary water governance mechanism that will protect existing groundwater resources while promoting conjunctive use of both surface and groundwater in the Lake Chad basin.

www.borntraeger-cramer.de
2012/0279/$ 0.25

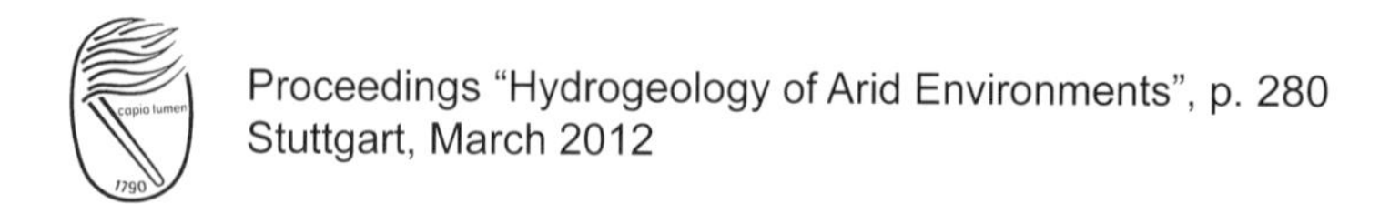
Proceedings "Hydrogeology of Arid Environments", p. 280
Stuttgart, March 2012

Poster Abstract

Using Simplified Water Budgeting to Estimate Environmental Impacts in Life Cycle Assessment (LCA)

Francesca Verones[1], Karin Bartl[2], Stephan Pfister[1], Stefanie Hellweg[1]

[1] ETH Zurich, Institute of Environmental Engineering, 8093 Zurich, Switzerland,
email: francesca.verones@ifu.baug.ethz.ch
[2] ETH Zurich, Institute of Agricultural Sciences – Animal Nutrition, 8092 Zurich, Switzerland

Key words: water budget, Life Cycle Assessment, Peru, environmental impacts

Life Cycle Assessment (LCA) is a methodology for quantifying the environmental impacts of a product. The life cycle includes all processes related to a product, from resource extraction and manufacturing over use to disposal and the results can be used for estimating the potential for improvement. Impacts which are generally accounted for include climate change, acidification, resource depletion, human toxicity and ecotoxicity. The consideration of water use related impacts in LCA is relatively new and many open questions remain. Environmental impacts from groundwater use, for example, have so far been tackled only for selected regions (van Zelm et al. 2011). Also the interaction between surface and groundwater are not explicitly addressed. As a first step towards the solution of this problem we derived characterization factors (CFs) for a groundwater-fed coastal wetland in Peru for different scenarios of agricultural water use, thereby distinguishing between surface water and groundwater use and consequent impacts. The basis for the calculation of the characterization factors is a water balance for the arid coastal region of Chancay-Huaral. In the region by far the largest share of water is used for irrigation, thus we focus on agricultural water use. The only wetland in the region, called Santa Rosa, is one of the most biodiverse coastal wetlands in Peru and therefore of considerable regional importance. We show that depending on the water source used for irrigation (surface water or groundwater) and as a function of irrigation efficiency, there is considerable difference in the environmental impact or benefit for the wetland. The use of surface water is beneficial for the wetland, if surplus irrigation water is recharging the aquifer, which in turn feeds the groundwater-fed wetland. Pumping of groundwater, on the other hand, has a negative impact on the wetland, since the wetland is deprived of some of its water source. This highlights the importance of taking the interaction between surface water and groundwater into account. Due to the increase in irrigation in the area, mainly with surface water, the wetland is currently growing, which confirms our finding that surface water use exerts a positive impact on this ecosystem. The results feed into a complete Life Cycle Assessment study of agricultural production in the region and are used for decision-making and planning of future agricultural development by local stakeholders and are disseminated through local stakeholder workshops.

References

van Zelm, R., Schipper, A. M., Rombouts, M., Snepvangers, J. & Huijbregts, M. A. J., 2011: Implementing groundwater extraction in Life Cycle Impact Assessment: Characterization Factors based on plant species richness for the Netherlands. – Environ. Sci. Technol. **45**: 629–635.

www.borntraeger-cramer.de
2012/0280/$ 0.25

Proceedings "Hydrogeology of Arid Environments", p. 281
Stuttgart, March 2012

Poster Abstract

Tademait Plateau: A Regional Groundwater Recharge Area in the Centre of the Algerian Sahara

K. Udo Weyer, James C. Ellis

WDA Consultants Inc., 4827 Vienna Drive NW, Calgary, AB, T3A 0W7; email: weyer@wda-consultants.com

Key words: groundwater flow systems, Hubbert's force potential, Tademait Plateau, groundwater recharge, In Salah, CO_2 sequestration

Traditionally, groundwater recharge in desert areas is considered to be of minor importance; hence any substantial fresh groundwater encountered in these regions is often labeled as fossil groundwater which supposedly rests at depth. Naturally, due to slow flow velocities of groundwater in most geological environments and due to the often considerable lateral extent (several hundred kilometers) of groundwater flow systems, groundwater ages of several ten thousands or possibly even hundred thousands of years may occur in active groundwater flow systems. This fact does not make that groundwater fossil, or make saline water within Mesozoic or Proterozoic layers connate water from early geological times.

Based on the physics of Hubbert's (1940) force potential and Tóth's (1962) subsequent Groundwater Flow Systems Theory, there exist regionally-extended groundwater flow systems even and in particular in arid regions and deserts like the Sahara. One of the indicators of substantial groundwater discharge areas are the oases prevalent in many areas of arid countries. For several thousand years, foggaras (fugharsas, qanats) were constructed in arid areas of Africa, Asia, and on the Arabian Peninsula. They were tunnels dug into talus slopes of mountains traced at the surface by lines of vertical shafts every 50 or 60 m. In present times these foggaras have been supplemented or replaced by deep well pumpage, as for example in In Salah located at the outer fringe of the Tademait Plateau and supplying water for a settlement area of approximately 45,000 people.

In fact there exists around the Tademait Plateau a continuous belt of oasis in the depressions to the south, the south-west and the west. All these oases are located on the edge of an elongated and curved topographical depression on the outer fringe of the Tademait Plateau. In the south-western area of this belt there is a parallel cluster of oases which, judged by their proximity to the Atlas Mountain chain, most likely get much of their water from recharge of artesian layers in the Atlas Mountains located to the northwest.

Within the Tademait Plateau, the Krechba gas field, a part of the In Salah cluster of gas fields, moved into the centre of international attention due to the testing of CO_2 sequestration in the very same field. Unexpectedly the CO_2 supercritical fluid, injected into the very same layer from which natural gas is produced and down dip from the natural gas, does not migrate up dip and towards the pressure sink as expected. Instead it moved in the opposite direction. This migration into the opposite direction can only be explained by the action of groundwater flow systems recharged on the predominately flat Tademait Plateau. The lateral flow distances from the centre of the regional recharge area Tademait Plateau to the discharge areas at the oases is approximately 100 to 300 km. The elevation differences are up to 550 m.

Proceedings "Hydrogeology of Arid Environments", p. 282
Stuttgart, March 2012

Poster Abstract

Managed Artificial Recharge into a Karst Aquifer – Wala Dam, Jordan

J. Xanke[1], A. Sawarieh[1], N. Seder[2], W. Ali[1], N. Goldscheider[1], H. Hötzl[1]

[1] KIT – Karlsruhe Institute of Technology, Institute of Applied Geosciences, Division of Hydrogeology, Adenauerring 20b, Building 50.40, 76131 Karlsruhe, email: julian.xanke@kit.edu
[2] Jordan Valley Authority (JVA)

Key words: managed artificial recharge, karst aquifer, hydrogeology

Groundwater is the main fresh water resource in Jordan but is limited in quantity and not always sufficient in quality. Natural recharge takes place by infiltration from rainwater or streams. Rare but intensive rainfall events can generate fast runoff in the mountains where surface water is conducted through wadis to the Jordan Valley, but only a small amount infiltrates naturally into the ground. To optimize the use of floodwaters, several storage dams were constructed in Jordan during the last decades along the slopes of the Jordan Valley. One of these is the Wala dam in Madaba region in Jordan. Its main function is to store water during the winter season and recharge it into the underlying Wadi Es Sir Limestone aquifer. Groundwater from this aquifer is used for the drinking water supply to nearby communities (Sawarieh et al. 2010).

Since its completion in 2002, the Wala reservoir fills during winter season and recharges the underlying aquifer. This intervention in natural processes has caused challenges in terms of control of the reservoir and its influence on the groundwater flow and quality. High amounts of suspended material are flushed into the reservoir and deposited, which results in a reduction of storage capacity and infiltration rate. This has caused overflow events in the past that cannot be avoided or controlled in the current situation, because the bottom gate of the dam is already covered with sediments. Nevertheless, the recharge still occurs but is reduced (Margane et al. 2009, Sawarieh et al. 2010).

The present study includes a geological and hydrogeological investigation of the aquifer to obtain a better understanding of the hydraulic conditions, which are influenced by faults, fractures and karst features. Observations of water level fluctuations in the reservoir and the aquifer intend to characterise the effect of the stored floodwaters on groundwater quality and quantity. Lack of rainfall data and runoff monitoring make it difficult to determine the exact amount of runoff for individual precipitation events, or an entire hydrological year. An approach is to calculate this runoff with water level records and the volume of the reservoir, but the latter is subject of constant changing due to sedimentation. Therefore, reliable data of infiltration rates are difficult to obtain. Since the beginning of operation of the dam, recurrent turbidity was detected in groundwater and could not be related to specific events, such as the overflow of the dam or changes of infiltration conditions. Identifying the source of this water quality degradation is part of the investigation.

"Managed Artificial Recharge" (MAR) is a concept, among others, to be used in the implementation of the Integrated Water Resources Management (IWRM) in the Lower Jordan Rift Valley, within the framework of the SMART II (Sustainable Management of Available Water Resources with Innovative Technologies) project (BMBF funding No.: FKZ 02WM1079).

References

Margane, A., Borgstedt, A., Hamdan, I., Subah, A. & Hajali, Z., 2009: Delineation of Surface Water Protection Zones for the Wala Dam. – Technical report no 12. BMZ-No.: 2005.2110.4, 126 p., Amman.

Sawarieh, A., Wolf, L., Ali, W. & Hoetzl, H., 2010: Quantity and Quality Pre-Assessments of Wala Reservoir impact on Groundwater system. – IWRM Conference, Karlsruhe, 2010.

www.borntraeger-cramer.de
2012/0282/$ 0.25

Proceedings "Hydrogeology of Arid Environments", p. 283
Stuttgart, March 2012

Poster Abstract

Estimation of Spatially Distributed Groundwater Recharge for Gaza Strip, Palestine

Z. Zomlot[1], M. Elbaba[1], O. Batelaan[1, 2]

[1] Department of Hydrology and Hydraulic Engineering, Vrije Universiteit Brussel, Brussels, Belgium
[2] Department of Earth and Environmental Sciences, Katholieke Universiteit Leuven, Leuven, Belgium

Key words: semi-arid, recharge, INFIL.03, Gaza

The Gaza Strip, is the second most "water-poor" region in the world, with 52 m^3 available per person each year (International Atomic Energy Agency Fact Sheet 2003). It is considered as a semi-arid area where groundwater is the only significant source of water, which is replenished by rain water infiltration. Long-term overexploitation has resulted in a decreasing water table, accompanied by degradation of its water quality. This leads to high levels of salinity and nitrate pollution, most of the groundwater is inadequate for both domestic and agricultural consumption. The spatial and temporal variability of groundwater recharge are key factors that need to be quantified for efficient groundwater management in Gaza strip. Few investigations have tried to estimate natural groundwater recharge for Gaza strip. Here in this paper a different approach using the INFIL3.0 model has been chosen to estimate the groundwater recharge. INFIL3.0 is a grid-based, distributed-parameter, deterministic water-balance model, which was developed for arid to semiarid environments. Daily climatic records of precipitation and air temperature for the period (1980–2009) are being used together with spatially distributed drainage-basin grid maps describing topography, geology, soils, and vegetation to simulate daily net infiltration and recharge. The presented results identify the spatial-temporal natural variability of the recharge for the Gaza strip. It is compared to previous study results and it will allow future improved groundwater management.

www.borntraeger-cramer.de
2012/0283/$ 0.25